UTAH OIL SHALE

Science, Technology, and Policy Perspectives

UTAH OIL SHALE

Science, Technology, and Policy Perspectives

EDITED BY

Jennifer P. Spinti

CRC Press
Taylor & Francis Group
Boca Raton London New York

CRC Press is an imprint of the
Taylor & Francis Group, an **informa** business

CRC Press
Taylor & Francis Group
6000 Broken Sound Parkway NW, Suite 300
Boca Raton, FL 33487-2742

First issued in paperback 2019

© 2017 by Taylor & Francis Group, LLC
CRC Press is an imprint of Taylor & Francis Group, an Informa business

No claim to original U.S. Government works

ISBN-13: 978-0-4987-2172-1 (hbk)
ISBN-13: 978-0-367-87298-4 (pbk)

Library of Congress Cataloging-in-Publication Data

Names: Spinti, Jennifer P, editor.
Title: Utah oil shale : science, technology, and policy perspectives /
Jennifer Spinti, editor.
Other titles: Oil shale
Description: Boca Raton : CRC Press, 2017. | Includes bibliographical
references and index.
Identifiers: LCCN 2016021017 | ISBN 9781498721721 (alk. paper)
Subjects: LCSH: Oil-shales--Utah. | Oil shale reserves--Utah. | Shale oils.
Classification: LCC TN858 .U83 2017 | DDC 553.2/8309792--dc23
LC record available at https://lccn.loc.gov/2016021017

Visit the Taylor & Francis Web site at
http://www.taylorandfrancis.com

and the CRC Press Web site at
http://www.crcpress.com

Contents

List of Figures

List of Tables

Preface

The idea for this book came from a desire to archive in one scholarly work the results from nearly a decade of research at the University of Utah's Institute for Clean and Secure Energy (ICSE) relating to the creation of an industry for unconventional oil production in the United States. Research funding came from the U.S. Department of Energy (DOE) with research direction from DOE and from the Energy Policy Act of 2005. ICSE was to serve as an ongoing source of unbiased information to the nation surrounding technical, economic, legal, and environmental aspects of developing oil sands and oil shale resources. While ICSE research included work on both oil sands and oil shale, the material in this book is focused on oil shale with a specific emphasis on Utah's Uinta Basin. The Uinta Basin contains one of the largest oil shale resources in the United States, and much of the research was conducted on oil shale samples from there.

The ICSE research program was very broad, with researchers from four colleges and one bureau: the College of Engineering, the College of Law, the College of Science, the College of Mines and Earth Sciences, and the Utah Bureau of Economic and Business Research. The chapters in this book reflect the breadth of the research. Each chapter is authored by ICSE researchers who are experts in the chapter topic. In addition, collaborators from the Utah Geological Survey and Brigham Young University are coauthors on several of the chapters. The book is loosely organized by topic. The first chapter provides an introduction to the ICSE research program. Chapters 2 and 3 address legal and policy considerations associated with oil shale development. Chapter 4 presents a basin-scale evaluation of the Uinta Basin oil shale resource. Chapters 5 through 9 discuss a series of experiments that were conducted on oil shale samples from the Skyline 16 oil shale core and other cores from the upper Green River Formation in the Uinta Basin. These experiments included sample characterization, pyrolysis and analysis of pyrolysis products, analysis of porosity and permeability evolution, and measurement of geomechanical properties under representative in situ conditions. Chapter 10 incorporates models and data from previous chapters to produce a suite of simulation scenarios that study the effect of well arrangement on energy ratio for an in situ process located in the Uinta Basin. Chapters 11 and 12 build on the analysis in Chapter 10. Chapter 11 attempts to answer the "how much will it cost" question for every in situ heating scenario discussed in Chapter 10, while Chapter 12 analyzes each scenario's carbon footprint and ozone precursor emission and discusses these emissions in the context of Uinta Basin air quality and broader energy policy.

The authors of this book hope that the information contained herein will provide a stepping stone for the next generation of oil shale research.

Acknowledgments

Funding for the work contained in this book came principally from the Department of Energy's National Energy Technology Laboratory (NETL) under award number DE-FE0001243. The authors thank NETL for their generous support and Robert Vagnetti, project director at NETL, for his guidance over the years. They also thank Dr. Olayinka Ogunsola at the Department of Energy for his leadership role and for his contributions to programmatic reviews.

Editor

Jennifer P. Spinti earned her PhD in chemical engineering from the University of Utah, Salt Lake City, Utah in 1997. Her doctoral work focused on the fate of nitrogen that remains in the char during pulverized coal combustion. From 1998 to 2009, she was a postdoctoral fellow and then a research associate in the Institute for Clean and Secure Energy (ICSE) at the University of Utah. During this time, she performed simulations of buoyancy-driven flows, including pool fires and flares, using high-performance computing resources as part of the Center for Simulation of Accidental Fires and Explosions.

She is currently a research associate professor in the Department of Chemical Engineering at the University of Utah. She served for six years as the assistant director of the Clean and Secure Energy from Domestic Oil Shale and Oil Sands Resources program within ICSE. She also organized the University of Utah Unconventional Fuels Conference and was engaged in research projects related to the development of oil shale and oil sands resources in Utah's Uinta Basin.

Contributors

Lauren P. Birgenheier
Department of Geology and
  Geophysics
University of Utah
Salt Lake City, Utah

Milind Deo
Department of Chemical
  Engineering
University of Utah
Salt Lake City, Utah

Julio C. Facelli
Department of Biomedical
  Informatics
University of Utah
Salt Lake City, Utah

Thomas H. Fletcher
Department of Chemical
  Engineering
Brigham Young University
Provo, Utah

Michal Hradisky
Department of Chemical
  Engineering
University of Utah
Salt Lake City, Utah

Kerry E. Kelly
Department of Chemical
  Engineering
University of Utah
Salt Lake City, Utah

Chen-Luh Lin
Department of Metallurgical
  Engineering
University of Utah
Salt Lake City, Utah

Charles L. Mayne
Department of Chemistry
University of Utah
Salt Lake City, Utah

John D. McLennan
Department of Chemical
  Engineering
University of Utah
Salt Lake City, Utah

Jan D. Miller
Department of Metallurgical
  Engineering
University of Utah
Salt Lake City, Utah

Anita M. Orendt
Center for High Performance
  Computing
University of Utah
Salt Lake City, Utah

Ronald J. Pugmire
Department of Chemical
  Engineering
and
Department of Chemistry
University of Utah
Salt Lake City, Utah

Terry A. Ring
Department of Chemical
    Engineering
University of Utah
Salt Lake City, Utah

John C. Ruple
S.J. Quinney College of Law
and
Wallace Stegner Center for Land,
Resources and the Environment
University of Utah
Salt Lake City, Utah

Philip J. Smith
Institute for Clean and Secure
    Energy
and
Department of Chemical
    Engineering
University of Utah
Salt Lake City, Utah

Mark S. Solum
Department of Chemistry
University of Utah
Salt Lake City, Utah

Jennifer P. Spinti
Institute for Clean and Secure
    Energy
and
Department of Chemical
    Engineering
University of Utah
Salt Lake City, Utah

Josh Staten
Department of Chemical
    Engineering
University of Utah
Salt Lake City, Utah

Thang Q. Tran
Department of Chemical
    Engineering
University of Utah
Salt Lake City, Utah

Pankaj Tiwari
Department of Chemical
    Engineering
Indian Institute of Technology
    Guwahati
Guwahati, India

Michael D. Vanden Berg
Utah Department of Natural
    Resources
Utah Geological Survey
Salt Lake City, Utah

Jonathan E. Wilkey
Department of Chemical
    Engineering
University of Utah
Salt Lake City, Utah

Units

μm	Micron
μs	Microsecond
Å	Angstrom
AF	Acre-feet
amu	Atomic mass unit
Atm	Atmosphere
bbl	Barrel
BPD	Barrels per day
Btu/gal	British thermal units per gallon
Btu/lb-°F	British thermal units per pound degree Fahrenheit
Btu/scf	British thermal units per standard cubic foot
g CO_{2e}/MJ	CO_2 equivalent per megajoule
cm	Centimeter
°C/min	Degrees celcius per minute
°C/h	Degrees celcius per hour
eV	Electron volts
ft	Feet
g	Grams
g VOC/MJ	Grams volatile organic carbons per megajoule
g/cm^3	Grams per cubic centimeter
GPa	Gigapascals
GPT	Gallons per short ton
GWh/day	Gigawatt-hours per day
gal/ton	Gallons per ton
h	Hour
in.	Inches
K	Kelvin
K/min	Kelvin per minute
kg	Kilogram
kg/m^3	Kilograms per cubic meter
kJ/mol	Kilojoules per mole
kWh	Kilowatt-hour
L/t	Liters per metric ton
L/min	Liters per minute
m	Meter
MCF	Thousand cubic feet
mD	Millidarcy
mg	Milligram
mg/mL	Milligrams/milliliter

mL/min	Milliliters per minute
MHz	Megahertz
min	Minute
MJ/kg	Megajoules per kilogram
mL	Milliliter
mm	Millimeter
mol	Mole
MPa	Megapascals
ms	Millisecond
MW	Megawatt
MWh/day	Megawatt-hour per day
nm	Nanometer
ppb	Parts per billion
ppm	Parts per million
psi	Pounds per square inch
psig	Pounds per square inch gauge
°R	Degrees Rankine
s	Seconds
ton/year	Short tons per year
W/m-°C	Watts per meter degree Celsius
wt%	Weight percent

Acronyms

2D	Two-dimensional
3D	Three-dimensional
AMSO	American Shale Oil
ANL	Argonne National Laboratory
ASTM	American Society for Testing and Materials
ATP	Alberta Taciuk process
BLM	Bureau of Land Management
C	Carbon
^{13}C	Carbon-13
CAA	Clean Air Act
CH	Methine
CH$_2$	Methylene
CH$_3$	Methyl
CH$_4$	Methane
CO	Carbon monoxide
CO$_2$	Carbon dioxide
CP	Cross-polarization
CPD	Chemical percolation devolatilization
CT	Computed tomography
DAEM	Distributed activation energy model
daf	Dry, ash-free
DCM	Dichloromethane
DEPT	Distortionless enhancement by polarization transfer
DOE	Department of Energy or design of experiments
DOGM	Division of Oil, Gas, and Mining
DSC	Differential scanning calorimeter
EER	External energy ratio
EER$_{n,y}$	Normalized energy ratio
EIA	Energy Information Administration
EIS	Environmental Impact Statement
EPA	Environmental Protection Agency
ESA	Endangered Species Act
FEHM	Finite element heat and mass
FID	Free induction decay
FIP	Federal Implementation Plan
FLPMA	Federal Land Policy and Management Act
FTIR	Fourier transform infrared spectroscopy
FWHM	Full width at half maximum
FWS	U.S. Fish and Wildlife Service
GC	Gas chromatography

GHG	Greenhouse gas
GR	Green River
GRF	Green River Formation
H	Hydrogen
^{1}H	Solution state proton (^{1}H)
HAP	Hazardous air pollutants
He	Helium
HHV	Higher heating value
HPTGA	High-pressure thermogravimetric analyzer
HRXMT	High-resolution x-ray microtomography
ICP	In situ conversion process
ICSE	Institute for Clean and Secure Energy
IRR	Internal (or investor's) rate of return
L	Lean
LBM	Lattice Boltzmann method
LCFS	Low-carbon fuel standard
LH	Latin hypercube
LHS	Latin hypercube sampling
LVDT	Linear variable displacement transducers
MAS	Magic angle spinning
MGPA	Most geologically prospective area
MS	Mass spectrometry
N	Nitrogen
N_2	Nitrogen (molecular)
N_2O	Nitrous oxide
NAAQS	National Ambient Air Quality Standards
NER	Net energy ratio
NMR	Nuclear magnetic resonance
NOE	Nuclear Overhauser effect
NO_X	Nitrogen oxides
NPV	Net present value
O	Oxygen
O_2	Oxygen (molecular)
O_3	Ozone
ODEQ	Oregon Department of Environmental Quality
PDF	Pair distribution function or probability distribution function
PM	Particulate matter
PSS	Production, separation, and storage
QEMSCAN	Quantitative evaluation of minerals by scanning electron microscopy
R	Rich
RD&D	Research, development, and demonstration
RHF	Restricted Hartree–Fock
RMP	Resource Management Plan
RuO_2	Ruthenium oxide

S	Sulfur
S/N	Signal-to-noise
SD	Standard deviation
SEM	Scanning electron spectroscopy
SIP	State Implementation Plan
SITLA	Utah School and Institutional Trust Lands Administration
SLBLM	Salt Lake Base Line and Meridian
SO_2	Sulfur dioxide
SP	Single pulse
TGA	Thermogravimetric analyzer
TMS	Tetramethylsilane
TOC	Total organic content
UGS	Utah Geological Survey
USD	U.S. dollars
USGS	United States Geological Survey
VOC	Volatile organic compound
WTI	West Texas Intermediate
WTP	Well-to-pump
WTW	Well-to-wheel
XANES	X-ray absorption near edge structure
XMT	X-ray microtomography
XNT	X-ray nanotomography
XPS	X-ray photoelectron spectroscopy
XRD	X-ray diffraction
XRF	X-ray fluorescence

1

A Decade of Oil Shale Research (2006–2015)

Jennifer P. Spinti and Philip J. Smith

CONTENTS

The Utah Heavy Oil Program within the University of Utah's Institute for Clean and Secure Energy (ICSE) was established in June 2006 to provide research support to federal and state constituents for addressing the wide-ranging issues surrounding the creation of an industry for unconventional oil production in the United States. Additional funding from the U.S. Department of Energy (DOE) allowed ICSE to continue its unconventional fuels research under the Clean and Secure Energy from Domestic Oil Shale and Oil Sands Resources Program.

The objectives for these programs, based on direction from the Energy Policy Act of 2005 and from DOE, were threefold: (1) update the 1987 technical and economic assessment of domestic heavy oil resources that was prepared by the Interstate Oil and Gas Compact Commission (Dowd et al. 1988), (2) develop an online repository for information, data, and software pertaining to heavy oil (specifically oil shale and oil sands) resources in North America, and (3) perform research that addressed challenges for managing and utilizing these natural resources. Additionally, ICSE was to serve as an ongoing source of unbiased information to the nation surrounding technical, economic, legal, and environmental aspects of developing oil sands and oil shale resources.

The first projective objective was accomplished in 2007 with the publication of "A Technical, Economic, and Legal Assessment of North American Heavy Oil, Oil Sands, and Oil Shale Resources" (UHOP 2007). The product of the second objective, the digital repository, is housed in the USpace Institutional Repository of the University of Utah (USpace 2015). Repository materials can be accessed through the USpace search engine. The entire collection can be viewed using the search term "ICSE." This book is the result of research work performed as part of the third objective.

1.1 Introduction

In 2005, the price of West Texas Intermediate (WTI), the standard U.S. benchmark for crude oil, was $66.88/bbl in 2015 U.S. dollars (USD) (or $56.64 in 2005 USD) (EIA 2015a), the average daily consumption of oil in the United States was 20.8 million barrels (EIA 2015b), and the United States was importing an average of 13.7 million barrels per day (BPD) (EIA 2015c). Concern about U.S. dependence on oil imports, especially from hostile countries, was rising along with oil consumption and price. On August 8, 2005, the Energy Policy Act of 2005 was signed into law by President George W. Bush (EPAct 2005). The purpose of the act was "to ensure jobs for our future with secure, affordable, and reliable energy." Section 369 declared the strategic importance of domestic oil shale and oil sand (e.g., tar sand) resources, encouraged the development of said resources in an environmentally sound manner, and called for a "national assessment of oil shale and tar sands resources for the purposes of evaluating and mapping oil shale and tar sands deposits." It also directed the Secretary of the Interior to make geologically prospective public lands within the states of Colorado, Utah, and Wyoming available for a research and development leasing program and to "complete a programmatic environmental impact statement for a commercial leasing program for oil shale and tar sands resources on public lands" (EPAct 2005). Interest in domestic oil shale surged as significant political effort and support led to the completion of a foundational Programmatic Environmental Impact Statement; finalization of federal oil shale leasing regulations; the 2006 and 2007 issuance of six initial Research, Development, and Demonstration (RD&D) leases for oil shale in Colorado and Utah (Crawford et al. 2013); the 2012 issuance of two additional RD&D leases in Colorado resulting from a second round of leasing (BLM 2012); and the reestablishment of the Oil Shale Symposia at the Colorado School of Mines beginning in 2006. Against this backdrop, the ICSE oil shale research program commenced in 2006.

1.2 What Is Oil Shale?

Oil shale is a fine-grained sedimentary rock bound with an organic material known as kerogen (Pugmire 2009); see Figure 1.1. Kerogen has not been exposed to sufficient heat and pressure to be converted into oil and gas.

FIGURE 1.1
Green River oil shale. (Photo courtesy of Michael Vanden Berg, UGS, Salt Lake City, UT.)

All known processes for disengaging kerogen from its mineral matrix and for converting kerogen to oil require thermal input.

Shale oil is a confusing term as it can refer to the oil being produced from places such as North Dakota's Bakken or to the oil that is produced when oil shale is heated. The former definition of shale oil, sometimes referred to as tight oil, refers to a light crude oil that is produced from low-permeability formations. Using techniques such as horizontal drilling and hydraulic fracturing. The latter definition of shale oil refers to the product of pyrolyzed (heated in the absence of air) oil shale. The properties of oil produced from oil shale depend on the type of kerogen and the thermal treatment method used. When Shell used an in situ (e.g. in the ground) process to produce shale oil from the Mahogany zone, an organic-rich oil shale layer of the Green River Formation in Colorado's Piceance Basin, the resulting liquid fuel had properties similar to those of jet or diesel fuel (Nair et al. 2008).

1.3 Uinta Basin Oil Shale Resource

1.3.1 Resource Size

The largest oil shale resource by volume in North America lies in the Eocene age deposits of the Green River Formation in Colorado, Utah, and

Wyoming; see Figure 1.2. Resource assessments by the U.S. Geological Survey (USGS) estimate total in-place oil for all oil shale zones in the Piceance Basin to be 1.53 trillion barrels, in the Greater Green River Basin to be 1.44 trillion barrels, and in the Uinta Basin to be 1.32 trillion barrels (Johnson et al. 2010a,b, 2011).

The Uinta Basin oil shale zone, discussed in detail in Chapter 4, consists of a series of rich (R) and lean (L) layers identified by a numbering system that decreases with depth. The richest layer is the R-7, also known as the Mahogany zone. The lean layers above and below the Mahogany zone are referred to as the A-groove and B-groove, respectively.

In 2008, the Utah Geological Survey (UGS) published a new assessment of Uinta Basin oil shale resources that investigated the entire basin. The study utilized data from oil shale wells and from "hundreds of geophysical logs from oil and gas wells drilled over the last two decades" (Vanden Berg 2008). These widespread data were used to map oil shale thickness and richness and to create isopach maps delineating oil yields of 15, 25, 35, and 50 gallons of shale oil per short ton (GPT) of rock. Thicknesses were centered around the extremely rich Mahogany bed of the Mahogany zone (R-7) within the Parachute Creek Member of the Green River Formation. From these isopach maps, new basin-wide resource numbers were calculated for each richness grade. A continuous interval of oil shale averaging 50 GPT contains an in-place oil resource of 31 billion barrels in a zone ranging up to 20 ft thick. An interval averaging 35 GPT, with a maximum thickness of 55 ft, contains an in-place oil resource of 76 billion barrels. The 25 GPT zone and the 15 GPT zone contain unconstrained resources of 147 and 292 billion barrels, respectively. The report also computed a potential economic resource in the Uinta Basin at 77 billion barrels based on criteria that included a resource grade of at least 25 GPT, a deposit thickness of at least 5 ft with less than 3000 ft of overburden, and a location on Bureau of Land Management (BLM), state, private, or tribal land that was not in conflict with current conventional oil and gas resources.

1.3.2 Skyline 16 Core

In May 2010, ICSE and UGS teamed up to drill 1000 ft of 4 in. diameter core in the upper Green River Formation oil shale deposit in the eastern Uinta Basin, Utah. The well was named Skyline 16 and its location was Uintah County, Utah, T11S, R25E, Sec. 9, UTM E 661444, UTM N 4415107. The purpose of the coring was to recover nearly the entire oil shale zone (Parachute Creek Member), providing "fresh" samples for the variety of geochemical and geomechanical tests that are discussed in this book.

During drilling, special care was taken to preserve the core for future testing. Starting at 260 ft and continuing down to about 700 ft, the core was slid into thick plastic sleeves and sealed with duct tape to help preserve moisture (see Figure 1.3). In addition, 12 1 ft sections of core (1 from the A-groove, 8 from the Mahogany zone, 1 from the B-groove, and 2 from

Oil shale deposits in the three-state area

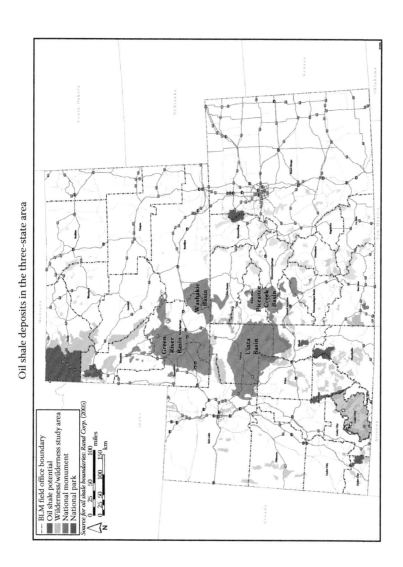

FIGURE 1.2
Location of Green River Formation oil shale deposits in Utah, Colorado, and Wyoming. The main basins are labeled. (Reprinted from 2012 Oil Shale and Tar Sands Programmatic EIS Information Center, U.S. Department of the Interior, Bureau of Land Management.)

(a)   (b)

FIGURE 1.3
Skyline 16 core drilling. (a) Marking footages as part of the process of logging the core. (b) Preserving the core in plastic sleeves. (Photo courtesy of Michael Vanden Berg, UGS, Salt Lake City, UT.)

the upper R-6) were wrapped in plastic wrap and sealed in ProtecCore, a special aluminum sleeve designed to preserve core in an in situ state. The core was slabbed with a one-third to two-thirds cut and stored at the UGS Core Research Center. The one-third slab was placed in display boxes and archived for future research and viewing (Figure 1.4), while the two-thirds section was placed back in the protective sleeves and preserved for future sampling and testing. Testing on samples from the Skyline 16 core is described in the subsequent chapters of this book.

1.4 Commercial Development of Oil Shale

When the BLM issued six RD&D leases in 2006 and 2007, five were located in Colorado. The leases were held by Chevron Oil Shale Company, EGL Resources, Inc. (now American Shale Oil or AMSO), and Shell Frontier Oil and Gas (which held three lease sites) (BLM 2006). The only lease in Utah (Uinta Basin) was held by Oil Shale Exploration, LLC (later purchased by Enefit American Oil) (BLM 2007). In a second round of leasing that began

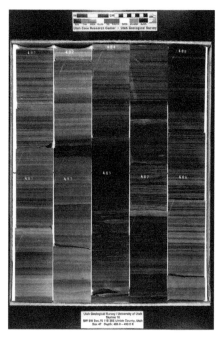

FIGURE 1.4
Photograph of box 47 of Skyline 16 core. The depth range for this box is 480.0–490.0 ft. The GR2 sample, discussed in subsequent chapters, was taken from the 486 to 487 ft interval. (Photo courtesy of Michael Vanden Berg, UGS, Salt Lake City, UT.)

in 2010, applicants included Aurasource, Inc., ExxonMobil Exploration Company, and Natural Soda Holdings, Inc. Only ExxonMobil and Natural Soda Holdings were eventually awarded leases (BLM 2012); both leases were in Colorado. Aurasource did not develop a cost reimbursement agreement or give the BLM any definitive time frame for the commencement of project operations, so BLM closed their application (BLM 2013). The eight leases in Colorado all involved in situ development (heating the oil shale in the ground), while the Utah project involved surface mining and retorting (e.g., heating) of the mined shale. Also during this time, Red Leaf Resources began development of its EcoShale® process, described as a modified in situ process that is part mining and part underground heating, on state-owned school trust lands in the Uinta Basin (Red Leaf 2015a).

In 2012, Chevron divested itself of its Colorado lease (*The Daily Sentinel* 2012) with Shell following suit in 2013 (Webb 2013). Nevertheless, other companies operating in Colorado continued moving forward with development plans. In 2014, the BLM approved RD&D lease plans of development that had been submitted by ExxonMobil and Natural Soda Holdings (ExxonMobil 2012, Natural Soda 2013, Webb 2014). During this time, AMSO continued both its activities at the RD&D lease site and its research program; recent papers discuss various issues associated with downhole heaters (Burnham et al. 2015, Reiling et al. 2015).

Across the border in Utah, development activity was increasing. In 2013, Enefit American Oil Chief Executive Officer Rikki Hrenko-Browning announced that the company hoped to have permitting completed by 2017 with first oil production in 2020. Production would then ramp up to the level of 50,000 BPD in the 2020s (O'Donoghue 2013). In May 2014, a press release from Red Leaf Resources announced, "The Utah Division of Water Quality has issued a Construction Permit to oil shale development company Red Leaf

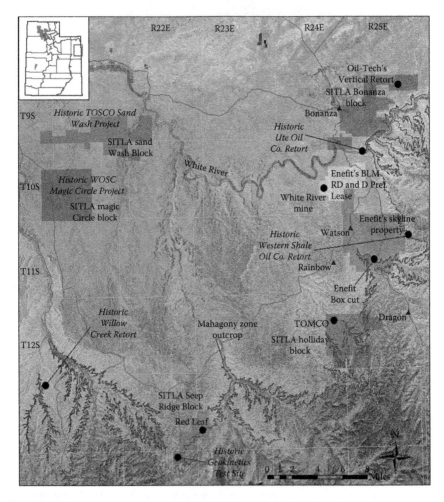

FIGURE 1.5
Map showing the location of projects and lands associated with oil shale development in Uintah County, Utah, including Red Leaf and Enefit. (Reprinted with permission from Aho, G.D., History of Utah's oil shale industry, in: Utah Geological Association Publication 44, October 2015.)

Resources for its EcoShale production capsule. Red Leaf now has all required permits in place and is poised to launch a 300,000 barrel commercial demonstration project in Uintah County, Utah" (Red Leaf 2014). These commercialization pathways were clear indicators that the mining and surface retorting processes were closer to commercial reality than were the in situ processes.

After the announcement of the Shell pullout from Colorado in 2013, Glen Vawter, president of the National Oil Shale Association, stated that "It looks like the future of oil shale right now is over there in Utah" (Webb 2013). This view is supported by the enthusiasm of local and state government officials for oil shale development. On a trip to Estonia in 2011, Uintah County officials toured Estonian oil shale mines and processing plants; Uintah County is the county in the Uinta Basin where Enefit and Red Leaf have their operations. A newspaper article published at the time stated that, "They believe the company that developed them will deliver on promises of up to 2,000 jobs doing the same thing in eastern Utah" (Loomis 2011). Utah Governor Herbert's 10-Year Strategic Energy Plan states, "Utah's energy portfolio should include fossil fuels, alternative fuels, renewable resources, and energy efficiency" (Herbert 2011). The plan emphasizes the size of the oil shale resource in Utah and references the Oil Shale Symposium as a source of information about newer and cleaner technologies to produce liquid transportation fuels from unconventional resources.

Figure 1.5 shows the location of Enefit and Red Leaf properties as well as other oil-shale-related activities in Uintah County, Utah. Additional information regarding the history of oil shale in Utah can be found in "History of Utah's Oil Shale Industry" by Aho (2015).

1.5 Future of Oil Shale in the Uinta Basin

With the constraints that there be less than 2000 ft of overburden and a stripping ratio of less than 10, there are 178.7 billion barrels of oil in-place associated with 15 GPT or greater oil shale source rock in Colorado's Piceance Basin compared with 31.7 billion barrels in the Uinta Basin (USGS 2015). Nevertheless, a convergence of political, regulatory, economic, and technological issues has focused the oil shale development spotlight on the Uinta Basin. The debate about whether to develop the resource continues. Vastly differing opinions are evident with oil shale; either it will be the energy source that will lead to the next oil production boom, similar to what is currently happening with tight oil, or it is a resource that will lead to environmental catastrophe if produced and as such is better left in the ground. Ultimately, economics will drive societal decisions about energy in general and development of oil shale in particular. Support for oil shale development will rest on the balance of the difference between its social benefits and social costs.

The price of WTI peaked in 2008 at $141.47/bbl ($154.84 in 2015 USD). By the beginning of the fourth quarter in 2015, the price of WTI had dropped to $46/bbl (EIA 2015a), U.S. oil consumption had dropped to a 2015 yearly average of 19.4 million BPD (EIA 2015b), oil imports had decreased to an average of 9.2 million BPD (2014 data) (EIA 2015c), and the United States and Saudi Arabia were competing for the title of the world's top oil producer (Smith 2014) with the Saudis having increased production to 9.8 million BPD in an effort to put U.S. shale oil producers out of business (Critchlow 2015).

In this difficult economic climate, the commercialization of oil shale in the Uinta Basin of Utah has been slow-tracked. In a press release dated March 26, 2015, Red Leaf announced "an adjustment to the construction schedule of its commercial demonstration project in anticipation of technological improvements that will result from design optimization work already underway... In part, this decision is the result of the low commodity oil price environment as well as recently identified efficiencies and engineering advances that may be utilized in the commercial demonstration project" (Red Leaf 2015b). Enefit, whose shares are owned by the government of Estonia, has been reassessing its business case, including options to mine a narrower (and richer) band of oil shale, to make a faster transition to underground mining that would result in less surface disturbance, to relocate the plant site to a more cost-beneficial location, and to reduce the preproduction investment by reducing the size of the processing unit from 550 to 280 tons/h of oil shale (Hrenko-Browning 2015). The project continues to move forward with a recent announcement about the draft Environmental Impact Statement for the utility corridor (Enefit 2015).

1.6 ICSE Research Contributions

In subsequent chapters, engineering, science, economic, legal, and policy research is presented relative to the development of the Uinta Basin oil shale resource. The purpose of this book is not to jump into the political fray surrounding resource development but rather to provide unbiased information as well as research results and analysis from the ICSE oil shale research program.

References

Aho, G.D. 2015. History of Utah's oil shale industry, in Vanden Berg, M.D., Ressetar, R., and Birgenheier, L.P., editors, Geology of Utah's Uinta Basin and Uinta Mountains: Utah Geological Association Publication 44, p. 319–335.

Burnham, A., C. Sandberg, K. Thomas et al. 2015 (October). Qualification of high-power-density mineral-insulated electric heater cables. Paper presented at the *35th Oil Shale Symposium*, Salt Lake City, UT.

Crawford, P. M., C. Dean, J. Stone, and J. C. Killen. 2013. Assessment of plans and progress on US Bureau of Land Management oil shale RD&D leases in the United States. Washington, DC: U.S. Department of Energy. http://energy.gov/sites/prod/files/2013/04/f0/BLM_Final.pdf. Accessed October 12, 2015.

Critchlow, A. 2015 (January 27). Saudi Arabia increases oil output to crush US shale frackers. http://www.telegraph.co.uk/finance/newsbysector/energy/11372058/Saudi-Arabia-increases-oil-output-to-crush-US-shale-frackers.html. Accessed July 1, 2015.

Dowd, W. T., V. A. Kuuskraa, and M. L. Godec. 1988 (January). A technical and economic assessment of domestic heavy oil: Final report. DOE/BC/10840-1. Oklahoma City, OK: Interstate Oil Compact Commission.

Enefit. 2015. EIS status report. http://enefitutah.com/2015/03/eis-status-report/. Accessed October 12, 2015.

Energy Policy Act of 2005 (EPAct). 2005. Public Law No. 109–58, 119 Stat. 594. http://energy.gov/sites/prod/files/2013/10/f3/epact_2005.pdf. Accessed August 15, 2015.

ExxonMobil Exploration Company. 2012 (November 8). RD&D lease plan of development. http://www.blm.gov/style/medialib/blm/co/information/nepa/white_river_field/fy14_scoping_and_comment.Par.81300.File.dat/XOM%20Oil%20Shale%20RD&D%20POD_final.pdf. Accessed October 12, 2015.

Herbert, G. R. 2011 (March 2). Energy initiatives and imperatives: Utah's 10-year strategic energy plan. http://www.utah.gov/governor/docs/10year-stragegic-energy.pdf. Accessed July 1, 2015.

Hrenko-Browning, R. 2015 (October). Enefit American Oil: A proven, efficient and environmentally sound means to help Utah become energy independent. Paper presented at the *35th Oil Shale Symposium*, Salt Lake City, UT.

Johnson, R. C., T. J. Mercier, M. E. Brownfield, M. P. Pantea, and J. G. Self. 2010a. An assessment of in-place oil shale resources in the Green River Formation, Piceance Basin, Colorado. In *Oil Shale and Nahcolite Resources of the Piceance Basin, Colorado*, U.S. Geological Survey Digital Data Series 69–Y, Chapter 1. Reston, VA: U.S. Geological Survey.

Johnson, R. C., T. J. Mercier, M. E. Brownfield, and J. G. Self. 2010b. Assessment of in-place oil shale resources in the Eocene Green River Formation, Uinta Basin, Utah and Colorado. In *Oil Shale Resources of the Uinta Basin, Utah and Colorado*, U.S. Geological Survey Digital Data Series 69–BB, Chapter 1. Reston, VA: U.S. Geological Survey.

Johnson, R. C., T. J. Mercier, R. T. Ryder, M. E. Brownfield, and J. G. Self. 2011. Assessment of in-place oil shale resources of the Eocene Green River Formation, Greater Green River Basin, Wyoming, Colorado, and Utah. In *Oil Shale Resources of the Eocene Green River Formation, Greater Green River Basin, Wyoming, Colorado, and Utah*, U.S. Geological Survey Digital Data Series 69–DD, Chapter 1. Reston, VA: U.S. Geological Survey Oil Shale Assessment Team. http://pubs.usgs.gov/dds/dds-069/dds-069-bb/.

Loomis, B. 2011 (July 9). Salt Lake Tribune. E. Utah officials fired up about oil shale after Estonia visit. http://archive.sltrib.com/article.php?id=15976325&itype=storyID/. Accessed August 15, 2015.

Nair, V., R. Ryan, and G. Roes. 2008. Shell ICP—Shale oil refining. Paper presented at the *28th Oil Shale Symposium*, Golden, CO. http://ceri-mines.org/documents/28thsymposium/presentations08/PRES_3-3_Nair_Vijay.pdf.

Natural Soda Holdings, Inc. 2013. 2013 plan of development: Oil shale research, development and demonstration (RD&D) tract COC 74299. http://www.blm.gov/style/medialib/blm/co/information/nepa/white_river_field/fy14_scoping_and_comment.Par.26594.File.dat/2013%20NSHI%20POD_Public_sm.pdf. Accessed April 6, 2016.

O'Donoghue, A. J. 2013 (July 13). *Deseret News*. Estonia company wants to pull 2.6 billion barrels of oil from Utah. http://www.deseretnews.com/article/865583090/Estonia-company-wants-to-pull-26-billion-barrels-of-oil-from-Utah.html?pg=all. Accessed July 1, 2015.

Pugmire, R. 2009. 13C NMR spectroscopy: A key contributor for development of the molecular structure of coal and other kerogens. ISCE seminar, University of Utah, Salt Lake City, UT. http://content.lib.utah.edu/cdm/singleitem/collection/ir-eua/id/3796/rec/1.

Red Leaf Resources, Inc. 2014. Press center. Now fully permitted, Red Leaf prepares for commercial production of oil shale (May 30). http://www.redleafinc.com/press-center. Accessed August 15, 2015.

Red Leaf Resources, Inc. 2015a. http://www.redleafinc.com. Accessed October 12, 2015.

Red Leaf Resources, Inc. 2015b (March 26). Press center. Red Leaf announces acceleration to technology optimization and adjustment to construction schedule for oil shale commercial demonstration project. http://www.redleafinc.com/press-center. Accessed October 12, 2015.

Reiling, V., J. Fauli, A. Papineau, R. LaFollette, L. Switzer, and V. Saubestre. 2015 (October). Surface qualification system for downhole heaters. Paper presented at the *35th Oil Shale Symposium*, Salt Lake City, UT.

Smith, G. 2014 (July 4). Bloomberg Business. U.S. seen as biggest oil producer after overtaking Saudi. http://www.bloomberg.com/news/articles/2014-07-04/u-s-seen-as-biggest-oil-producer-after-overtaking-saudi. Accessed July 1, 2015.

The Daily Sentinel. 2012 (February 28). Chevron halting oil shale effort. http://www.gjsentinel.com/news/articles/chevron-halting-oil-shale-effort. Accessed July 1, 2015.

U.S. Department of the Interior, Bureau of Land Management (BLM). 2006 (December 15). Interior department issues oil shale research, development and demonstration leases for public lands in Colorado. News release. http://www.blm.gov/wo/st/en/info/newsroom/2006/december/blm_announces_waiver.html. Accessed July 1, 2015.

U.S. Department of the Interior, Bureau of Land Management (BLM). 2007 (June 28). Interior Department issues oil shale research, development and demonstration lease for public lands in Utah. News release. http://www.blm.gov/ut/st/en/info/newsroom/2007/06/interior_department.html. Accessed July 1, 2015.

U.S. Department of the Interior, Bureau of Land Management (BLM). 2012 (August 30). BLM approves two Colorado oil shale RD&D leases to test innovative heat extraction technologies—Second round of leases encourage promising, innovative techniques. http://www.blm.gov/co/st/en/BLM_Information/newsroom/2012/blm_approves_two_colorado.html. Accessed July 1, 2015.

U.S. Department of the Interior, Bureau of Land Management (BLM). 2013 (March). Approved land use plan amendments/record of decision (ROD) for allocation of oil shale and tar sands resources on lands administered by the Bureau of Land Management in Colorado, Utah, and Wyoming and final programmatic environmental impact statement.

U.S. Department of the Interior, U.S. Geological Survey (USGS). 2015 (February). In-place oil shale resources of the Mahogany zone, Green River Formation, sorted by grade, overburden thickness, and stripping ratio, Piceance Basin, Colorado, and Uinta Basin, Utah. Fact sheet 2015–3005. http://pubs.usgs.gov/fs/2015/3005/pdf/fs2015-3005.pdf. Accessed July 1, 2015.

U.S. Energy Information Administration (EIA). 2015a. Petroleum and other liquids. Spot prices. http://www.eia.gov/dnav/pet/pet_pri_spt_s1_d.htm. Accessed November 2, 2015.

U.S. Energy Information Administration (EIA). 2015b. Petroleum and other liquids. U.S. product supplied of crude oil and petroleum products. http://www.eia.gov/dnav/pet/hist/LeafHandler.ashx?n=pet&s=mttupus1&f=a. Accessed November 2, 2015.

U.S. Energy Information Administration (EIA). 2015c. Petroleum and other liquids. U.S. imports of crude oil and petroleum products. http://www.eia.gov/dnav/pet/hist/LeafHandler.ashx?n=PET&s=MTTIMUS1&f=A. Accessed November 2, 2015.

USpace Institutional Repository (USpace), University of Utah. 2015. http://uspace.utah.edu. Accessed November 2, 2015.

Utah Heavy Oil Program (UHOP), University of Utah. 2007. A technical, economic, and legal assessment of North American heavy oil, oil sands, and oil shale resources: In response to Energy Policy Act of 2005 Section 369(p). http://content.lib.utah.edu/cdm/singleitem/collection/ir-eua/id/2778/rec/1. Accessed August 15, 2015.

Vanden Berg, M. 2008. Basin-wide evaluation of the uppermost Green River Formation's oil-shale resource, Uinta Basin, Utah and Colorado. Special Study 128. Salt Lake City, UT: Utah Geological Survey. http://files.geology.utah.gov/online/ss/ss-128/ss-128txt.pdf.

Webb, D. 2013 (September 24). The daily sentinel. Shell shock: Rio Blanco oil shale project axed. http://www.gjsentinel.com/news/articles/shell-shock-8232rio-blanco-oil-shale-project-axed. Accessed July 1, 2015.

Webb, D. 2014 (March 11). The daily sentinel. Back in oil shale. http://www.gjsentinel.com/news/articles/back-in-oil-shale. Accessed July 1, 2015.

2

Legal and Policy Considerations Involving Oil Shale–Bearing Lands and the Resources They Contain

John C. Ruple

CONTENTS

This chapter focuses on ownership and control of oil shale resources within Utah, the conflicting mandates and policies guiding the various resource owners and managers, how conflicting direction may affect access to oil shale resources and technology development, and possible ways of addressing these conflicts. Section 2.1 summarizes the land and resources involved, land and resource ownership, and how divergent management mandates could impact oil shale development. Section 2.2 summarizes key legal considerations impacting access to oil shale resources, as well as the social context within which development would occur. Section 2.3 discusses the implications for industry and for the federal government.

2.1 Resource Characterization and Ownership

2.1.1 Resource Characterization

The world's largest known oil shale deposits are contained in the Green River Formation, which covers portions of Colorado, Utah, and Wyoming (see Figure 2.1). Estimates put the share of these oil shale resources in Utah at approximately 1.32 trillion barrels,[1] though much of this is likely undevelopable due to physical or economic constraints. Commercially producible resources represent an estimated 147.4 billion barrels of oil equivalent[2] and are shown in Figure 2.2. The shaded area in Figure 2.2 reflects the thickness of shale with the potential to produce at least 25 gallons of oil equivalent per short ton (GPT) of shale.[3] The Most Geologically Prospective Area (MGPA), which is outlined in purple, was defined by the Bureau of Land Management (BLM) and reflects the same 25 GPT resources, but with a 25 ft minimum thickness and no more than 3000 ft of overburden.[4] The MGPA reflects the BLM's assessment of technical recoverability and is used to determine which federal lands are available for leasing.

To put these potential supplies in perspective, the Prudhoe Bay Oil Field contains 13.5 billion barrels of oil, and the mean estimate of recoverable oil from the coastal plains of the Arctic National Wildlife Refuge is 10.4 billion barrels.[5] Domestic petroleum demand is about 20 million barrels per day. Oil shale resources in Utah could therefore, in theory, supply all U.S. domestic oil needs for more than 20 years.

2.1.2 Resource Ownership

A commercial oil shale industry cannot develop without access to oil shale resources. The threshold question, with regard to resource access, is who owns the oil shale resources? Ownership of the oil shale resources within Utah is split between four primary entities: the federal government, the State of Utah, the Ute Tribe of Indians, and private property owners. Ownership of the land surface overlying oil shale resources within Utah is shown in Figure 2.3.

Ownership of the land surface[6] overlying oil shale resources is quantified at two geographic scales. The area shown in Figure 2.2 was divided into zones with 500 ft or less of overburden (ex situ lands) and zones with 501–3000 ft of overburden (in situ lands) in order to better understand development

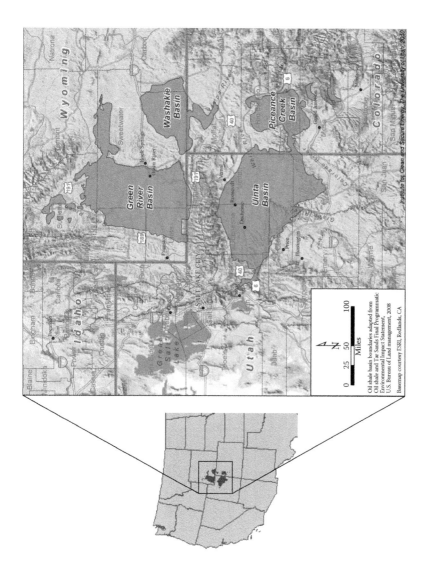

FIGURE 2.1
Oil shale deposits of the Green River Formation. (Adapted from the Oil Shale and Tar Sands Final Programmatic Environmental Impact Statement, U.S. Bureau of Land Management, 2008.)

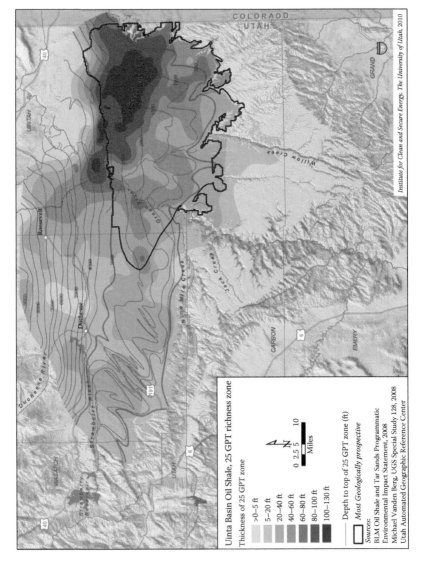

FIGURE 2.2
Oil shale and overburden thickness within the Uinta Basin.

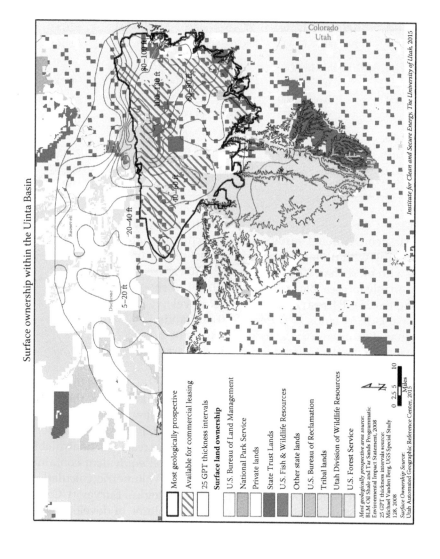

FIGURE 2.3
Surface ownership and available federal lands within the Uinta Basin.

TABLE 2.1

Surface Ownership of Developable Oil Shale Bearing Lands within the Uinta Basin by Development Constraints (Acres)

	BLM	Private	SITLA	Tribal	Unavailable	Total
MGPA						
Ex situ lands	50,200	37,700	30,600	24,500	45,500	188,400
In situ lands	242,700	17,700	77,700	82,900	230,700	651,800
Total	292,900	55,400	108,300	107,400	276,200	840,200
Oil shale area outside of the MGPA						
Ex situ lands	0	2,100	800	7,600	5,500	16,000
In situ lands	0	24,600	5,500	53,400	100,700	184,100
Total	0	26,700	6,300	61,000	106,200	200,100
Total oil shale area						
Ex situ lands	50,200	39,800	31,400	32,100	51,000	204,500
In situ lands	242,700	42,300	83,200	136,300	331,400	835,900
Total	292,900	82,100	114,600	168,500	382,400	1,040,400

possibilities. Consistent with BLM assumptions, ex situ lands could be mined using conventional surface mining techniques, and in situ lands are accessible only with underground mining techniques or in situ processing. Oil shale with more than 3000 ft of overburden was treated as unrecoverable with reasonably foreseeable technologies. Table 2.1 shows surface ownership overlaying the 25 GPT resource under these development constraints.

Developable BLM lands include only those lands within the MGPA that are designated as open for oil shale leasing as of 2013. Lands identified as "Unavailable" in Table 2.1 include Utah State Parks, state wildlife management areas, state lands underlying navigable waters, and Utah Department of Transportation rights-of-way. National Forest System lands and BLM lands that are outside of the MGPA are treated as unavailable because they could be leased only if management plans are amended to allow for oil shale development. Similarly, BLM-managed lands within the MGPA that are not specifically designated as available for oil shale leasing are treated as unavailable.

Table 2.2 quantifies the approximate volume of oil equivalent contained beneath lands held by each category of owners. Volumetric calculations were derived from data provided by the Utah Geological Survey (UGS) by dividing the project area into 160-acre cells. The volume of developable oil shale for each cell was computed using the 160-acre surface area and the thickness of the 25 GPT layer within each cell. Oil shale deeper than 3000 ft and with a 25 GPT layer less than 25 ft thick was excluded. Cell volumes were then converted into barrels of oil equivalent based on conversion factors provided by UGS.

Within Utah, the BLM controls roughly 292,900 acres of the surface overlying oil shale resources that are available for leasing and developable

TABLE 2.2

Surface Ownership of Developable Oil Shale Bearing Lands within the Uinta Basin by Development Constraints (Millions of Barrels of Oil Equivalent)

	BLM	Private	SITLA	Tribal	Unavailable	Total
MGPA						
Ex situ lands	3,724	3,174	2,315	1,215	3,105	13,533
In situ lands	28,603	2,266	9,012	9,495	23,438	72,814
Total	32,327	5,440	11,327	10,710	26,543	86,347
Oil shale area outside of the MGPA						
Ex situ lands	0	119	55	365	356	895
In situ lands	0	1,306	323	3,071	5,652	10,351
Total	0	1,425	378	3,436	6,008	11,246
Total oil shale area						
Ex situ lands	3,724	3,293	2,370	1,579	3,462	14,428
In situ lands	28,603	3,571	9,334	12,565	29,090	83,164
Total	32,327	6,864	11,704	14,144	32,552	95,592

with in situ thermal treatment or underground mining. The BLM controls approximately 50,200 acres of leasable land overlying resources that are available for surface mining. The Utah School and Institutional Trust Lands Administration (SITLA) controls roughly 114,600 acres of land overlying oil shale resources subject to in situ recovery or underground mining and 31,400 acres of land overlying resources subject to surface mining. Tribal lands encompass 168,500 acres overlying oil shale resources subject to in situ recovery or underground mining and 32,100 acres of land overlying resources subject to surface mining. Private interests control less total acreage, 82,100 developable acres, but a higher percentage of private holdings are located along the Mahogany outcrop, where surface mining operations would dominate. The Mahogany zone is the richest oil shale zone in the Green River Formation, as discussed in Chapter 4.

Volumetrically, BLM-managed lands that are potentially available for oil shale leasing contain approximately 32.3 billion barrels of oil equivalent, while SITLA and tribal lands contain approximately 11.7 and 14.1 billion barrels of oil equivalent, respectively. Private interests control almost 6.9 billion barrels of oil equivalent, roughly half of which is recoverable using surface mining technologies, making them particularly powerful movers with respect to those technologies.

2.1.2.1 Divergent Management Objectives

Once control of oil shale resources is established, the question becomes whether owners will allow development, and if so, under what conditions. Land management requirements, which vary by landowner, dictate which lands can be developed and under what conditions development can occur. Divergent federal and state objectives have important implications for both the type and extent of development. As can be seen in Figure 2.3, land ownership within the Uinta Basin is highly fragmented, with SITLA lands frequently surrounded by federal land. If development does occur, failure to reconcile inconsistent management directives and objectives could engender conflict.

The BLM manages the vast majority of federal lands within the Uinta Basin and operates under a multiple-use, sustained-yield mandate, balancing resource extraction and preservation.[7] Applying this mandate, the BLM determined that the highlighted portion of the 2 million acres within the MGPA are appropriate for development, as shown in Figure 2.3.[8] The remaining BLM acreage is unavailable for leasing because of overriding resource considerations and management objectives.

SITLA manages the vast majority of state-owned lands within the Uinta Basin. In contrast to the BLM, SITLA must manage trust lands in the most "prudent and profitable manner possible" to support public schools and institutions.[9] By law, SITLA must "obtain the optimum values from the use of trust lands and maximize revenues for the trust beneficiaries."[10]

2.2 Legal Considerations

2.2.1 Oil Shale Leasing

2.2.1.1 Federal Commercial Leases

On November 28, 2008, the BLM programmatically amended resource management plans in Colorado, Utah, and Wyoming to make approximately 1.9 million acres of public lands available for application for oil shale leasing. Roughly 670,000 of the available acres were in Utah. Less than 3 years later, pursuant to a litigation settlement, the BLM decided to take a "fresh look" at its 2008 decision in light of new information regarding the greater sage-grouse, which was then the subject of a pending petition for Endangered Species Act listing, new administrative direction regarding management of lands with wilderness characteristics, and new assessments of in-place oil shale resources completed by the U.S. Geological Survey. In March of 2013, the BLM finalized its review, reducing the total amount of federal land available for oil shale development to approximately 678,000 acres as shown in Figure 2.3. As part of its 2013 decision, the BLM decided to "allow commercial development only after a technology has been proven to be viable on, for example, a [Research, Development, and Demonstration, or] RD&D lease."[11] In this new decision, approximately 296,000 acres of federal land surface and subsurface minerals and 64,000 acres of federal minerals beneath nonfederally owned surface resources in Utah are potentially available for oil shale development.

In 2008, the Secretary of Interior also issued rules establishing federal commercial oil shale lease terms. These rules allow for issuance of exploration licenses covering up to 25,000 acres and the leasing of tracts up to 5,760 acres in size. However, leaseholders are limited to no more than 50,000 acres in any one state. Leases are subject to a $2.00 per acre annual rental charge. Production royalties start at 5% and increase to 12.5% over time. An environmental impact review is required before issuing a lease or exploration license and prior to approving a plan of development.

Under the 2008 rules, an application to lease must include information regarding the technologies that would be used to develop the tract and a description of sensitive resources known to exist within the lease area. The application must also include a "description of how the proposed lease development would avoid, or, to the extent practicable, mitigate impacts on species or habitats protected by applicable state or federal law or regulations, and impacts on wildlife habitat management" before a lease can be offered for bid.[12]

These final BLM leasing rules were challenged in court, and as part of a settlement agreement, the BLM agreed to consider certain regulatory revisions. On March 27, 2013, the BLM published a proposed rule to amend its commercial oil shale regulations.[13] As part of the revision, the BLM is considering whether a single royalty rate or rate structure should be set in advance to provide greater certainty to potential lessees, or whether some administrative flexibility may be retained to make adjustments to royalty terms after more is known about the costs and resource impacts associated with emerging oil shale technologies.

The BLM is also considering denying a commercial lease application, a postlease plan of development, or a conversion of an RD&D lease into a commercial lease unless it determines that oil shale operations could occur without "unacceptable environmental risk." Risk would be assessed in a qualitative manner in light of the nascent character of the oil shale industry. As an alternative to the proposed unacceptable environmental risk standard, the BLM is considering whether "unacceptable environmental consequences" might be a more appropriate standard for issuance of commercial leases. The unacceptable environmental consequence standard grew out of National Environmental Policy Act compliance and, though not explicitly defined, already applies for conversion of existing RD&D leases to commercial operations. The BLM requested comments on how to define both standards as part of its rule making efforts.

As of the writing of this chapter, the final oil shale leasing rule has not been issued and the 2008 commercial leasing rule remains in effect until the proposed amendments are finalized. No commercial oil shale leases have, however, been offered or issued under the 2008 rules, and the BLM is unlikely to entertain commercial oil shale leasing until the federal commercial leasing rules are updated.

2.2.1.2 Federal RD&D Leases

Over the past decade, the BLM has twice issued RD&D oil shale leases in order to encourage technology development. If technologies prove to be commercially viable, the lessee can convert the RD&D lease into a commercial lease, securing the right to develop both the primary lease tract and an adjacent "preference area." During the first round of RD&D leasing, the BLM issued five RD&D leases in Colorado and one in Utah. As of 2013, operators had ceased work on four of the six RD&D leases, and none of the remaining leases had proceeded to commercial development. The leases each cover 160-acre primary lease tracts and 4960-acre preference areas.[14]

Enefit American Oil, which acquired Oil Shale Exploration Company, holds the only RD&D lease in Utah. The Enefit RD&D lease tract includes "Tract Ua" and the existing White River Mine, which were development candidates under the Prototype Oil Shale Leasing Program in the 1970s. Enefit also holds title or options to significant adjacent private land and intends to develop the existing mine and RD&D lease tract in conjunction with adjacent nonfederal oil shale resources.[15] In addition to the estimated 545 million barrels of oil equivalent that Enefit controls on BLM lands, Enefit controls an estimated additional 2 billion barrels of oil equivalent on nearby state and private leases and holdings.[16]

In November 2009, the BLM invited nominations for a second round of oil shale RD&D leases.[17] AuraSource, Inc. proposed to mine oil shale within Utah and to use aboveground retorting technology to retort oil shale.[18] AuraSource's nomination was canceled when it failed to initiate the required environmental reviews.[19]

2.2.1.3 Nonfederal Oil Shale Leasing

The State of Utah controls sizeable oil shale resources, actively promotes oil shale development,[20] and can lease state lands independent of federal oil shale leasing rules. The State of Utah's support for oil shale development is reflected in royalty reductions to encourage oil shale development, a 10-year exemption from severance taxes for oil shale development, and tax exemptions for motor vehicle fuels derived from Utah oil shale.[21]

Virtually all of the state's oil shale resources are located on SITLA lands. SITLA leases have traditionally conveyed development rights for up to a 10-year period that continues (1) if oil shale is being produced in paying quantities or (2) if the tract is subject to diligent operations reasonably calculated to advance or restore production, and the operator pays annual minimum royalties. Rental rates are no less than $1 per acre or fraction thereof per year and no less than $500 per tract per year. Leases are subject to an 8% production royalty. SITLA has discretion to increase the royalty rate by up to 1% annually to a maximum of 12.5% after the first 10 years of the lease.[22]

SITLA's regulations require submission and approval of an operating plan prior to any ground-disturbing activity. The operating plan must include access and infrastructure locations as well as a site reclamation plan.[23] SITLA may also require cultural, paleontological, and biological surveys of leased land, reasonable mitigation of impacts to other resources occasioned by surface or subsurface operations on the lease, and surface use or right-of-way agreements as necessary for development.[24] The Utah Division of Oil, Gas, and Mining's reclamation bonding requirements are applicable to SITLA leases, and SITLA may require supplemental bonding.[25] As of March 2015, SITLA was overseeing 51 active oil shale leases, covering over 68,469 acres of land, down considerably from the 99 active leases covering 97,848 acres identified in 2009.[26]

2.2.1.4 Access to Nonfederal Inholdings

As can be seen in Figure 2.3, SITLA and private lands are often scattered across the landscape, and access issues may occur where federal public lands surround private- or SITLA-managed lands. Holders of SITLA leases are legally entitled to reasonable access across BLM-managed lands in order to access SITLA lands, even if the surrounding federal lands are not open for development.[27] Crossing BLM lands, however, requires a BLM-granted right-of-way.[28] Rights-of-way must include conditions to "minimize damage to scenic and esthetic values and fish and wildlife habitat and otherwise protect the environment."[29]

Striking a balance between reasonable access and reasonable protection of competing resource values involves complex discretionary decisions on the part of the BLM. If these decisions represent "major federal actions significantly affecting the quality of the human environment"[30] under the National Environmental Policy Act, federal environmental impact reviews are triggered.[31] Such reviews may increase the time and expense required to obtain rights-of-way across federal lands and to access oil shale resources.

In an effort to reduce management conflicts, rationalize fragmented landscapes, and minimize access challenges, the State of Utah and the federal government have on multiple occasions traded lands. Under such land exchanges, the federal government conveys lands to the state that are appropriate for development and that allow the state to consolidate its lands into more manageable blocks. In return, the state conveys lands to the federal government that are at risk of development, and where development would negatively impact sensitive resources or interfere with landscape-scale management considerations. Such exchanges are authorized by the Federal Land Policy and Management Act (FLPMA), provided that the exchange is in the public interest and the lands involved are equal in value.[32] Complex procedural requirements and high transaction costs, however, make such exchanges difficult to consummate.[33] Today, most federal–state exchanges are accomplished through project-specific legislation sidestepping the FLPMA process.

2.2.2 Land Use Planning Considerations

2.2.2.1 BLM Resource Management Plans

As noted earlier, BLM lands are subject to a multiple-use, sustained-yield management mandate. In order to address and balance the range of possible uses contained within this mandate, the BLM must inventory all public lands and the resources they contain. Based on these inventories, the BLM then develops Resource Management Plans (RMPs) for the public lands it administers. The RMPs function as zoning plans for public lands administered by the BLM, determining what uses and protections are appropriate based on existing conditions and statutory requirements.

In 2008, the BLM released the Vernal, Utah RMP, which covers the oil shale–bearing region and includes management stipulations addressing at least 86 separate resource considerations. In 2013, the BLM programmatically amended 10 RMPs in Colorado, Utah, and Wyoming, including the Vernal RMP, to determine which areas would be available for oil shale development. These programmatic amendments do not replace individual RMPs, and management requirements applicable to other resource uses or values that are contained in the Vernal RMP remain in effect.

The BLM also recently finalized programmatic RMP amendments to protect the greater sage-grouse.[34] Greater sage-grouse habitat overlaps oil shale resources, and the provisions contained in the programmatic amendments are likely to directly impact activities within the Uinta Basin. Consequently, those considering development of oil shale resources on federal public lands within Utah will need to pay close attention to the 2008 Vernal RMP, the 2013 programmatic RMP amendments allocating lands for oil shale leasing, and the programmatic RMP amendments relating to the greater sage-grouse.

2.2.2.2 Multiple Mineral Development

According to the BLM, "[c]ommercial oil shale development... is largely incompatible with other mineral development activities and would likely preclude these other activities while oil shale development and production are ongoing."[35] Depending on the technologies used, developing oil shale prior to conventional oil and gas, or vice versa, may also affect the subsequent extraction of the other resource. The potential for conflicts over mineral development is significant as large portions of the Uinta Basin are already undergoing oil and gas development. A 2008 Congressional Research Service report states that 83% of the MGPA oil shale area in Utah is already leased for oil and gas development.[36] In the Uinta Basin, UGS paints a more detailed picture of conflicting mineral rights:

> A significant portion of the Uinta Basin's oil-shale resource, approximately 25% for each grade, is covered by conventional oil and gas fields.... In particular, the extensive Natural Buttes gas field covers a significant portion of land underlain by oil shale averaging 25 GPT, ranging to 130 ft thick, and under roughly 1500 to 4000 ft of cover. Furthermore, this field is expected to expand in size and cover more oil-shale rich lands to the east. Of the 18.4 billion barrels contained in 25 GPT rock having thicknesses between 100 and 130 ft, 7.8 billion barrels, or 42%, are located under existing natural gas fields.
>
> However, lands where the oil-shale deposits are under less than 1000 ft of cover currently do not contain significant oil and gas activity (except the Oil Springs gas field) as compared to lands with deeper oil-shale resources. The majority of planned oil-shale operations will be located on lands having less than 1000 ft of cover. This does not mean that oil-shale deposits located within oil and gas fields will be permanently off limits. In fact, most of the conventional oil and gas reservoirs are located far below the Mahogany zone. It simply demonstrates that regulators will need to recognize that resource conflicts exist and plan their lease stipulations accordingly.[37]

Despite these challenges, federal oil shale leasing rules do not preclude the BLM from leasing the same lands for deposits of other minerals, but the "lessee or the licensee must make all reasonable efforts to avoid interference with other such authorized uses."[38] The BLM's oil shale leasing regulations recognize the potential for conflict but state only that the oil shale lessee or licensee is "required to conduct operations in a manner that will not interfere with the established rights of existing lessees or licensees."[39]

The terms contained in leases for oil, natural gas, or other minerals that occupy the same land as oil shale must therefore be considered, as these lease terms will dictate the rights held by the competing mineral lessee. Between 1968 and 1989, federal oil and gas leases within Utah's Uinta Basin contained stipulations protecting future oil shale development. These stipulations generally prevent oil and gas drilling that would result in undue waste of oil shale resources or otherwise interfere with oil shale development. Subsequent leases lack this provision and, as the BLM recognizes, "[w]here these oil shale stipulations do not exist in oil and gas leases, without some accommodation being made between oil shale developers and prior lease holders, oil shale development may not be able to proceed."[40]

On SITLA land, SITLA reserves "the right to enter into mineral leases and agreements with third parties covering minerals other than the leased substances, under terms and conditions that will not unreasonably interfere with operations under this Lease in accordance with Lessor's regulations, if any, governing multiple mineral development."[41] SITLA also reserves the right to designate Multiple Mineral Development Areas and impose additional terms and conditions necessary to integrate and coordinate multiple mineral development.

2.2.3 Environmental and Jurisdictional Considerations

2.2.3.1 Wildlife and the Endangered Species Act

Uintah County, which includes the vast majority of oil shale resources within Utah, is home to 9 species protected under the Endangered Species Act (ESA),[42] 18 species designated as state species of concern, and 5 species subject to special management agreements intended to preclude the need for ESA protection. The region also includes important habitat for elk, mule deer, and pronghorn antelope.

The ESA protects and aids in the recovery of "listed" species and the ecosystems upon which they depend by prohibiting the "take" of listed animals, except under federal permit. "Take" is interpreted expansively to include "harm." "Harm," in turn, includes "significant habitat modification or degradation." The prohibition against a "take" applies to any "person," regardless of land ownership, and sizeable penalties can apply for Act violations.

If a federal action, including authorization or funding of an otherwise private activity, is likely to adversely impact a listed species, Section 7 of the ESA requires federal agencies to consult with the U.S. Fish and Wildlife Service (FWS). The FWS must then issue a "biological opinion" stating whether the activity would jeopardize the species or adversely modify the species' critical habitat. If the lead federal agency minimizes impacts to the maximum extent practicable and the proposed action will not appreciably reduce the likelihood of survival or recovery of a listed animal species, the FWS issues an "incidental take statement" protecting the agency and their permittees from liability for the incidental take of a listed species that occurs as a result of otherwise lawful activities.

Section 10 of the ESA can provide relief to operators who want to develop property inhabited by listed species but lack the "federal nexus" created by federal land use, a federal permit, or federal funding. Nonfederal landowners can obtain a permit from the FWS, protecting them from liability for the incidental "take" of a listed species that results from otherwise legal activities, provided they first develop an approved habitat conservation plan. Habitat conservation plans include an assessment of the likely impacts on the species resulting from the proposed action, the steps that the permit holder will take to minimize and mitigate the impacts, and the funding available to carry out the steps.

As applied to federal oil shale leasing, the ESA may require Section 7 consultation at the leasing phase and additional consultation at the development phase, depending on the level of detail available from and considered for each preceding phase. Consultation will require an assessment of the lease site as well as an overall evaluation of the indirect and cumulative effects of commercial development on listed species and their critical habitat. Plants are handled somewhat differently under the ESA and the following discussion of greater sage-grouse and endemic plants indicates the types of challenges sensitive species are likely to pose for commercial oil shale development on public lands. A discussion of how water use could impact ESA-listed fishes is contained in Chapter 3.

Greater sage-grouse habitat overlies significant oil shale resources within the Uinta Basin. Between 1999 and 2005, the FWS received 8 ESA listing petitions for the greater sage-grouse. In 2010, the FWS concluded that the bird warranted listing, but that listing was precluded by other higher priority actions. This determination placed the greater sage-grouse on the ESA candidate species list. As part of a subsequent litigation settlement, the FWS agreed to publish ESA listing rules or not-warranted findings for 251 species, including the greater sage-grouse, that were on the candidate list.

In an effort to avoid an ESA listing, the State of Utah developed a *Greater Sage Grouse Conservation Plan,* and in September 2015, the BLM and U.S. Forest Service completed amendments to 98 separate federal land use plans to address sage-grouse habitat loss, fragmentation, and other threats to the species. Revised plans include the Vernal Field Office RMP, which addresses management of oil shale-bearing lands within the Uinta Basin. The 98 plan amendments, together with state conservation efforts, reduced threats to the greater sage-grouse across 90 percent of the species' breeding habitat and enabled the FWS to conclude that the bird did not require ESA protection.[43]

The programmatic plan amendments preclude most new surface-disturbing activities in leks (breeding grounds) and in the highly valued sagebrush ecosystem areas identified as "Sagebrush Focal Areas." Plan amendments also limit or avoid new surface disturbance in "Priority Habitat Management Areas," of which Sagebrush Focal Areas are a subset, and minimize surface disturbance in "General Habitat Management Areas." Additionally, the plan amendments include mating, rearing, and wintering season disturbance restrictions, as well as buffers around sage grouse leks. For roads and infrastructure related to energy development, the buffer extends 3.1 miles from the edge of the lek. Most actions must be avoided within buffers around Priority Habitat Management Areas, unless a lesser buffer distance is shown to provide the same or greater level of sage grouse protection. Actions must be avoided within buffers around General Habitat Management Areas unless a lesser buffer distance is shown to provide the same or greater level of sage grouse protection, or the project will cause only minor disturbance and any residual impacts are addressed through compensatory mitigation.[44] Programmatic plan amendments are the subject of ongoing multi-party litigation.

While it is possible that programmatic plan amendments may be struck down through litigation, such an outcome could force the FWS to revisit its decision of not warranted for listing. Accordingly, the safest course of action for oil shale developers is to completely avoid leks and associated buffers.

In addition to wildlife, the most geologically prospective oil shale area is home to several protected plant species. Although the ESA's Section 9 prohibition against "taking" listed species does not apply to plants, it is illegal under the ESA to damage, destroy, or remove ESA-protected plants from lands that are under federal jurisdiction. Section 7 consultation requirements also still apply, and all federal agencies must insure that any action authorized, funded, or carried out by such agency does not jeopardize the continued existence of any ESA-listed species or result in the destruction or adverse modification of its critical habitat.

Several plants protected under the ESA are found in close proximity to oil shale. Shrubby reed-mustard is endangered, and the Uinta Basin hookless cactus and clay reed-mustard are both listed as threatened under the ESA. Graham beardtongue and White River beardtongue are endemic to the Uinta Basin in Utah and in immediately adjacent Rio Blanco County, Colorado. The FWS identifies key threats to these beardtongues as including "oil shale mining." In 2014, the FWS proposed listing both species as threatened under the ESA. The proposed listing was withdrawn later the same year following formalization of a Conservation Agreement between the FWS, the BLM, Uintah County, the Utah Department of Natural Resources, and SITLA. The Conservation Agreement effectively increases the level of protection afforded to these species because protections now apply on state and private land as well as federal lands. Oil shale developers will therefore need to review the Conservation Agreement and ensure their operations comply with its requirements.

One of the unique challenges posed by federally protected endemic plants is that protection is predicated on avoidance. With respect to conventional oil and gas, roads and well pads can be shifted to avoid localized plant populations. However, the plant species noted above are found only in areas where oil shale is located close to the surface and where surfacing mining is likely to occur. Avoidance may prove difficult given the extensive surface disturbances associated with surface mining.

While space constraints preclude discussion of the full range of land use-related activities that could potentially impact oil shale development, prospective developers should note that avoidance challenges similar to those noted for ESA-listed plants will be encountered with archaeological and paleontological resources. The lack of comprehensive survey data documenting the occurrence of archaeological and paleontological resources may further complicate planning and avoidance efforts.

2.2.3.2 Working in Indian Country[45]

Tribal lands pose unique jurisdictional and regulatory issues. "Indian Country" can include current and past Indian reservation lands, dependent Indian communities, and Indian allotments—all of which can extend well beyond current reservation boundaries. Indian Country within eastern Utah includes most of the region's oil shale and other energy resources.

While states can normally assume jurisdiction to implement provisions of several key federal environmental laws, including the Clean Air Act (discussed in Chapter 12), states generally cannot assert regulatory control within Indian Country. Until tribes assume responsibility for delegable federal environmental programs, the U.S. Environmental Protection Agency (EPA) retains regulatory primacy. Because the Ute Tribe of Indians has not assumed regulatory jurisdiction for most delegable federal environmental laws, the EPA will continue to administer these laws within the Uinta Basin. Approximately 70% of Utah's oil and natural gas production occurs within Indian Country, and energy developers will therefore need to work with the EPA to obtain appropriate environmental permits for their operations.

While the EPA will have regulatory primacy with respect to federal environmental laws, the State of Utah will almost certainly want to coordinate with the EPA to ensure that federal actions and authorizations harmonize with the state's ongoing efforts to manage air, water, range, mineral, and wildlife resources. Early engagement with the state and the EPA by prospective developers will minimize the risks associated with potentially confusing jurisdictional boundaries and different regulatory requirements.

With respect to mineral resources underlying tribal lands, the Department of the Interior, in association with the tribe, administers minerals on tribal or Indian lands that are held in trust by the federal government. Portions of the Uintah and Ouray Reservation, which is inhabited by the Ute Tribe of Indians, is held in trust, while the tribe owns other parts of the reservation free of trust obligations. Tribes have the sole authority for leasing mineral rights on tribal lands owned in fee. The nature of land ownership must therefore be considered if developers pursue oil shale development on tribal or non-tribal lands within Indian Country.

2.2.4 Trust and Social License

Nonlegal factors can greatly influence project success, and recent events have shown that oil shale development can generate strong opposition. Where opposition exists, developers will need to acquire a social license to operate in addition to the legal approvals, the most significant of which were discussed above. A social license reflects "the demands on and expectations for a business enterprise that emerge from neighborhoods, environmental groups, community members, and other elements of the surrounding civil society. In some instances, the conditions demanded by 'social licensors' may be tougher than those imposed by regulation."[46]

Trust is at the heart of social licensing and is essential to avoiding stakeholder opposition and prolonged litigation. The absence of adequate information regarding development plans and resource impacts breeds fear and suspicion and invites speculation that can shift the debate over oil shale development from realistic impacts to worst-case scenarios. With respect to oil shale development, uncertainty and suspicion are rooted in history. In 1980, Exxon announced its plans for the Colony Project—a massive oil shale development in Western Colorado that would employ an estimated 22,000 workers and produce 8 million barrels of oil per day. The region exploded, with the town of Parachute's population quadrupling in three short years. Grand Junction invested millions preparing for the influx of workers, building a new airport, a shopping mall, and five new schools to keep pace with the growing population.

Then, on May 2, 1982, also known as "Black Sunday," Exxon's board of directors announced that they would shutter the Colony Oil Shale Project. Overnight, 2100 people employed at the project became unemployed, most without warning. Within a week, a thousand people fled Garfield County. Between 1983 and 1985, nearly 24,000 people left Garfield and Mesa counties. Regional unemployment climbed from near 0% to 9.5%. Approximately $85 million in annual payroll rapidly disappeared from the regional economy.[47] "[F]oreclosures in Mesa County increased from 98 in 1980 to over 1,600 in 1985.... Businesses folded by the score. Banks that had survived the Depression now went under.... The oil shale bust triggered a regionwide financial collapse on a scale not seen even in most Midwestern steel towns."[48]

Researchers from the Rand Corporation caution that in light of energy sector volatility, the uncertainties associated with oil shale development, local community concerns, and a host of other factors, "an attempt to rush or shortcut development is likely to generate significant opposition at the local, state, and even national levels."[49] Researchers from the University of Colorado are more blunt, warning that "[a]ctions perceived by local residents as careless or hasty will call up the specter of Black Sunday and reinforce animosity, if not create outright opposition, among community members."[50] Adopting an open, no-surprises approach to community engagement will help reestablish trust between the nascent oil shale industry and the public.

2.3 Management Implications

2.3.1 Implications for Industry

Given the challenges associated with obtaining a federal oil shale lease, industry interest will naturally gravitate to nonfederal lands. As the largest consolidated blocks of state and private land occur along the Mahogany outcrop, where surface mining technologies are likely to predominate, developers of these technologies may have a strategic advantage, at least within Utah.

As the oil shale industry proceeds, the breadth and complexity of environmental laws will impact development efforts. Endangered species protection and BLM land management requirements pose two of the most pressing challenges for prospective developers. While Indian Country jurisdictional issues complicate matters, none of these challenges are unprecedented or insurmountable.

Prospective oil shale developers should begin by crafting a legally and politically viable framework to ensure adequate resource protection while addressing the realistic impacts of commercial oil shale development. Good decisions require good information, and industry may see significant rewards in terms of permitting time if they proactively consult with regulatory agencies to develop the baseline data needed for informed decision making. Barrick Gold's recent efforts with respect to the greater sage-grouse may provide a model. In 2015, Barrick, the FWS, the Nature Conservancy, and the BLM agreed to set up a sage-grouse "conservation bank." Barrick will improve sage-grouse habitat on private lands, earning credits that Barrick can later use to offset impacts elsewhere, when it seeks to expand mining operations on federal lands.

> The conservation bank concept commits Barrick and other land users to achieve 'net conservation benefits' for the greater sage-grouse by encouraging greater gains in functional sage-grouse habitat through preservation and restoration than what is lost through development activities. Over time, the application of this concept should result in significant, landscape-scale improvements to habitat conditions throughout the region. Through implementation of this conservation bank, Barrick will obtain assurance that the voluntary compensatory mitigation measures taken by the company, when sufficient to provide a net conservation gain to the species, will be accounted for by BLM and the FWS as the agencies review the company's future proposed mining operations.[51]

The Nature Conservancy played a critical role in brokering this agreement, developing the methodology used to define credits accrual and to determine when credits will be available for use. Of equal importance, The Nature Conservancy stood as a credible intermediary among industry, government, and the environmental community—facilitating dialogue, improving trust, and lending credibility to the effort.

The challenges inherent in reconciling energy development and resource protection are not new. As the preceding example demonstrates, benefits can accrue to broad constituents when resource protection and development are pursued as part of a coordinated agenda. But collaborative successes are unlikely to occur unless the participants receive support from broad constituencies, including local residents and the environmental community. Industry should begin fostering the relationships needed to collaboratively address these multifaceted challenges well in advance of development. Early and effective engagement by industry will be especially important in light of the strong feelings oil shale development can engender.

The threat of piecemeal and haphazard oil shale development impairing valuable resources may be enough to bring diverse stakeholders to the table.

But even the best group will founder unless the process fosters understanding and trust. Efforts to build trust should begin with industry as it seeks to overcome a trust deficit. Trust requires that industry be more forthcoming so that the public understands industry intentions, the scope of likely development, and the trade-offs involved. In the absence of full disclosure and good information, those that oppose oil shale development may use uncertainty to their advantage.

2.3.2 Implications for Government

Federal officials engaged in policy development should weigh carefully the indirect consequences of federal inaction. Significant oil shale resources are available on nonfederal lands, and federal inaction will not prevent oil shale development. Red Leaf Resources is the only oil shale developer within Utah to make significant progress toward commercial operations, and, as of 2015, Red Leaf held active oil shale leases for almost 56,000 acres of SITLA land.

Shifting oil shale development onto nonfederal lands may have two important consequences. First, because a much higher percentage of nonfederal resources are located near the surface, incentivizing nonfederal development may disproportionately drive surface mining and associated technologies. This technological shift may be unavoidable, at least within Utah, because deeper oil shale resources are often located in areas already undergoing significant oil and gas development. Second, indirectly incentivizing nonfederal development reduces federal involvement in energy and environmental policy development. The question for the federal government should be how to engage most effectively.

As it moves forward, the federal government should seek to reduce uncertainty, both for oil shale operators and the public. Clarifying federal energy, environmental, and greenhouse gas policies would provide valuable guidance for prospective oil shale developers. Likewise, engaging in landscape-level planning that promotes coordinated management of multiple resources across jurisdictional boundaries could improve the decision-making climate while addressing the competing mandates that can drive conflict. Rationalizing the landscape through federal–state land exchanges also represents an opportunity to advance mutual interests while reducing uncertainty and conflict surrounding management of a fragmented landscape.

Incentivizing efforts to answer important factual questions about oil shale development is in the best interest of all stakeholders. RD&D leasing provides one tool for answering these questions. Finalizing the commercial oil shale leasing rule will also create certainty for industry and environmental stakeholders by clarifying what is expected of the nascent industry.

Regardless of the leasing mechanism, resource managers must address a unique set of challenges with respect to rare plants, archaeological resources, and paleontological resources. These resources are unique because management invariably focuses on avoidance. Effective avoidance, however, requires detailed knowledge of resource locations, which, throughout much of eastern Utah, is lacking. Absent detailed information about the location and extent of these resources, regulators will have a difficult time determining the context or intensity of potential resource impacts. Further, the breadth of surface disturbance associated with oil shale development will make avoidance more difficult than it would be with conventional oil and gas development, where wells and road locations can be adjusted to avoid sensitive sites. Accordingly, resource surveys should precede leasing, first to prevent state or federal agencies from inadvertently leasing lands that are subject to significant constraints on development, and second, to provide potential lessees with an accurate assessment of constraints on development prior to bidding on an oil shale lease.

Policy makers have a unique and unprecedented opportunity to create policies incentivizing development that advances broad social goals. Because technology is immature, the federal government could link access to federal oil shale resources to development of technologies with lighter environmental footprints. The federal government could also provide price support for shale oil produced using technologies that are capable of producing fuels with significantly reduced impacts to air, water, land, and wildlife.

Incentivizing breakthrough technologies could benefit from the establishment of an environmental benchmark for conventionally produced hydrocarbons and from linking federal support for oil shale development to achievement of a significant reduction in per-unit impact over that benchmark. By utilizing such an approach, the federal government could incentivize the development of technologies that carry a lower social cost, rather than technologies with the lowest internalized economic cost. That could, at least in theory, produce a more desirable outcome—provision of abundant domestic liquid transportation fuels, the development of which involves substantially lower environmental impacts than those that would result from continued reliance on conventional hydrocarbon resources.

Notes

1. Ronald C. Johnson et al., U.S. Geological Survey, Assessment of In-Place Oil Shale Resources in the Eocene Green River Formation, Uinta Basin, Utah and Colorado 1 (2010) *available at* http://pubs.usgs.gov/dds/dds-069/dds-069-bb/REPORTS/69_BB_CH_1.pdf.
2. Based on resources capable of producing at least 25 GPT of shale and less than 3000 ft below the surface. If shales bearing 15 GPT and subject to the same overburden constraints were developed, available resources increase to 292.3 billion barrels. Michael D. Vanden Berg, Utah Geological Survey, Basin-Wide Evaluation of the Uppermost Green River Formation's Oil-Shale Resource, Uinta Basin, Utah and Colorado 7 (2008).

3. *See* Vanden Berg, *supra* note 2, at 6, 10.
4. Bureau of Land Mgmt, U.S. Dep't of the Interior, Approved Land Use Plan Amendments/Record of Decision (ROD) for Allocation of Oil Shale and Tar Sands Resources in Lands Administered by the Bureau of Land Management in Colorado, Utah, and Wyoming and Final Programmatic Environmental Impact Statement 6-7 (2013) (hereinafter, 2013 Record of Decision) *available at* http://ostseis.anl.gov/documents/docs/2012_OSTS_ROD.pdf.
5. U.S. Energy Info. Admin., Analysis of Crude Oil Production in the Arctic National Wildlife Refuge, Report no. SR-OIAF/2008-03 (May 2008) *available at* http://files.geology.utah.gov/online/ss/ss-128/ss-128txt.pdf.
6. Surface ownership may differ from resource ownership, as the surface owner may not own the underlying minerals. Spatial information regarding mineral estate ownership is, however, unavailable, making surface ownership the best available indicator.
7. 43 U.S.C. § 1701(a)(7)) (2012).
8. 2013 Record of Decision, *supra* note 4 at 4–5.
9. Utah Code Ann. § 53C-1-102(2)(b) (2014).
10. Utah Code Ann. § 53C-1-302(1)(b)(iii) (2014).
11. 2013 Record of Decision, *supra* note 4 at 4.
12. 43 C.F.R. §§ 3900—3936.40 (2014).
13. 78 Fed. Reg. 18547 (March 27, 2013).
14. 70 Fed. Reg. at 33753 (June 9, 2005).
15. INTEK, Inc., *Secure Fuels from Domestic Resources: Profiles of Companies Engaged in Domestic Oil Shale and Tar Sands Resource and Technology Development* 22–23 (5th ed. 2011) *available at* http://energy.gov/sites/prod/files/2013/04/f0/SecureFuelsReport2011.pdf.
16. Peter M. Crawford et al., Assessment of Plans and Progress on US Bureau of Land Management Oil Shale RD&D Leases in the United States 20 (2012) *available at* http://energy.gov/sites/prod/files/2013/04/f0/BLM_Final.pdf.
17. 74 Fed. Reg. 567867-69 (November 3, 2009).
18. *See* AuraSource, Inc., *Technology Summary,* http://www.aurasourceinc.com/AuraFuel.htm.
19. 78 Fed. Reg. 18547, 18549 (March 27, 2013).
20. Utah Code Ann. § 63M-4-301(1)(b)(i) (2014).
21. Utah Code Ann. §§ 53C-2-414, 59-5-120, and 59-13-201(3)(a)(iii) (2014).
22. Utah Admin. Code R850-22-500 (2015).
23. Utah Admin. Code R850-22-700(1) (2015).
24. Utah Admin. Code R850-22-700(2) (2015).
25. Utah Admin. Code R850-22-800(1) (2015).
26. Robert B. Keiter et al., Analysis of Environmental, Legal, Socioeconomic and Policy Issues Critical to Development of Commercial Oil Shale Leasing on the Public Lands in Colorado, Utah, and Wyoming under the Mandate of the Energy Policy Act of 2005 A-25 (2009) *available at* http://content.lib.utah.edu/cdm/ref/collection/utlawrev/id/6522.
27. Utah v. United States, 486 F. Supp. 995, 1002 (D. Utah 1979).
28. 43 U.S.C. §§ 1761-70 (2012).
29. 43 U.S.C. § 1765 (a)(ii) (2012).
30. 42 U.S.C. § 4332(2)(C) (2012).

31. *See* Sierra Club v. Hodel, 848 F.2d 1068, 1090-91 (10th Cir. 1988) (holding that BLM actions to insure county road construction proposal did not exceed the scope of its right-of-way through public lands and did not constitute "major federal action," but the BLM's duty to prevent unnecessary degradation of adjoining wilderness study areas elevated situation to one of major federal action).

32. 43 U.S.C. § 1716 (2012).

33. John Ruple and Robert Keiter, *The Future of Federal—State Land Exchanges*, University of Utah College of Law Research Paper No. 84 (2014), *available at*: http://ssrn.com/abstract=2457272 or http://dx.doi.org/10.2139/ssrn.2457272.

34. Bureau of Land Mgmt., U.S. Dep't of the Interior, Record of Decision and Approved Resource Management Plan Amendments for the Great Basin Region, Including the Greater Sage-Grouse Sub-Regions of Idaho and Southwestern Montana, Nevada and Northeastern California, Oregon, Utah, 2015, *available at* http://www.blm.gov/style/medialib/blm/wo/Communications_Directorate/public_affairs/sage-grouse_planning/documents.Par.44118.File.dat/GB%20ROD.pdf.

35. Bureau of Land Mgmt., U.S. Dep't of the Interior, Proposed Land Use Plan Amendments for Allocation of Oil Shale and Tar Sands Resources in Lands Administered by the Bureau of Land Management in Colorado, Utah, and Wyoming and Final Programmatic Environmental Impact Statement, 4–19 (2012) (hereinafter 2012 Final PEIS).

36. Anthony Andrews, Cong. Research Serv., *Developments in Oil Shale* 15-16 (2008).

37. Vanden Berg, *supra* note 2, at 10 (internal references omitted).

38. 43 C.F.R. § 3900.40(a) (2015).

39. 43 C.F.R. § 3900.40(b) (2015).

40. 2012 Final PEIS, *supra* note 34, at 4–18.

41. *See e.g.*, Utah State Mineral Lease for Oil Shale, ML-49953 § 2.2 (Aug. 30, 2005).

42. 16 U.S.C. §§ 1531 through 1544 (2014). For a comprehensive overview of the Endangered Species Act, *see* George Cameron Coggins and Robert L. Glicksman, 3 Pub. Nat. Resources L. § 29 (2nd ed.).

43. 80 Fed. Reg. 59858-942 (Oct. 2, 2015).

44. Bureau of Land Management, U.S. Dep't of the Interior, Utah Greater Sage-Grouse Approved Resource Management Plan Amendment, Attachment 4 (2015), *available at* http://www.blm.gov/style/medialib/blm/ut/natural_resources/SageGrouse/ARMPA_appendices.Par.31778.File.dat/Utah_ARMPA.pdf.

45. For a summary of Indian Country jurisdictional issues within Utah, *See* Heather J. Tanana and John C. Ruple, *Energy Development in Indian Country: Working Within the Realm of Indian Law and Moving Towards Collaboration*, 32 Utah Envtl. L. Rev. 1 (2012).

46. Neil Gunningham, Robert A. Kagan, and Dorothy Thornton, *Social License and Environmental Protection: Why Businesses Go Beyond Compliance*, 29 Law & Social Inquiry 307, 308 (2004).

47. For an in-depth review of the oil shale boom and of "Black Sunday," *see* Andrew Gulliford, Boomtown Blues: Colorado Oil Shale (Rev. Ed. 2003), and Jason L. Hanson and Patty Limerick, Univ. of Colorado Center for the Am. West, What Every Westerner Should Know About Oil Shale: A Guide to Shale Country, 16–17 (2009).

48. Hanson and Limerick, *supra* note 47 at 16–17.
49. James T. Bartis et al., Rand Corp., Oil Shale Development in the United States: Prospects and Policy Issues, 43 (2005).
50. Hanson and Limerick, *supra* note 47 at 27.
51. Press Release, Interior Department, Barrick Gold of North America & The Nature Conservancy Announce Partnership to Protect Sagebrush Habitat, March 26, 2015, *available at* http://www.doi.gov/news/pressreleases/interior-department-barrick-gold-of-north-america-and-the-nature-conservancy-announce-partnership-to-protect-sagebrush-habitat.cfm.

3

Legal and Policy Considerations Involving Water for Oil Shale Development

John C. Ruple

CONTENTS

This chapter summarizes the legal and policy issues impacting water availability for oil shale development within Utah. The discussion begins with a summary of the legal system used to allocate Utah's finite water resources, endangered species issues impacting water availability, and the social license needed to develop water-intensive projects efficiently. With this framework in place, the discussion turns to the demand for water resulting from both oil shale development and other water uses that will compete for the region's scarce water resources. This chapter then discusses potential sources of supply, focusing on physical availability and existing water allocations. Reflecting upon supply and demand, this chapter concludes by addressing the implications for the nascent oil shale industry and the government entities shaping water and energy policy.

3.1 Water Resource Allocation

3.1.1 Law of Water Allocation within Utah[1]

In Utah, water belongs to the public and is available to all for appropriation and beneficial use. The state administers a water right permitting system to ensure that water is used in accordance with the public interest, and water users must obtain state authorization to use water prior to initiating appropriations. Water rights are obtained through application to the Office of the State Engineer. If the application is approved, the applicant is allowed time to develop the proposed water source and to put the water to a beneficial use. When the water is put to a beneficial use, the applicant files proof of development and beneficial use with the State Engineer. Upon verification of development and the amount of beneficial use, the State Engineer issues a Certificate of Appropriation "perfecting" the water right. Each approved application and perfected water right contains provisions governing the source of supply, the point of diversion, the nature of use, the quantity of water appropriated, the rate of diversion/withdrawal, and the season of use.

Each water right also has a priority date that normally coincides with the date upon which the water right application was filed. When demand for water exceeds the available supply, those with senior rights can require full or partial curtailment of junior water users' diversions, leaving junior priority water users with less than their allotted amount of water, or with no water at all. Sharing of water during times of shortage is not required. Therefore, the more senior the water right, the more valuable that water right becomes during times of drought.

Wasteful uses of water are not protected, and appropriators are generally unable to hold water rights for speculative future needs. If a water right is not put to a beneficial use within the time period specified by law, the water right reverts to the state and becomes available to the public for appropriation. Extensions of time to perfect a water right may be available where the applicant exercises diligence in developing water rights, but the development is reasonably delayed. Where multiple extensions have been granted, perfection of long dormant rights could displace rights with junior priority dates that were developed earlier because the priority date associated with a water right generally relates back to the date of the original water right application.

Changes in the use authorized by a water right are allowed subject to the general rule that they do not result in an injury to other water users. Indeed, water rights reflect a property right that may be sold or leased independent of the land upon which the water is used. The Office of the State Engineer, however, must review and approve changes to the point of diversion, place of use, or other elements of a water right before the water right owner can undertake a change.

It follows that when insufficient water is available to satisfy the needs of all prospective users, markets develop and water rights are conveyed to economically more profitable uses. Historically, conversion of agricultural water rights to municipal and industrial rights has facilitated significant western expansion.[2]

3.1.2 Endangered Species Act[3]

State-issued water rights are but part of the water availability picture. While states have primacy with respect to water resource allocation, state laws generally fall to federal law when conflicts arise.[4] The Endangered Species Act (ESA),[5] accordingly, may represent the single greatest legal hurdle impacting water availability. The ESA prohibits the "take" of listed animals, except under federal permit. "Take" is interpreted expansively to include "harm." "Harm," in turn, includes "significant habitat modification or degradation." The take prohibition's reach is broad, applying regardless of land ownership.

If a federal action is likely to adversely affect a listed species, section 7 of the ESA requires federal agencies to formally consult with the U.S. Fish and Wildlife Service. Following consultation, the service issues a "biological opinion" stating whether the activity would jeopardize the species or adversely modify the species' critical habitat. Federal leases, permits, and funding all can trigger section 7 consultation requirements. If the applicant reduces impacts to the maximum extent practicable and the proposed action will not appreciably reduce the likelihood of survival or recovery of a listed animal species, the service issues an "incidental take statement" authorizing the take of a listed species incidental to otherwise lawful activities.

Section 10 of the ESA provides relief to nonfederal landowners who want to develop property inhabited by listed species. Nonfederal landowners can obtain an "incidental take permit" from the U.S. Fish and Wildlife Service provided that the landowner has an approved Habitat Conservation Plan. Habitat Conservation Plans include an assessment of the likely impacts on the species resulting from the activity, the steps that the permit holder must take to minimize and mitigate the impacts, and the funding available to carry out those steps. Habitat Conservation Plan development and approval, however, can take years and involve considerable expense.

Water right utilization is subject to the ESA's prohibition against the "take" of a listed species. The ESA may, therefore, require projects consuming water that would have otherwise been available to a listed species to maximize species protection, effectively subordinating state water rights and federal water delivery contracts to the ESA.

The White and Green rivers are the most likely sources of water for oil shale development within Utah. Figure 3.1 shows the location of these rivers within the Most Geologically Prospective Area (MGPA) for oil shale development in the Uinta Basin; see also Figures 2.1 through 2.3 in Chapter 2 for an overview of the area. The White River, which is approximately 160 miles long, drains approximately 5120 square miles in western Colorado and eastern Utah. The White River flows through an area containing the region's richest oil shale resources and has been described as the "first-choice source of water" for oil shale development.[6] The Green River, while significantly larger, is farther from Utah's oil shale resources.

All of the White River within Utah is designated as critical habitat for one or more ESA-listed fish. The White River is uniquely important to the Colorado pikeminnow (an ESA-listed fish) because of quality in-channel habitat and linkage to other important habitats. Also, "as one of the least altered major tributaries to the Green River, it makes biological, physical, and chemical contributions to the Green that are similar to its historic contributions."[7] In 2011, researchers documented spawning by razorback sucker (also an ESA-listed fish) in the White River for the first time, adding to the river's importance.[8]

Revised instream flow recommendations for the White River are being developed by an interagency advisory committee and will be finalized as part of the White River Management Plan. The Management Plan will identify historic and most likely future depletion scenarios, identify the effects of past and future water development on endangered fish habitat, develop final flow recommendations for the White River, and identify recovery actions needed to offset depletion effects.[9]

Endangered fish also occupy the Green River within Utah, from Dinosaur National Monument to the Green River's confluence with the Colorado River. Instream flow recommendations for the Green River were developed in response to ESA listings and are incorporated into the operating plan for Flaming Gorge Dam.[10] In addition, appropriations from the Green River below Flaming Gorge Dam with a priority date of December 2, 1994, or later are subject to seasonal base flow requirements.[11] Water rights applications and change applications approved after September 21, 2009, are also subject to year-round bypass flow requirements, including spring peak flow requirements intended to mimic spring runoff.[12]

Because the White and Green rivers are the two main surface water sources in proximity to oil shale resources within Utah, management of the instream flows required to protect and restore ESA-listed fish species that inhabit these rivers will directly impact water resource availability, including water for oil shale development.

3.1.3 Social License to Operate

As noted in Chapter 2, a social license to operate is a practical necessity above and beyond both the legal license to use water obtained from the state and

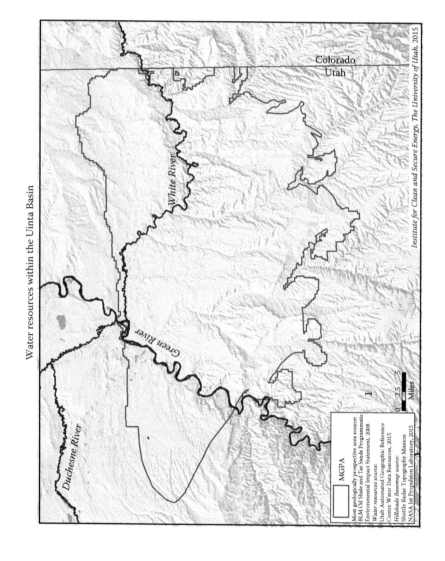

FIGURE 3.1

Water resources within the MGPA for oil shale development in the Uinta Basin.

the permitting requirements contained in the ESA; see Section 2.2.4. Water used to extract, process, refine, and upgrade oil shale is water that will not be available for agricultural, environmental, or municipal and industrial purposes—a significant concern given the arid nature of Utah's Uinta Basin and the region's fully allocated water supplies. Activities that increase the demand for finite existing water supplies will impact every aspect of social and economic activity.

Westerners can be fiercely protective of their water resources and distrustful of those competing to use the same finite resource. Oil shale developers may need to purchase or lease water rights from existing water users; doing so in the face of distrust and public opposition will complicate those efforts. While the prospect of economic opportunity is appealing to many, industry should not assume that economic opportunities would overshadow local or regional concerns over water resources. Many residents of eastern Utah and western Colorado remember "Black Sunday," the collapse of the last oil shale boom that was discussed in Chapter 2, and the devastating social consequences that followed.[13]

Trust is essential for avoiding prolonged litigation and social opposition. Unfortunately, much of the discourse over water for oil shale development does more to fuel conflict than create opportunities or illuminate the choices ahead. The oil shale industry has not been as forthcoming as many would like with respect to its development plans or the resource inputs that would be required if development proceeds, fueling speculation that development threatens the environment in rural communities.

3.2 Water Demand

3.2.1 Oil Shale Industry Demand

Water demand should be considered both in terms of the needs of the nascent oil shale industry and in terms of demands by competing users, as both groups of users will be vying for the same finite supply of water. Quantifying water demand associated with commercial oil shale production is difficult at best as there are presently no commercial-scale oil shale developments in the United States. Based on a review of eighteen prior studies, a mean water consumption estimate of 2.5 gallons per gallon of oil produced was computed.[14] However, as noted in the Rand Report, "[r]eliable estimates of water requirements will not be available until the technology reaches the scale-up and confirmation stage."[15]

Quantification of water requirements depends on consumption at each phase of the production process. For mining and surface retorting, water is needed for dust control during materials extraction, crushing, transport, storage, and disposal; for cooling, reclaiming, and revegetating spent shale; for upgrading raw shale oil into a pumpable oil suitable for refinery feedstock; and for various plant uses including sanitary waste systems and environmental controls such as exhaust gas scrubbing.

In situ retorting eliminates or reduces a number of these water requirements, but considerable volumes of water may be required for oil and synthesis gas extraction, postextraction cooling, product upgrading and refining, environmental control systems, power production, and postproduction site reclamation and revegetation. Use and consumption at each phase depend on the technologies utilized.

This conceptual understanding of the process, however, does little to help quantify future needs. The most frequently cited water use estimates are based on dated technologies and assumptions that have become questionable with the passage of time.[16] More recent estimates are often based on small-scale tests, the results of which may not be scalable. Furthermore, assertions by those touting new low-water technologies have not been vetted in the public arena or at a commercial scale. Additionally, the scope of activities considered in various estimates is often inconsistent or unclear, making evaluation and comparison across estimates problematic.

The uncertainty associated with demand forecasting is compounded by the uncertain rate of industrial growth and the size of a mature domestic oil shale industry. Total water use also depends on whether oil shale displaces or supplements conventional hydrocarbon production. If oil shale development displaces conventional hydrocarbon production, then water use associated with ongoing oil and gas production may offset some or all oil shale development–related demand. Given the compounding nature of these assumptions, water demand estimates should be treated with skepticism.

3.2.2 Water Demand Independent of Oil Shale Development

Water for oil shale is not a question that can be answered in isolation. Utah's Uinta Basin also contains conventional oil and natural gas. Within Utah, the Environmental Impact Statement (EIS) for the Bureau of Land Management's Vernal Field Office Resource Management Plan anticipates that 4345 new natural gas wells will be drilled between 2009 and 2023, most of which will be drilled in the Monument Butte–Red Wash area along the White River. The EIS also anticipates 130 additional coalbed methane wells and 2055 new oil wells over this time period.[17] New wells will be in addition to the 5785 active wells existing within the Uinta Basin as of 2009.[18] In sum, water demand is likely to increase as the total number of oil and gas wells is anticipated to more than double.

Independent of energy production, the Uinta Basin's population was estimated at just over 52,000 persons in 2010. The Governor's Office of Planning and Budget estimates that the basin's 2030 population will increase to just over 63,000 people and reach more than 80,000 in 2060. Approximately 213,000 acres support irrigated agriculture, and, unlike other regions of the state, the acreage under irrigation within the Uinta Basin is expanding. Municipal and industrial water use in the basin totals 25,582 acre-feet (AF) annually and is projected to increase by approximately 15% between 2010 and 2060.[19] These projections do not include water to support oil shale development. "Aside from the concerns addressed in Section 3.3.6, municipal and industrial supplies appear adequate to meet anticipated demand. However, when considered in the aggregate," local shortages may occur and those local shortages could be exacerbated by commercial-scale oil shale development.

Accurately estimating population growth that would result from commercial oil shale development is also challenging. Aggregate population change will depend upon the number and size of facilities, the number of workers required to support various technologies, the ratio of permanent to temporary workers, the indirect employment created by workers flowing into the development area, and whether workers bring their families. How that population increase will be distributed across the landscape depends on facility location as well as the availability of permanent and temporary housing and other services. A 2007 study suggested a total population increase of 43,250 to 51,900 for a 100,000 barrel per day in situ oil shale operation—almost doubling the Uinta Basin's 2010 population. This estimate was developed by using productivity levels from the conventional oil and gas industry as a proxy for oil shale industry needs.[20] Population increases and associated water demand will affect most profoundly those communities proximate to development, such as Vernal and Roosevelt.

3.3 Water Supply

3.3.1 Regional Climate

Within the Uinta Basin, precipitation averages just 15.9 in annually, and average precipitation is skewed by heavy mountain snowfall at the edge of the basin.[21] Furthermore, the Upper Colorado River Basin, which includes the Uinta Basin, is subject to significant fluctuations in precipitation and extended periods of drought that must be considered in planning for future water use. Based on reconstructed Colorado River flow records, the National Academy of Sciences concludes "extended drought episodes are a recurrent and integral feature of the basin's climate," and more pronounced than those observed over the past century. Limited and unreliable precipitation pose a significant challenge for the fledgling oil shale industry. As the Congressional Research Service observed:

> Adjusting the demand for water as supplies shrink during droughts is difficult. Federal, state, and local authorities make water resource

decisions within the context of multiple and often conflicting laws and objectives, competing legal decisions, and entrenched institutional mechanisms, including century-old water rights and long-standing contractual obligations (i.e., long-term water delivery and power contracts).[22]

Climate change will compound drought-related challenges. Mean temperatures in the Colorado River Basin have been increasing steadily and have warmed more than any other region in the United States. Temperature increases cause fall and winter precipitation to occur as rain rather than snow and accelerate spring snow melt. A smaller snowpack and earlier run-off translate into decreased summer and fall streamflow in the Colorado River and its tributaries and to less water availability in summer and fall. Temperature increases also result in a growing season that begins earlier and lasts longer, increasing evapotranspiration and compounding stress on available water resources. Modeling by researchers from NASA, Columbia University, and Cornell University found that:

> [T]he mean state of drought in the late 21st century over the Central Plains and Southwest will likely exceed even the most severe mega-drought periods of the Medieval era in both high and moderate future [GHG] emissions scenarios, representing an unprecedented fundamental climate shift with respect to the last millennium. . . . Our results point to a remarkably drier future that falls far outside the contemporary experience of natural and human systems in Western North America, conditions that may present a substantial challenge to adaptation.[23]

A large commercial oil shale industry would increase pressure on finite water supplies during periods of prolonged drought. If shortages were to persist, junior water right holders, including industry, could see their access to scarce water resources reduced significantly. At a minimum, the cost associated with water acquisition would be sure to increase.

3.3.2 Physical Availability

Utah's oil shale resources are located within the Uinta Basin, which, as already noted, receives scant precipitation. The Green River is the largest river within the basin; the Duchesne and White rivers are the major tributaries to the Green River. The White River is the surface water source closest to the oil shale resources, flowing west from its headwaters in Colorado through the heart of Utah's oil shale country before joining the Green River. Because of its size and proximity, the White River is the most appealing source of water for oil shale development.[24] Notably, the vast majority of Colorado's most geologically prospective oil shale area also drains to the White River, and development upstream in Colorado could impact water availability downstream in Utah.

The Green and Duchesne river systems are also potential sources of supply, though diversions from either system would involve a more extensive system of pipelines and pumping that would increase costs compared to those associated with withdrawals from the White River. Expanding out, the Colorado River, while south of Utah's oil shale resources, is important primarily because withdrawals from its tributaries, including the White and Green rivers, are regulated under two interstate compacts and a complex body of law commonly known as "the Law of the River."

3.3.3 Allocation of the Colorado River and Its Tributaries[25]

The Colorado River Compact apportions surface water among the seven states that drain to the Colorado River. These states are Arizona, California, Colorado, Nevada, New Mexico, Utah, and Wyoming. The compact divides the watershed into upper and lower basins based on whether lands drain to the Colorado River at points above or below Lee Ferry, Arizona.[26] In total, the compact allocates 17,500,000 AF annually—much more water than is physically available during most years.

Under the compact, both the upper and lower basins are entitled to annual consumptive use of up to 7,500,000 AF of water. The lower basin is also "given the right to increase its beneficial consumptive use of such waters by one million acre-feet per annum" if surplus water is available. Mexico is also entitled to 1,500,000 AF annually pursuant to the treaty with Mexico, and this water is provided out of flows surplus to the upper and lower basins' entitlements. When surplus flows are unavailable, the obligation to Mexico is borne by an equal reduction in each basin's apportionment.

The upper basin's 7,500,000 AF entitlement is misleading. In all but the most severe droughts, the upper basin is obligated to deliver an average of 7,500,000 AF to the lower basin, regardless of precipitation. The upper basin's apportionment may therefore be reduced first to satisfy obligations to the lower basin and again to satisfy the upper basin's share of obligations to Mexico.[27] Increasing water use and the prospect of reduced precipitation combined with prolonged drought periods will only exacerbate water allocation challenges.

From 1906 through 2005, the Colorado River flow at Lees Ferry discharged, on average, 15,072,000 AF annually. More recent studies place average annual flow at between 13,000,000 and 14,700,000 AF, with significant year-to-year variability.[28] In light of these river flow estimates, evaporation estimates, and the upper basin states' obligations to the lower basin and to Mexico, the upper basin states are left with an average annual allocation of approximately 6,000,000 AF—possibly much less.

While the amount of water that will be available from the Colorado River in the future is uncertain, allocation of available resources is defined clearly.

The upper basin states' share of the Colorado River is divided according to the Upper Colorado River Compact, and Utah is entitled to 23% of the flow. Applying 23% to the 6,000,000 AF presumably available to the upper basin yields an average annual consumptive right from the Colorado River system of 1,369,000 AF for Utah. After accounting for current depletions, Utah has, on average, 361,500 AF of unused Colorado River apportionments—all of which is currently allocated.[29] Therefore, while Utah appears to have sufficient water rights to proceed with significant oil shale development, that development must rely on reallocation of existing water allocations.

With respect to the White River, neither the Colorado River Compact nor the Upper Colorado River Compact states how much water the State of Colorado must leave in the river for Utah's downstream users, and no other interstate agreement governs the river's apportionment. The lack of formal agreement has not been problematic to date because of limited development within the White River basin, but conflict is possible if demand increases.

3.3.4 Existing Allocations and Potential Sources of Supply

Within Utah, surface waters are fully appropriated throughout the oil shale–bearing region. Groundwater resources are also scarce and generally limited to domestic or temporary supplies, when they are available at all. Unless oil shale developers already hold valid water rights, they must acquire existing rights.

While prospective oil shale developers have sought to secure water rights for oil shale development in numerous ways over the past several decades, Red Leaf Resources, Inc., is the only oil shale developer in Utah that is known to currently hold an approved water right or water right application.[30] Other prospective developers may hold valid water rights or water right applications under the name of a corporate subsidiary that was not identified for this analysis. Prospective oil shale developers may also have secured an adequate source of supply by contracting with an existing water right holder. A prospective producer may, for example, have secured a contractual option to purchase or lease water at a later date, when they are ready to begin production. Alternatively, a prospective producer may have contracted to obtain water from a purveyor such as the Uintah Water Conservancy District. For those that have not secured a source of water, some water may be available from the Uintah Water Conservancy District, which holds a large water right on the White River. This right allows for "mining, drilling, retorting, steam generation, cooling, and related uses."

The White River, while the most convenient source of supply, is not the only option. In 1958, the Bureau of Reclamation filed to appropriate water from the Green River for the Flaming Gorge Dam; this application has been segregated into four separate state water rights, the most significant of which involves 447,500 AF and is held by Utah's Board of Water Resources.

Although the state has conveyed much of this water to other users, water may be available from remaining state water rights or from the subsequently segregated rights.[31]

3.3.5 Produced Water

When oil and gas are developed, water is often withdrawn along with the targeted hydrocarbons. Water produced as a byproduct of oil and gas production represents a potential source of supply for the oil shale industry. Likewise, oil shale development may produce water as a by-product of the retorting process that will need to be disposed of or utilized in the shale oil production process. Either scenario, however, involves unique legal challenges.

Produced water management poses a challenge to western appropriative water law because water withdrawn from the hydrocarbon-bearing formation is a depletion of the source aquifer, but unlike groundwater withdrawals for agricultural or domestic uses, primary production of oil and gas does not consume the water withdrawn. Furthermore, unlike more conventional water uses where excess water can be returned to the source of supply, returning produced water to the source aquifer can impede hydrocarbon production and is therefore counterproductive unless carefully controlled to enhance hydrocarbon recovery.

Water rights were traditionally not required for water produced as a by-product of hydrocarbon extraction because this water was considered a waste product. Increases in produced water generation and concern over impacts to other water resources resulting from produced water generation and disposal have precipitated efforts to revise water law to better address produced water. Produced water generators should be attentive to this changing legal landscape and recognize that produced water minimization may make those permitting processes easier to navigate.

Furthermore, where produced water contains even small amounts of hydrocarbon residue or other pollutants, discharge to surface waters or injection into a drinking water aquifer is likely to trigger complex treatment and disposal regulations. Discharges to surface waters will be regulated under the Clean Water Act, while discharges to groundwater must comply with the Safe Drinking Water Act.

Like oil shale operators, oil and gas operators have a strong incentive to reduce the volume of produced water generated as part of their development operations. Reducing the produced water volume means fewer disposal wells, infiltration galleries, evaporation ponds, associated pumps and pipelines, water quality treatment works, and the Clean Water Act and Safe Drinking Water Act permitting requirements they entail. When produced water generation cannot be avoided, operators have an incentive to use the water produced as a by-product of their operations for their own operational requirements. While a water right and treatment are likely required for reuse, reuse moderates the need to obtain additional scarce and potentially expensive external water sources, moderates disposal costs and permitting requirements, and reduces the amount of water subject to the beneficial use requirements discussed in Section 3.1.1. Recognizing that elimination of by-product water generation is unlikely and that reuse will not completely eliminate disposal requirements, operators should look for disposal options that avoid waste and that protect the quality of other water resources. Where produced water quality is compromised, operators will need to obtain state or federal discharge permits.

Efforts by oil and gas operators to reduce produced water generation and to increase produced water reuse may limit produced water as a source of supply for prospective oil shale developers. However, even with the most aggressive reduction and reuse programs, some level of excess produced water generation is likely. The oil shale industry may benefit, as it would represent a market for a product that is of limited use to conventional oil and gas operators and which would otherwise represent a disposal challenge. Availability to third-party users will likely require a water right change authorization or a water right if one was not required for formation dewatering.

Produced water generators, prospective produced water users, and government regulators alike must act proactively and be flexible in adapting to site-specific issues and constraints, a rapidly evolving legal framework, and a resource that may change over time. Foreseeable increases in energy production will likely drive more stringent disposal and appropriations requirements and all involved must be flexible and proactive in responding to these challenges.

3.3.6 Indian Reserved Water Rights[32]

American Indians are in a uniquely powerful position to affect water resource development and use, including water for oil shale production. Establishment of federal Indian reservations impliedly reserves to the Indians inhabiting those reservations the water needed to meet the needs of the reservation, even if water rights are not expressly discussed or quantified in the treaty or executive order creating the reservation. The priority date associated with Indian reserved rights is generally the date upon which the reservation was created. Unlike water rights granted under state law, reserved rights cannot be lost due to nonuse.

With regard to potential oil shale development within Utah, the Uintah and Ouray Indian Reservation, which is home to the Ute Tribe of Indians, is important because of its proximity to oil shale resources and its expansive size; see Figure 2.3. As noted in Chapter 2, the Ute Tribe of Indians directly control over 168,500 acres of land overlying oil shale resources that are estimated at 11.7 billion barrels of oil equivalent. The Utes could also, theoretically, support off-reservation oil shale development.

Negotiations during the 1980s and 1990s resulted in the Ute Indian Rights Settlement, which was contained in the federal government's Reclamation Projects Authorization and Adjustment Act of 1992. Believing that an agreement was at hand, the State of Utah codified the Ute Indian Water Compact into state law, subject to ratification by the parties.[33] The Utes, however, did not ratify the compact.[34]

The stumbling blocks to ratification have thus far involved water right administration rather than the quantity, seniority, or potential use of the Utes' water. The compact, therefore, is a reasonable starting point for discussing the tribe's rights. Under the compact, the Utes would obtain the right to consume up to 248,943 AF annually, including the right to consume up to 32,880 AF annually from the White River and its tributaries. The remaining water would be taken from the Duchesne and Green river systems. The priority date for these rights would be 1861 or 1882, except when water is supplied from storage in the Central Utah Project.

Under the compact, the Utes' water "shall not be restricted to any particular use but may be used for any purpose selected by the Tribe," including "sale, lease, or any other use whatsoever."[35] Furthermore, the Ute's water rights would be senior to all but a handful of water rights within the basin and therefore not subject to call during times of shortage.

The tribe is therefore in a unique position to provide water for commercial oil shale development, either on or off their reservation, should they so choose. Conversely, a decision to utilize water rights for other purposes would make water for oil shale development harder to obtain. Regardless of how the Ute's water rights are put to use, these water rights could displace many junior water rights and upset what has long been a fairly stable allocation of resources. Accordingly, resolution of tribal reserved rights and clarification of water development plans should be a high government priority, and prospective oil shale developers should pay close attention to these efforts.

> *The discussion of the Ute's water rights is not meant to imply that the Tribe should or should not support oil shale development. The discussion is included only to highlight the significance of the Tribe's water right holdings and the transformative effect Tribal decisions about water development could have on all water users within the region.*

3.4 Implications

3.4.1 Implications for Industry

Water demand for oil shale development involves significant uncertainty. This uncertainty plays out against a backdrop of water scarcity, finite and fully allocated water resources, and the "Black Sunday" oil shale bust. Consequently, the prospect of oil shale development is of great concern to many members of the public. Industry can respond to these concerns by being more forthcoming and transparent about the pace and scale of development anticipated as well as the water needs associated with various technologies. Increased transparency should include more details on when, where, how, and how much water would be used as well as whether estimates reflect an increase in domestic and municipal use attributed to industrial growth and water needed for site reclamation and revegetation.

In discussing water use, the nascent oil shale industry could benefit by putting its needs into context, such as comparing the water required to produce liquid transportation fuels from oil shale to water used to produce comparable fuels from more traditional hydrocarbon resources.[36] Demonstrating that producing liquid transportation fuels from oil shale requires less water than producing comparable fuels from conventional hydrocarbons may make oil shale development more appealing. Water use concerns also will take a very different form depending on whether oil shale development displaces more water-intensive oil and gas development or represents demands in addition to those associated with ongoing production.

Unless they already hold valid water rights—and those rights are senior enough to provide reliable supplies during shortages associated with normal fluctuations in precipitation, climate change, and population growth—operators will need to acquire rights to existing sources of supply. Some water is likely available from the state, which holds extensive water rights on the White River and Green River systems. Agricultural users may also be willing to sell or lease their supplies. Additionally, if the Ute Tribe of Indian's reserved rights claims are resolved, and the tribe decides to support additional energy development, the tribe could potentially provide access to water resources. Oil shale developers may also gain by forging cooperative relationships with conventional energy producers and by exploring their ability to utilize water generated as a by-product of conventional oil and gas production.

Regardless of the supplier, prospective oil shale developers will benefit from early and open dialogue with these entities and the trust such dialogue could help establish. As plans proceed, operators will also need to engage with regulators, as early engagement will be needed to identify permitting requirements and to navigate complex regulatory processes, including ESA compliance.

3.4.2 Implications for Government

Policy makers should avoid expending significant efforts on predicting the amount of water needed for oil shale development. Such predictions depend on untested assumptions regarding the mix of technologies deployed, the water use requirements associated with each of those technologies, as well as both the scale and pace of development. These assumptions are difficult to support given the nascent nature of the industry. Furthermore, uncertainty compounds as these assumptions are linked together. Finally, the onus of obtaining necessary water resources is on industry, and existing legal structures provide sufficient mechanisms for prospective developers to lease or purchase water. Dedicating substantial government resources to demand quantification, therefore, will provide limited benefit and divert scarce resources away from water resource–related planning and management efforts that are likely to provide greater benefits.

Resources expended on quantifying demand would be better spent on efforts to reduce systemic uncertainty regarding water resource allocation and authorized uses. Such efforts would benefit not just prospective oil shale users, but all who rely on scarce water resources and who must allocate resources and make decisions in an atmosphere of uncertainty. Interstate apportionment of the White River would go a long way toward resolving water availability in downstream Utah. Resolving this uncertainty would benefit not just prospective oil shale developers, but all water users within the White River basin. Interstate compacts provide a valuable tool to resolving allocation dilemmas and represent an attractive alternative to costly and protracted litigation. Given the time and effort required to finalize interstate agreements, it makes sense to begin negotiations now, before a development boom occurs.

Settlement of the Ute Indians' reserved right claims would also help clarify the relative value of competing water rights. Clarity regarding the nature and extent of competing water rights will facilitate the market reallocation of water rights that will inevitably accompany commercial oil shale development. While prior efforts to resolve both interstate apportionment of the White River and the Utes' reserved water right claims have been unsuccessful, energy production may provide the spark needed to move forward. Importantly, resolving each of these challenges provides utility that extends well beyond the oil shale industry.

Federal and state policies should provide certainty with respect to water resource availability, water resource allocation, and the trade-offs involved in oil shale development. Clarifying water allocations, determining the weight given to competing resource values, and articulating clear national energy and environmental strategies would provide the sideboards needed to refine assessments of both the water needed by, and available for, oil shale development.

Policy makers must also recognize the limits inherent in current information about the nascent oil shale industry, the resources that industry will require, and the effluent streams it will produce. A commercial oil shale industry will take decades to develop, if it develops at all. Decisions made today should focus on resolving uncertainty and maintaining flexibility to adapt to the changes ahead. Policies should drive development of an industry consistent with carefully articulated national energy and environmental objectives, including water resource management, emphasizing transparency and innovation while avoiding irretrievable resource commitments.

Because oil shale development is yet to occur at a commercial scale, policy makers have a unique and potentially unprecedented opportunity to develop policies incentivizing development that advances broad social goals. As noted in Chapter 2, carefully developed policy initiatives could incentivize production of domestically produced fuels with lower water consumption and lighter environmental footprints—fuels that represent a significant improvement over the status quo.

Notes

1. Utah's water allocation system is summarized in Steven E. Clyde, *Utah Water and Water Rights, in* 6 Water and Water Rights (Amy K. Kelley, ed., 3rd ed. 2015).
2. *See generally,* Mark Reisner, Cadillac Desert: The American West and Its Disappearing Water (Rev. ed. 1993) (discussing water development in California), and Patricia Nelson Limerick and Jason L. Hanson, A Ditch in Time: The City, the West, and Water (2012) (discussing water development in Colorado).
3. For a comprehensive overview of the Endangered Species Act, *see* George Cameron Coggins and Robert L. Glicksman, 3 Pub. Nat. Resources L. § 29 (2nd ed.).
4. U.S. Const. art VI, § 2.
5. 16 U.S.C. §§ 1531-1544 (2012).
6. Utah Energy Office, Utah Dep't of Natural Res. and Energy, An Assessment of Oil Shale and Tar Sands Development in the State of Utah, Phase II: Policy Analysis 27 (1982).
7. Leo D. Lentsch et al., Utah Div. of Wildlife Res., The White River and Endangered Fish Recovery: A Hydrological, Physical and Biological Synopsis 37 (2000).
8. U.S. Fish and Wildlife Service, Dep't of the Interior, *Final 2013-2014 Assessment of "Sufficient Progress" under the Upper Colorado River Endangered Fish Recovery Program in the Upper Colorado River Basin, and of Implementation of Action Items in the January 10, 2005, "Final Programmatic Biological Opinion on the Management Plan for Endangered Fishes in the Yampa River Basin"* 13 (Sept. 20, 2013) *available at* http://www.coloradoriverrecovery.org/documents-publications/section-7-consultation/sufficientprogress/2013SufficientProgressMemo.pdf.
9. *Id.* at 22.
10. Bureau of Reclamation, Dep't of the Interior, Record of Decision, Operation of Flaming Gorge Dam Final Environmental Impact Statement (2006).
11. Div. of Water Rights, Utah Dep't of Natural Res., *Policy Regarding Applications to Appropriate Water and Change Applications Which Divert Water From the Green River Between Flaming Gorge Dam, Downstream to the Duchesne River* (Nov. 30, 1994) *available at* http://waterrights.utah.gov/wrinfo/policy/topics/green.pdf.

12. Div. of Water Rights, Utah Dep't of Natural Res., *2009 Proposed Water Rights Policy Regarding Applications to Appropriate Water and Change Applications Which Divert Water From the Green River Between Flaming Gorge Dam and the Duchesne River* (Aug. 20, 2009) (*hereinafter* 2009 Green River Policy) *available* at http://www.waterrights.utah.gov/meetinfo/m20090820/policy-upcorviMC09L.pdf.

13. *See* Andrew Guilliford, Boomtown Blues: Colorado Oil Shale (2003 rev. ed.).

14. John Ruple and Robert Keiter, Inst. for Clean and Secure Energy, Topical Report: Policy Analysis of Water Availability and Use Issues For Domestic Oil Shale and Oil Sands Development, [DOE Award No.: DE-FE0001243] 6 (2010) *available at* http://papers.ssrn.com/sol3/papers.cfm?abstract_id=2483318.

15. Bartis et al., Rand Corp., Oil Shale Development in the United States: Prospects and Policy Issues 20 (2005).

16. Ruple and Keiter, *supra* note 14 at n.5.

17. Bureau of Land Mgmt., U.S. Dep't of the Interior, Vernal Field Office Proposed Resource Mgmt. Plan and Final Environmental Impact Statement 3-36 (2008) *available at* http://www.blm.gov/ut/st/en/fo/vernal/planning/rmp/proposed_rmp_eis/proposed_rmp_eis_document.html.

18. Div. of Oil, Gas, and Mining, Utah Dep't of Natural Res., Summary Production Report by County (2009) *available at* https://fs.ogm.utah.gov/pub/Oil&Gas/Publications/Reports/Prod/County/Cty_Oct_2009.pdf.

19. Div. of Water Res., Utah Dep't of Natural Res., Uintah Basin Planning for the Future ch. 3 (Feb. 2015) *available at* http://www.water.utah.gov/Planning/SWP/Unitah/UintahBasin2015.pdf. An acre-foot of water is the amount of water needed to cover one acre of land with twelve inches of water, or 325,851 gallons.

20. Inst. for Clean and Secure Energy, Univ. of Utah, A Technical, Economic, and Legal Assessment of North American Oil Shale, Oil Sands, and Heavy Oil Resources 6.31 (2007) *available at* http://content.lib.utah.edu/cdm/ref/collection/ir-eua/id/2778.

21. Div. of Water Res., Utah Dep't of Natural Res., Uintah Basin 2012 Inventory 3 (2013).

22. Peter Folger et al., Cong. Research Serv., Drought in the United States: Causes and Issues for Congress 12 (2009).

23. Benjamin I. Cook et al., *Unprecedented 21st Century Drought Risk in the American Southwest and Central Plains*, Science Advances 6 (Feb. 12, 2015).

24. Utah Energy Office, Utah Dep't of Natural Res. and Energy, An Assessment of Oil Shale and Tar Sands Development in the State of Utah, Phase II: Policy Analysis 27 (1982).

25. For a comprehensive overview of the law pertaining to the Colorado River System, *see* Lawrence J. MacDonnell, *Colorado River Basin, in* 6 Water and Water Rights (Amy K. Kelley, ed., 3rd ed. 2015).

26. "Lee Ferry" and "Lee's Ferry" are distinct locations along the Colorado river. Lee Ferry is the hydrologic divide between the upper and lower basins under the Colorado River Compact, and is used as the measurement point for the allocation between the two basins. Lee's Ferry, about a mile upstream of Lee Ferry, is the location of the U.S. Geological Survey's stream gauge. The Paria River enters the Colorado River between Lee's Ferry and Lee Ferry, and its gauged flow is added to the Lee's Ferry gauged flow to measure the upper basin's total delivery to the lower basin.

27. The upper basin states' delivery obligations can be reduced to as little as 7,0000,000 AF annually if Lake Powell's storage capacity falls below 5,900,000 AF (24% of capacity). U.S. Dep't of Interior, Record of Decision, Colorado River Interim Guidelines for Lower Basin Shortages and the Coordinated Operations for Lake Powell and Lake Mead 50 (Dec. 2007).
28. Committee on the Scientific Bases of Colorado River Basin Water Management, National Research Council, Colorado River Basin Water Management: Evaluating and Adjusting to Hydroclimatic Variability 104 (2007).
29. 2009 Green River Policy, *supra* note 12.
30. Water right 49-2330.
31. Uintah Basin Planning for the Future, *supra* note 19 at 68.
32. For a comprehensive overview of Indian reserved water rights, *see*, 4 Water and Water Rights § 37.03 (Amy K. Kelley, ed., 3rd ed. 2015).
33. Utah Code Ann. §§ 73-21-1 and -2 (2014).
34. *See* Daniel McCool, Native Waters: Contemporary Indian Water Settlements and the Second Treaty Era 177-82 (2002). For a detailed discussion of prior efforts to resolve the Ute's water rights, including an extensive discussion of concerns over state administration, *see* John Shurts, Indian Reserved Water Rights: The Winters DOCTRINE IN ITS SOCIAL AND LEGAL CONTEXT, 1880S-1930S (2000).
35. Utah Code Ann. § 73-21-2 (2014).
36. *See e.g.,* May Wu et al., Argonne National Laboratory, *Consumptive Water Use in the Production of Ethanol and Petroleum Gasoline* (2009) (comparing water requirements for liquid fuel production across multiple sources, including oil shale).

4

Evaluation of the Upper Green River Formation's Oil Shale Resource in the Uinta Basin, Utah

Michael D. Vanden Berg and Lauren P. Birgenheier

CONTENTS

The upper Green River Formation (GRF) of the Uinta Basin in Utah contains world-class source rock and hosts a prolific, thermally immature oil shale resource. As part of the larger oil shale research mission of the Institute for Clean and Secure Energy (ICSE) at the University of Utah, this geologic study characterized and mapped the thickness and richness of each individual oil shale horizon, from the R-4 zone up to the lower R-8. To compensate for the lack of core-based Fischer assay oil yield data in the study area, bulk density or sonic geophysical logs from oil and gas wells were used to estimate oil shale grade. In addition to oil shale thickness and richness, stratigraphic changes in inorganic mineralogy of individual oil shale horizons from the Skyline 16 research core were explored using x-ray fluorescence (XRF) and Quantitative Evaluation of Minerals by Scanning Electron Microscopy (QEMSCAN) analysis.

The oil shale zone with the most potential for development is the Mahogany zone (R-7), which contains oil shale grades up to 30 gallons of oil per short ton of rock (GPT) within an interval that is up to 140 ft thick. Since the organic-rich Mahogany zone outcrops on the east and southeast portions of the basin, surface/underground mining coupled with surface retorting, as well as surface-capsule retorting, are the only production methods currently being pursued in the Uinta Basin. Using an economic cutoff grade of 15 GPT, the only zones other than the R-7 containing grades that exceed this cutoff are the R-4 and the upper R-6 zones then only in small isolated areas. The R-5 and lower R-8 zones, due to a variety of geologic factors, fail to meet the minimum economic cutoff grade of 15 GPT. Rich (R) zones below the Mahogany, such as the upper R-6, contain more clay minerals and dolomite as compared to the Mahogany zone and the R-8. Mineralogical differences between oil shale zones should be considered in regard to potential impacts on retorting processes and products.

4.1 Introduction

In the 1960s, the U.S. Department of the Interior started an aggressive program to describe and estimate the GRF oil shale resource, kicking off the first major oil shale boom in the United States. The dramatic increase in petroleum prices resulting from the Organization of the Petroleum Exporting Countries oil embargo of 1973 triggered a second resurgence of oil shale research during the mid-1970s and early 1980s. When oil prices plummeted in the mid-1980s, so did research associated with oil shale. As crude oil prices again rose to new heights in the late 2000s, interest in truly unconventional fuel sources, such as oil shale, was once again renewed.

The largest known oil shale deposits in the world are in the Eocene GRF, which covers portions of Utah, Colorado, and Wyoming (Figure 4.1). The Uinta Basin alone is estimated to hold a resource of 1.32 trillion barrels of in-place oil shale (USGS 2010), with approximately 77 billion barrels as a potential economic resource (Vanden Berg 2008). Lacustrine sediments of the GRF were deposited in two large lakes that occupied a 25,000-square-mile area in the Piceance, Uinta, Green River, and Washakie sedimentary basins. Fluctuations in stream inflow caused large expansions and contractions of the lakes, as evidenced by widespread intertonguing of marly, open lacustrine strata with beds of land-derived sandstone and siltstone. During arid times, the lakes contracted in size and the water became increasingly saline and alkaline (Dyni 2003). The warm alkaline waters provided excellent conditions for the abundant growth of cyanobacteria

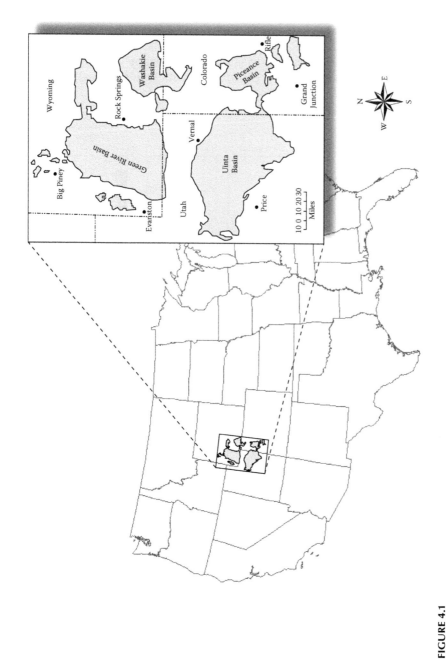

FIGURE 4.1

Oil shale resource areas of Utah, Colorado, and Wyoming. (Adapted from Bunger, J.W. et al., *Oil Gas J.*, 102(30), 16, 2004.)

(blue-green algae), which are thought to be the major source of the organic matter in the oil shale (Dyni 2003). The organic matter preserved in the shale is called kerogen, which when heated can be converted to crude oil and natural gas.

Figure 4.2 shows a stratigraphic section of the Parachute Creek Member of the upper GRF in the Uinta Basin, Utah, based on data from the Coyote Wash 1 core hole (section 22, T. 9 S., R. 23 E., Salt Lake Base Line and Meridian). The interval with the richest oil shale is named the Mahogany zone (R-7 in the nomenclature of Donnell and Blair (1970) and Cashion and Donnell (1972)),

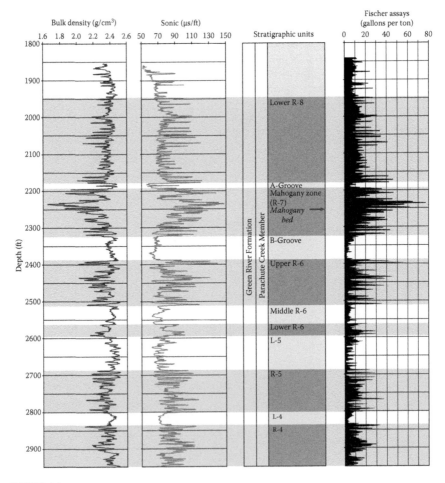

FIGURE 4.2

Stratigraphy of the Parachute Creek Member of the upper GRF illustrated by bulk density, sonic, and oil yield plots from the Coyote Wash 1 core hole. "R" refers to a rich oil shale zone and "L" refers to a lean oil shale zone; stratigraphic nomenclature for oil shale zones is derived from Donnell and Blair (1970) and Cashion and Donnell (1972).

where individual beds, such as the Mahogany bed, can contain more than 70 GPT and the entire zone is commonly over 100 ft thick.

The entire length of the Mahogany zone outcrop has been mapped at the 1:100,000 and/or 1:24,000 scale. This outcrop defines the southern and eastern boundary of the study area. The southeastern extent of the outcrop was digitized from fourteen 7.5 minute geologic quadrangles, and the remaining sections of outcrop were digitized from three 30 ft × 60 ft geologic quadrangle maps. The fourteen 7.5 minute quadrangles are Agency Draw NE (Pipiringos 1979), Agency Draw NW (Cashion 1984), Bates Knolls (Pipiringos 1978), Burnt Timber Canyon (Keighin 1977a), Cooper Canyon (Keighin 1977b), Davis Canyon (Pantea 1987), Dragon (Scott and Pantea 1985), Flat Rock Mesa (Pantea and Scott 1986), Nutters Hole (Cashion 1994), Rainbow (Keighin 1977c), Southam Canyon (Cashion 1974), Walsh Knolls (Cashion 1978), Weaver Ridge (Cashion 1977), and Wolf Point (Scott and Pantea 1986). The 30 ft × 60 ft quadrangles are the Huntington (Witkind 1988), Price (Weiss et al. 1990), and Westwater (Gualtieri 1988).

Roughly 180 oil shale exploration wells were drilled between 1954 and 1983. These wells were mostly located in central to southern Uintah County, typically near the well-mapped outcrop of the Mahogany zone. A few wells, drilled farther west and north, reached the Mahogany zone at more than 2000 ft below the surface. Their cores were analyzed for oil shale grade using the modified Fischer assay technique, as described by Stanfield and Frost (1949) and later adopted by the American Society for Testing and Materials (1980). This method was developed primarily for evaluating the Green River oil shale resources. Generally, the assays were made on crushed samples prepared from 1 to 2 ft lengths of quartered core. A complete database of Fischer assays for wells from the state of Utah can be found in Vanden Berg et al. (2006).

Fischer assays performed on rotary cuttings from oil and gas wells averaged over 10 ft intervals are also cataloged in Vanden Berg et al. (2006). However, these data are less reliable than core data due to contamination by mixing of cuttings, to borehole cave-ins, and to depth errors, which result when the samples are inaccurately lagged for travel time up the borehole. Because these data are generally unreliable and typically underestimate the resource quality, assays from rotary cuttings were not used in this study.

In contrast to the study described in this chapter, Vanden Berg (2008) evaluated the Uinta Basin's oil shale resource by looking at the thickness of continuous intervals of oil shale averaging 15, 25, 35, and 50 GPT, regardless of oil shale horizon. These intervals were determined for 186 oil and gas wells with calculated assay data, as well as 107 oil-shale-specific wells with assays derived from the core, for a total of 293 wells. These continuous zones were calculated starting at the Mahogany bed, adding assay values above or below until the desired average oil yield cutoff was reached.

As calculated in Vanden Berg (2008), a continuous interval of oil shale averaging 50 GPT contains an in-place oil resource of 31 billion barrels in a zone ranging up to 20 ft thick. An interval averaging 35 GPT, with a maximum thickness of 55 ft, contains an in-place oil resource of 76 billion barrels. The 25 GPT zone and the 15 GPT zone contain unconstrained resources of 147 billion barrels and 292 billion barrels, respectively. The maximum thickness of 25 GPT rock is about 130 ft, whereas the maximum thickness of 15 GPT rock is about 500 ft.

After placing several constraints on Utah's total in-place oil-shale resource, Vanden Berg (2008) determined that approximately 77 billion barrels of oil could be considered as a potential economic resource. This estimate is for deposits that are at least 25 GPT, at least 5 ft thick, under less than 3000 ft of cover, not in conflict with current conventional oil and gas resources, and located only on Bureau of Land Management (BLM), state, private, and tribal lands.

In order to better understand Utah's oil shale resource, ICSE and the Utah Geological Survey teamed up to drill 1000 ft of 4 in. diameter core in the upper GRF oil shale deposits in the eastern Uinta Basin, Utah. The purpose was to recover nearly the entire oil shale zone (Parachute Creek Member), providing "fresh" samples for a variety of geochemical and geomechanical tests.

The well was spudded on May 18, 2010, in the upper GRF, very near the Uinta/GRF contact (in fact, the boundary could be seen on the hillside next to the drill rig). Over the course of seven days, 986 ft of continuous core was recovered, starting at 20 ft below surface, down to a total depth of 1006 ft.

During drilling, special care was taken to preserve the core for future testing. Starting with samples taken at 260 ft and continuing down to about 700 ft, the core was stored in thick plastic sleeves and sealed with duct tape to help preserve the core's moisture. In addition, 12 1 ft sections (1 from the A-groove, 8 from the Mahogany Zone, 1 from the B-groove, and 2 from the upper R-6) of the core were wrapped in plastic wrap and sealed in ProtecCore, a special aluminum sleeve designed to preserve the core in an in situ state. Three of these samples (GR1, Mahogany zone "rich," 461.9–462.9 ft, ~60 GPT; GR2, Mahogany zone "lean," 485.9–486.9 ft, ~28 GPT; and GR3, upper R-6, 548.2–549.2 ft, ~22 GPT) were used for extensive testing by other researchers involved with ICSE (Table 4.1). The core was trucked to Salt Lake City and is stored at the Utah Geological Survey's Core Research Center. Later, the core was slabbed with a one-third to two-thirds cut; the one-third slab was placed in display boxes and archived for future research and viewing, while the two-thirds section was placed back in the protective sleeves and reserved for future sampling/testing.

Figure 4.3 displays a detailed lithologic log of the Skyline 16 core plotted next to geophysical logs. Fischer assay analyses were not run on this core, but the bulk density log can be used to estimate oil yield (oil yield is inversely proportional to bulk density). The core is also separated into the

TABLE 4.1

Summary of Semiquantitative XRF and QEMSCAN Analysis of Oil Shale Samples from the Skyline 16 Core

Sample	Depth	Oil Shale Zone	Estimated Oil Yield and TOC	XRF (Homogenized 1 ft Interval)		QEMSCAN Mineralogy
GR1	461.9–462.9 ft	Mahogany zone (rich)	~60 GPT ~23% TOC	Na_2O = 2.2% MgO = 5.0% Al_2O_3 = 5.8% SiO_2 = 31.2% P_2O_5 = 0.8% SO_3 = 2.4%	K_2O = 3.3% CaO = 39.1% TiO_2 = 0.9% MnO = 0.1% Fe_2O_3 = 8.8% SrO = 0.4%	Dolomite = 15.4% Calcite = 9.3% Illite ("clay" and "feldspar") = 31.4% Quartz = 16.7% Pyrite = 1.3% Other, incl. org. C = 26.0%
	QEMSCAN 462.75 ft					
GR2	485.9–486.9 ft	Mahogany zone (lean)	~28 GPT ~11% TOC	Na_2O = 1.7% MgO = 5.4% Al_2O_3 = 3.9% SiO_2 = 21.9% P_2O_5 = 0.3% SO_3 = 0.6%	K_2O = 1.8% CaO = 59.1% TiO_2 = 0.4% MnO = 0.1% Fe_2O_3 = 4.3% SrO = 0.6%	Dolomite = 25.5% Calcite = 56.6% Illite ("clay" and "feldspar") = 4.3% Quartz = 5.9% Pyrite = 0.0% Other, incl. org. C = 9.7%
	QEMSCAN 486.1 ft					
GR3	548.2–549.2 ft	Upper R-6	~22 GPT ~9% TOC	Na_2O = 0.8% MgO = 5.6% Al_2O_3 = 7.3% SiO_2 = 32.0% P_2O_5 = 0.3% SO_3 = 1.8%	K_2O = 8.6% CaO = 34.7% TiO_2 = 0.8% MnO = 0.1% Fe_2O_3 = 7.5% SrO = 0.4%	Dolomite = 33.1% Calcite = 7.1% Illite ("clay" and "feldspar") = 33.4% Quartz = 4.1% Pyrite = 0.8% Other, incl. org. C = 21.5%
	QEMSCAN 548.6 ft					

Notes: TOC is total organic carbon. XRF values are given in weight percent and QEMSCAN in area percent.

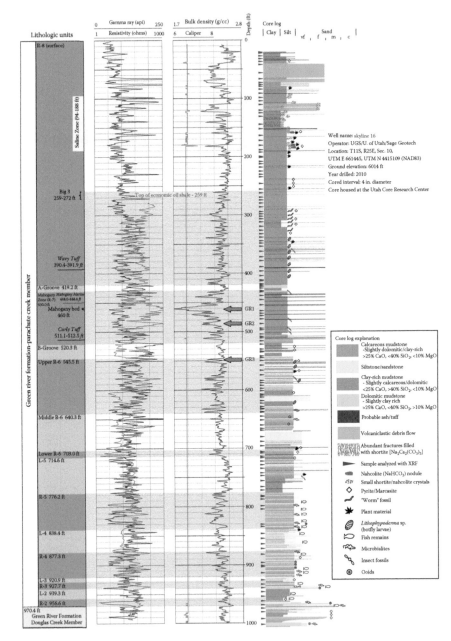

FIGURE 4.3
Detailed lithologic log for the Skyline 16 research core plotted with geophysical logs.

aforementioned rich and lean zones and the bottom of the core captures about 30 ft of the Douglas Creek Member.

4.2 Methods

Stratigraphic tops for 10 rich and lean oil shale zones were picked using a combination of Fischer assay data and geophysical logs. The sequence of horizons consists of the R-4 oil shale zone at the base and, in ascending order, the L-4, R-5, L-5, lower/middle R-6, upper R-6, B-Groove, Mahogany (R-7), A-Groove, and lower R-8 zones (Figure 4.2). The R-6 zone was subdivided to highlight the organic richness of the upper R-6 compared to the leaner lower/middle R-6. The top of the lower R-8 zone is defined in this study by the "Big 3" oil shale beds described by Donnell (1997). From these tops, stratigraphic thicknesses were determined for each zone and isopach maps were generated (Figures 4.4 and 4.5).

Included on the isopach maps are overburden contours at 1000 ft intervals. A structure contour map was generated in ArcGIS™ displaying the surface of each oil shale zone in feet above sea level. This structure contour map was then subtracted from a digital elevation model of the Uinta Basin providing accurate overburden thickness contours. A few estimated data points were added in areas having little or no data as a means to provide more geologically accurate overburden contours, particularly near the outcrop. Overburden thickness equals zero at the outcrop in the southern and eastern portions of the basin and gradually increases in thickness, up to 9000 ft, to the north.

After stratigraphic oil shale zone tops had been picked, each zone was assessed for variations in oil shale grade, in units of GPT. Oil grade was calculated using available core-based Fischer assay data or estimated from bulk density or sonic logs (Bardsley and Algermissen 1963, Tixier and Curtis 1967, Smith et al. 1968). The process for estimating oil shale grade from geophysical logs was the same as that used in Vanden Berg (2008).

Using ArcGIS software, richness maps for each oil shale zone were plotted using a spline fit algorithm with tension. In some cases, individual grade values were edited to remove spurious "bulls-eyes" from the maps. The extent of each map was constrained to only include the area with sufficient subsurface data for a certain zone. The richness maps were subdivided into 3 GPT intervals, with each interval having the same color across all maps for easy comparison (Figures 4.6 and 4.7). Grades below the economic cutoff of 15 GPT are in shades of gray, whereas grades above this cutoff are in shades of red (U.S. BLM 2008).

Nondestructive qualitative XRF analysis was performed on several selected intervals from the Skyline 16 research core to help determine inorganic mineralogy, which is difficult to detect based on visual core inspection alone;

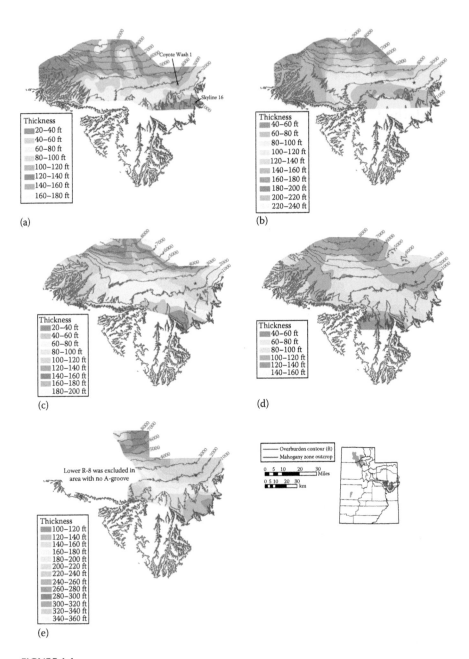

FIGURE 4.4
Isopach maps for each organic-rich oil shale interval within the upper GRF. The location of the Coyote Wash 1 core (Figure 4.2) and the Skyline 16 research core (Figure 4.3) are marked with red stars. (a) R-4, (b) R-5, (c) upper R-6, (d) Mahogany zone, and (e) lower R-8.

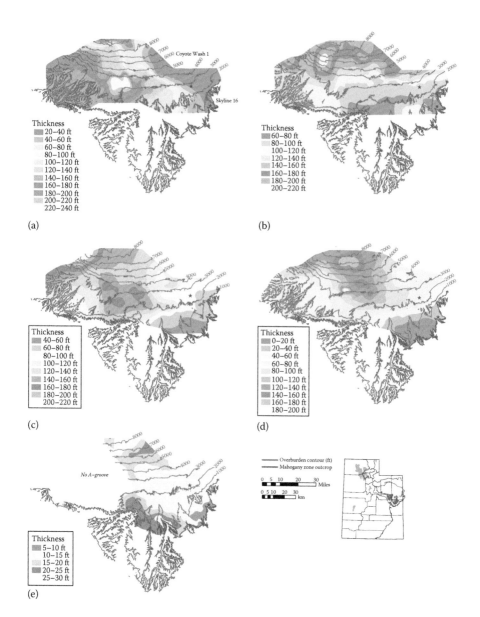

FIGURE 4.5

Isopach maps for each organic-lean oil shale interval within the upper GRF. The locations of the Coyote Wash 1 core (Figure 4.2) and the Skyline 16 research core (Figure 4.3) are marked with red stars. (a) L-4, (b) L-5, (c) lower–middle R-6, (d) B-groove, and (e) A-groove.

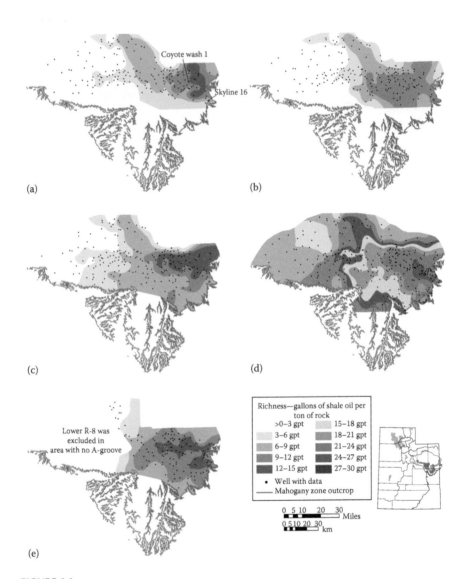

FIGURE 4.6
Richness maps for each organic-rich oil shale interval within the upper GRF. The locations of the Coyote Wash 1 core (Figure 4.2) and the Skyline 16 research core (Figure 4.3) are marked with red stars. (a) R-4, (b) R-5, (c) upper R-6, (d) Mahogany zone, and (e) lower R-8.

see Table 4.1. XRF analysis was performed using a Rigaku miniZSX XRF-WD machine on whole-rock samples at approximately 5 ft intervals with adjustments made according to key lithologic changes (red arrows/triangles on Figure 4.3). Specifically, CaO was used as a proxy for calcite abundance, MgO was used as a proxy for dolomite abundance, and SiO_2 was used as a proxy for clay/silt/sand (siliciclastics) abundance. The dominant inorganic

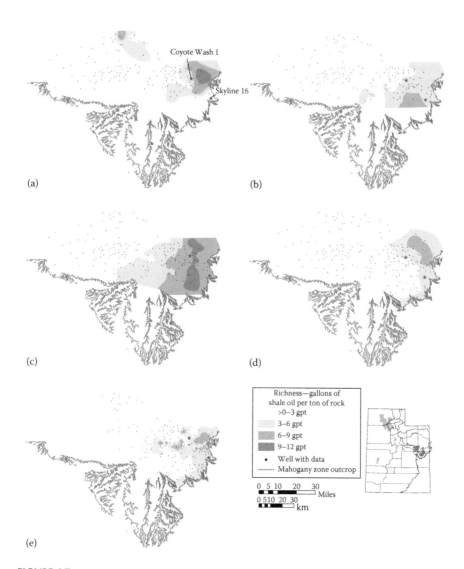

FIGURE 4.7

Richness maps for each organic-lean oil shale interval within the upper GRF. The locations of the Coyote Wash 1 core (Figure 4.2) and the Skyline 16 research core (Figure 4.3) are marked with red stars. (a) L-4, (b) L-5, (c) lower–middle R-6, (d) B-groove, and (e) A-groove.

mineralogy for the Skyline 16 core (Figure 4.3) was defined based on the following XRF criteria (in weight percent excluding any organic material):

1. Calcareous mudstone (blue), >25% CaO, <40% SiO_2, and <10% MgO
2. Dolomitic mudstone (light green), >25% CaO, <40% SiO_2, and >10% MgO
3. Clay-rich mudstone (brown), <25% CaO, >40% SiO_2, and <10% MgO

In addition, semiquantitative XRF analysis was performed on the GR1, GR2, and GR3 Skyline 16 samples (Figure 4.3; Table 4.1). The select samples were processed into pressed-powder pellets and analyzed using the same Rigaku miniZSX XRF-WD machine. To make the pellets, 4.5 g of the sample was combined with 0.5 g of powdered paraffin. This mixture was then placed in a container and mixed in a mechanical tumbler for approximately 30 min. The 5 g sample was then poured into an aluminum sample holder (Spex-cap), placed in a pellet die in a hydraulic press, and pressed with 41,370 kPa of pressure for approximately 2 min. The prepared pellets were then analyzed using a semiquantitative XRF application designed to specifically analyze oil shale samples. Semiquantitative analysis determines the approximate elemental component quantities as oxide compounds, totaling 100% (excluding any organic material), over a range of possible elemental oxide components (Table 4.1).

QEMSCAN mineralogy analysis was performed on three samples, each 1 cm^3 in size, that were carefully cut from the three 1 ft oil shale zones (GR1, GR2, and GR3) in the Skyline 16 research core (Figure 4.3; Table 4.1). The core samples were cut to fit a 1 cm diameter epoxy mold, with stratigraphic-up direction marked. The epoxy was mixed and allowed to impregnate the sample over 24 h in a vacuum chamber. Following this, the epoxy pots were removed and the epoxy cylinders were polished using progressively finer meshes. To finish the preparation process, the cylinders were coated in carbon.

The samples were run on a model 4300 Zeiss Eco 50 SEM platform QEMSCAN, with a tungsten filament and four light-element Brüker Xflash energy dispersive x-ray detectors, where the spectral analysis engine automatically identifies elemental composition at every predetermined spacing interval. Low-resolution scans with 10 μm spacing were performed on all samples. High-resolution scans with two-micron 2 μm spacing were also performed.

Once elemental composition was determined from the QEMSCAN analysis, software was used to calculate concentrations of elements and to assign a predefined mineral to each space on the grid. Conversion of spatial elemental data to mineralogy requires careful interpretation. In collaboration with the designers of the QEMSCAN technique and with clay mineralogists, the SMART method was developed specifically for shales and was equipped to deal with the complexity of clay minerals. It runs through several sets of mineral definitions and is outlined further in Haberlah et al. (2011). The QEMSCAN results were manipulated and examined in iDiscover (a program bundled with the QEMSCAN) to identify area percentages of numerous minerals. Additionally, QEMSCAN provided microfabric analysis, a visual map of the spatial distribution of mineral composition of oil shale laminations and beds.

The QEMSCAN and XRF datasets each have particular strengths and drawbacks. QEMSCAN analysis provides detailed microfabric mineralogic data, but only for a 1 cm² area of the core, making it difficult to upscale that data to larger rock volumes. It is important to note that each 1 ft oil shale core interval displays heterogeneity in richness and mineralogy within that interval and that each 1 cm² QEMSCAN sample is simply a best estimate of the properties of the 1 ft interval. Therefore, the QEMSCAN sample may not be a true average of richness and mineralogy of the entire 1 ft core interval. Whole-rock XRF provides a qualitative and less visual analysis as compared to QEMSCAN but can be performed quickly so that stratigraphic changes and trends over the entire thickness of the oil shale deposits can be evaluated.

4.3 Results

4.3.1 R-4 Oil Shale Zone

The R-4 oil shale zone ranges in thickness from about 40 ft in the southeast to over 170 ft in the north (Figure 4.4a). The thinner R-4 in the south-central portion of the basin (and subsequent thicker L-4) is the result of fluvial sands originating from the south that reduced the amount of profundal (oil shale) deposition in this area. In general, the R-4 records several roughly 2 ft thick, carbonate-dominated, shallowing-upward sequences, with profundal, dark-brown, organic-rich, carbonate mudstone gradually transitioning upward into sublittoral to littoral, light tan, organic-lean, dolomitic microbial carbonate (Figures 4.3 and 4.8). These alternations between organic-rich and organic-lean rocks greatly reduce the overall richness of the R-4 zone. There is only one small area within the R-4, in the central-eastern portion of the basin, which reaches the economic cutoff grade of 15 GPT (Figure 4.6a). However, a large area with grades averaging between 9 and 15 GPT is located in central Uintah County.

4.3.2 L-4 Oil Shale Zone

The L-4 organic-lean zone ranges in thickness from 25 ft in the east and north up to 240 ft in the central portion of the basin (Figure 4.5a). In general, the L-4 contains numerous roughly 2 ft thick, coarsening-upward, siliciclastic packages (Figures 4.3 and 4.8). A thicker L-4 in the central portion of the basin is the result of fluvial sands entering the ancient lake from the south. Average oil shale grade for the L-4 is very low, typically less than 2 GPT; however, small pockets of slightly higher grade (up to 10 GPT) are found to the east and far north (Figure 4.7a).

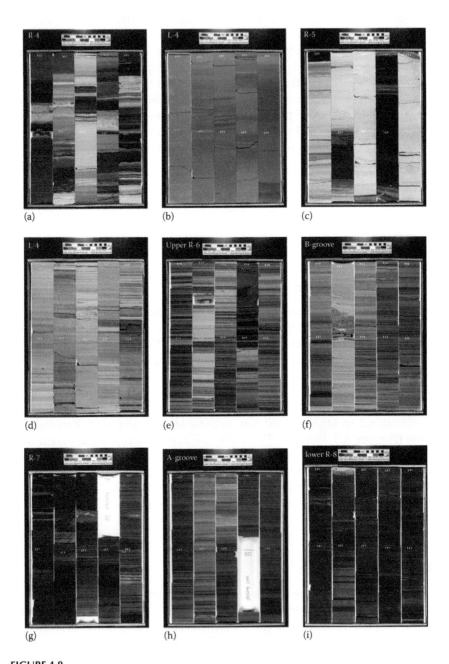

FIGURE 4.8

Examples of oil shale zones from the Skyline 16 research core. (a) R-4, 890–900 ft. (b) L-4, 850–860 ft. (c) R-5, 820–830 ft. (d) L-5, 740–750 ft. (e) Upper R-6, 570–580 ft. (f) B-groove, 530–540 ft. (g) Mahogany (R-7), 460–470 ft. (h) A-groove, 420–430 ft. (i) Lower R-8, 380–390 ft.

4.3.3 R-5 Oil Shale Zone

The R-5 oil shale zone ranges in thickness from about 60 ft in the southeast to over 200 ft in the far north (Figure 4.4b). Similar to the R-4 zone, the R-5 records several roughly 5 ft thick, carbonate-dominated, shallowing-upward sequences, with profundal, dark brown, organic-rich, carbonate mudstone transitioning upward into sublittoral to littoral, light tan, organic-lean, dolomitic microbial carbonate (Figures 4.3 and 4.8). As before, these alternations between organic-rich and organic-lean rocks greatly reduce the overall richness of the R-5 zone. No areas within the R-5 zone average the economic grade of 15 GPT (Figure 4.6b). However, a large area with grades averaging between 9 and 15 GPT is located in central Uintah County.

4.3.4 L-5 Oil Shale Zone

The L-5 organic-lean zone ranges in thickness from 60 ft in the southeast to over 200 ft in the northwest (Figure 4.5b). Similar to the L-4, the L-5 contains several roughly 2 ft thick, coarsening-upward, siliciclastic packages, but with more silt and clay and only rare occurrences of sandstone (Figures 4.3 and 4.8). Average oil shale grade for the L-5 is very low, typically less than 3 GPT; however, average grades reach up to 7 GPT in the southeast portion of the basin (Figure 4.7b).

4.3.5 Lower to Middle R-6 Oil Shale Zones

The R-6 oil shale zone is subdivided into two distinct units, the lower R-6 (organic-rich, but thin) to middle R-6 (lean) and the richer upper R-6 (Figures 4.2 and 4.3). The lower/middle R-6 grades from a thin zone of organic-rich carbonate (lower R-6) into a thicker lean zone (middle R-6) consisting of organic-lean marlstone to claystone and siltstone. The thickness of the lower/middle R-6 ranges from about 50 ft in the southeast to over 200 ft in the northwest (Figure 4.5c). Average oil shale grade for the thicker area of the lower/middle R-6, west of central Uintah County, is generally less than 5 GPT but averages up to 11 GPT in eastern Uintah County. However, the lower/middle R-6 never exceeds the 15 GPT economic cutoff (Figure 4.7c).

4.3.6 Upper R-6 Oil Shale Zone

The upper R-6 oil shale zone generally increases in thickness from the south (about 30 ft) to the north (190 ft) (Figure 4.4c). Somewhat similar to the R-4 and R-5 zones, the upper R-6 records several roughly 2 ft thick, carbonate-dominated, shallowing-upward sequences; however, the sequences represent relatively deeper facies (mostly laminated carbonate mud) (Figures 4.3 and 4.8).

As before, the rapid transitions between organic-rich and organic-lean intervals greatly reduce the overall richness of the upper R-6 zone. In central Uintah County, the upper R-6 averages slightly more than the economic cutoff grade of 15 GPT (Figure 4.6c). Significant acreage surrounding this small zone averages greater than 10 GPT, including along the eastern outcrop. Outside the Mahogany zone, the upper R-6 is the second most (albeit distant) economic oil shale zone.

QEMSCAN analysis of GR3, a relatively rich oil shale from the upper R-6, indicates that dolomite is the dominant mineral present, composing 33.1% of the sample with an equal amount of illite (illite plus feldspar) (33.4%) and relatively minor amounts of calcite (7.1%) and quartz (4.1%) (Table 4.1; Figures 4.9 and 4.10). QEMSCAN analysis classifies organic carbon and pore space as background, so these components are included in the "other" (26.0%) mineral category. Due to the low porosity of these samples, a large proportion of the material that is classified as "other" is interpreted as organic carbon along with any other accessory minerals that occur, such as apatite, smectite, and chlorite, all of which compose less than 1% of the sample. Black regions shown in the backscatter electron image gathered during QEMSCAN analysis highlight organic richness within individual laminations (Figure 4.9).

The semiquantitative XRF data largely agree with the QEMSCAN data, even though the sample sizes are different. GR3 contains 34.7% CaO, 5.6% MgO, and 32.0% SiO_2 by weight, indicating that the majority of the sample contains carbonate (calcite with significant dolomite), with a significant amount of silica, likely in the form of clay and feldspar (also indicated from the higher percentages of Al, K, and Fe) and possibly minor quantities of silt (Table 4.1).

4.3.7 B-Groove Oil Shale Zone

The B-groove organic-lean zone ranges in thickness from about 10 ft in the southeast to over 200 ft in eastern Duchesne County (Figure 4.5d). The B-groove correlates to Remy's (1992) S1 sandstone in the central Uinta Basin, the thickness of which greatly influences the overall thickness of the B-groove in this area. Overall, the B-groove consists of laminated, organic-lean carbonates, as well as claystone to sandstone (Figures 4.3 and 4.8). Average oil shale grade for the B-groove is very low, typically less than 3 GPT, except for a limited area in central Uintah County that averages up to 9 GPT (Figure 4.7d). The overall thickness of the B-groove might be the limiting factor in the economic viability of accessing the underlying upper R-6 rich zone.

4.3.8 Mahogany (R-7) Oil Shale Zone

The Mahogany (R-7) oil shale zone is well recognized as the most prospective oil shale interval in the Uinta Basin (Dyni et al. 1991). This zone consists

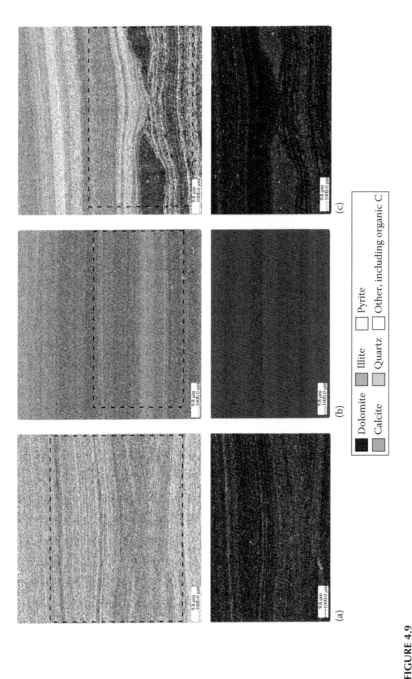

FIGURE 4.9

QEMSCAN and backscatter electron images of oil shale samples from the Skyline 16 core at 10 μm resolution. Area of BSE image highlighted with dashed black line on QEMSCAN image. (a) GR1 sample from a rich interval of the Mahogany zone. (b) GR2 sample from a lean interval of the Mahogany zone. (c) GR3 sample from a rich interval of the upper R-6.

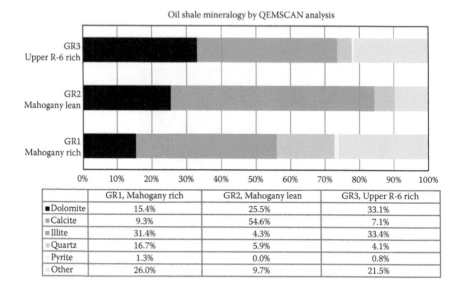

Oil shale mineralogy by QEMSCAN analysis

	GR1, Mahogany rich	GR2, Mahogany lean	GR3, Upper R-6 rich
■ Dolomite	15.4%	25.5%	33.1%
■ Calcite	9.3%	54.6%	7.1%
■ Illite	31.4%	4.3%	33.4%
■ Quartz	16.7%	5.9%	4.1%
Pyrite	1.3%	0.0%	0.8%
Other	26.0%	9.7%	21.5%

FIGURE 4.10
Mineralogy results of QEMSCAN analysis of oil shale samples from the Skyline 16 core.

almost entirely of thinly laminated, profundal, organic-rich marlstone, ranging in thickness from about 50 to 140 ft, and averaging 80 to 100 ft near the
outcrop to the east and southeast (Figures 4.4d and 4.8). The Mahogany zone
isopach displays a central thinner portion, trending northwest/southeast,
bracketed by thicker intervals to the south and north. The origin of this trend
is unclear (a slight topographic high in the lake creating less accommodation
space?), but it is similar to the trend of the regional gilsonite veins. The grade
of the Mahogany zone is dominated by the very rich Mahogany bed, which
reaches up to 80 GPT and represents the ancient lake's highest level (USGS
2010, Tänavsuu-Milkeviciene and Sarg 2012). The richness of the Mahogany
bed helps pull the overall average oil shale grade to a high of nearly 30 GPT
(Figure 4.6d). In fact, the majority of the Mahogany zone in Uintah County is
over the cutoff economic grade of 15 GPT, including significant areas along
the outcrop to the east and south.

QEMSCAN analysis of the GR2 Mahogany lean sample indicates that calcite is dominant with 54.6% calcite, 25.5% dolomite, 4.3% illite, and 5.9%
quartz (Table 4.1; Figures 4.9 and 4.10). The low abundance of "other" minerals (9.7%), which includes organic carbon, confirms the relatively lean
nature of the sample. In contrast, the GR1 sample contains a high percentage
of "other" minerals (30.1%), a large proportion of which is likely organic carbon (Table 4.1; Figures 4.9 and 4.10). Most of the calcite in the GR2 Mahogany
lean sample is replaced by organic matter in the GR1 Mahogany rich sample. The GR1 sample additionally contains 15.4% dolomite, 9.3% calcite, and

31.4% illite (Table 4.1; Figures 4.9 and 4.10). Semiquantitative XRF data from GR2 yields 59.1% CaO, 5.4% MgO, and 21.9% SiO_2, indicating more calcite than dolomite and slightly less clay/feldspar than the GR3 upper R-6 sample (Table 4.1). XRF data from GR1 yields 39.1% CaO, 5.0% MgO, and 31.2% SiO_2, indicating the sample has less calcite (and probably more dolomite) than GR2 similar to GR3 from the upper R-6) and contains more clay/feldspar (Table 4.1).

4.3.9 A-Groove Oil Shale Zone

The relatively thin A-groove ranges in thickness from about 10 ft in the east to about 20–30 ft in the north and south (Figure 4.5e). The A-groove consists of an organic-lean, dolomitic marlstone that pinches out to the west (Figures 4.3 and 4.8). Remy's (1992) S2 sandstone in the central to western Uinta Basin is roughly equivalent to the A-groove, but only the dolomitic lean zone is mapped in this study. Average oil shale grade for the A-groove is very low, typically less than 3 GPT, and does not exceed about 8 GPT (Figure 4.7e).

4.3.10 Lower R-8 Oil Shale Zone

The extensive R-8 zone includes all oil shale above the A-groove (or Mahogany zone on the west side of the basin where the A-groove is absent) to the inter-fingering GRF/Uinta Formation boundary. This study subdivides the R-8 into a lower, middle, and upper section. The top of the lower R-8 is defined by the "Big 3" oil shale beds, the top of the middle R-8 is defined by oil shale bed 76, and the upper R-8 is the zone between bed 76 and the top of the GRF (the "Big 3" and bed 76 are defined by Donnell (1997)). The top of the lower R-8 defines the extent of potentially economic oil shale, and the lower R-8 is the only interval evaluated in this study (Birgenheier and Vanden Berg 2011).

The lower R-8 oil shale zone generally increases in thickness from the southeast (about 115 ft) to the northwest (up to 350 ft) (Figure 4.4e). The lower R-8 represents a relatively stable lake level and consists almost entirely of marginally rich, profundal marlstone (Figures 4.3 and 4.8). Variations in organic richness could be caused by small fluctuations in lake level—small enough to consistently deposit profundal facies, but not significant enough to change the amount of organic productivity/preservation. Oil shale grade within the lower R-8 reaches 10–13 GPT in most of central Uintah County but greatly decreases to the west and north (Figure 4.6e). There are no areas where the grade is above the economic cutoff of 15 GPT.

4.3.11 Combined R-4 to Lower R-8 Oil Shale Interval

The entire assessed oil shale interval, from the R-4 to the lower R-8, increases in thickness from about 700 ft in the southeast portion of the basin to over 1600 ft in the northwest (Figure 4.11a). The average grade of the

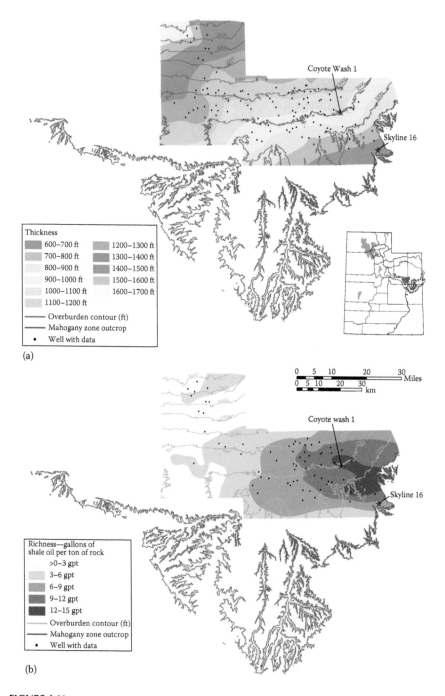

FIGURE 4.11

(a) Isopach and (b) richness map for the combined R-4 to lower R-8 oil shale interval within the upper GRF.

total interval reaches a high of 13.5 GPT in east-central Uintah County and gets significantly leaner to the west and north (Figure 4.11b). Qualitative XRF data from the Skyline 16 core, as well as several other oil shale cores (yet to be published), indicate that average MgO/CaO ratios decrease by 0.1 above the base of the Mahogany zone compared to the lower rich zones (Figure 4.12); this decreasing ratio can also be seen in Figure 4.3 as a shift from brown/green below the Mahogany to blues above. This increase in calcium carbonate within the Mahogany zone and R-8 interval could require slightly different retorting technology compared to the more dolomitic lower rich zones.

4.4 Summary and Discussion

The upper GRF oil shale resource was evaluated by mapping the thickness and richness of each oil shale zone from the R-4 to the lower R-8. As previously determined, the Mahogany zone is the interval with the most oil shale development potential with grades up to 30 GPT and typical thicknesses of 80–120 ft. In addition, the richest portion of the Mahogany zone is near outcrop, making it suitable for mining and surface retorting technologies. Outside the Mahogany, only the R-4 and upper R-6 zones have grades exceeding the minimum economic cutoff of 15 GPT and then only in small areas in east-central Uintah County. If in situ technologies advance to the point of becoming commercially viable, the most prospective areas in the Uinta Basin would be in east-central Uintah County, where overall oil shale grade of the upper GRF reaches its highest value of 13.5 GPT for a section of rock nearly 1000 ft thick. It is more likely that companies pursuing in situ techniques would focus their efforts in the Piceance Basin in Colorado, where the oil shale deposits are much richer and thicker. The exception might be if technologies can be developed using horizontal drilling techniques combined with in situ heating. In this case, the Mahogany zone oil shale deposits in the Uinta Basin would be an ideal horizontal, in situ resource target.

Detailed geologic interpretation of the Skyline 16 research core reveals why the lower rich zones in the Uinta Basin are not as economically attractive as equivalent zones in the Piceance Basin. These zones (R-4, R-5, and somewhat the R-6) contain several cycles that alternate between organic-lean carbonate (littoral deposits) and organic-rich oil shales (profundal deposits). The abundance of organic-lean, littoral carbonates in these zones greatly dilutes the overall oil yield of these lower rich zones. In addition, there seems to be a change in chemistry between the lower oil shale intervals and the Mahogany/R-8 zones. The lower zones are more dolomitic and the Mahogany and R-8 intervals are more calcite-rich.

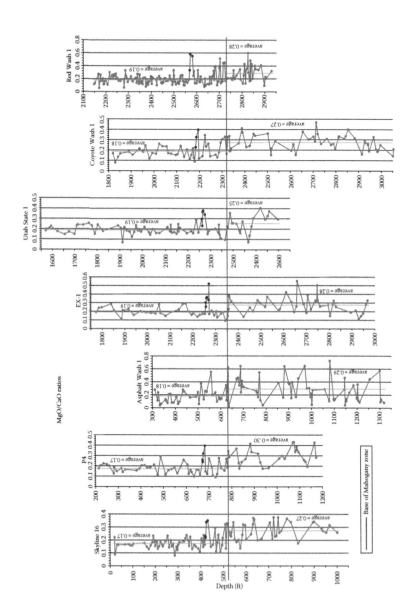

FIGURE 4.12

Plots of Mg/Ca ratios for several cores in the Uinta Basin. The horizontal line marks the base of the Mahogany zone and a transition from more dolomitic (MgO) rocks below (MgO) to more calcium carbonate-rich (CaO) rocks above (a shift of ~0.1). The darker dots are measurements taken from the A-groove, which is a thin dolomite zone above the Mahogany.

This transition seems to be present basin-wide and could have consequences for retorting technologies that might target the Mahogany as well as the lower rich zones.

___

References

American Society for Testing and Materials. 1980. Standard test method for oil from oil shale (resource evaluation by the USBM Fischer assay procedure), ASTM Designation D 3904-80. In *1980 Annual Book of ASTM Standards, Part 25.* Philadelphia, PA: American Society for Testing and Materials, pp. 513–525.

Bardsley, S. R. and S. T. Algermissen. 1963. Evaluating oil shale by log analysis. *Journal of Petroleum Technology* 15:81–84.

Birgenheier, L. and M. Vanden Berg. 2011. Core-based integrated sedimentologic, stratigraphic, and geochemical analysis of the oil shale bearing Green River Formation, Uinta Basin, Utah. Topical report submitted to the U.S. Department of Energy, National Energy Technology Laboratory, DOE Award No. DE-FE0001243. Salt Lake City, UT: University of Utah, Institute for Clean and Secure Energy.

Bunger, J. W., P. M. Crawford, and H. R. Johnson. 2004. Hubbert revisited: Is oil shale America's answer to peak-oil challenge? *Oil & Gas Journal* 102(30):16–24.

Cashion, W. B. 1974. Geologic map of the Southam Canyon quadrangle, Uintah County, Utah. Reston, VA: U.S. Geological Survey Miscellaneous Field Studies Map MF-579, scale 1:24,000.

Cashion, W. B. 1977. Geologic map of the Weaver Ridge quadrangle, Uintah County, Utah and Rio Blanco County, Colorado. Reston, VA: U.S. Geological Survey Miscellaneous Field Studies Map 824, scale 1:24,000.

Cashion, W. B. 1978. Geologic map of the Walsh Knolls quadrangle, Uintah County, Utah, and Rio Blanco County, Colorado. Reston, VA: U.S. Geological Survey Miscellaneous Field Studies Map MF-1013, scale 1:24,000.

Cashion, W. B. 1984. Geologic map of the Agency Draw NW quadrangle, Uintah County, Utah. Reston, VA: U.S. Geological Survey Miscellaneous Field Studies Map MF-1717, scale 1:24,000.

Cashion, W. B. 1994. Geologic map of the Nutters Hole quadrangle, Uintah and Carbon Counties, Utah. Reston, VA: U.S. Geological Survey Miscellaneous Field Studies Map 2250, scale 1:24,000.

Cashion, W. B. and J. R. Donnell. 1972. Chart showing correlation of selected key units in the organic-rich sequence of the Green River Formation, Piceance Creek Basin, Colorado, and Uinta Basin, Utah. Washington, DC: U.S. Geological Survey Oil and Gas Investigations, Chart OC 65.

Donnell, J. R. 1997. Correlation of individual oil-shale beds in the upper part of the Green River Formation, Piceance Creek Basin, Colorado, and Uinta Basin, Utah. Denver, CO: U.S. Geological Survey Open-File Report 97–714.

Donnell, J. R. and T. W. Blair, Jr. 1970. Resource appraisal of three rich oil-shale zones in the Green River Formation, Piceance Creek Basin, Colorado. *Colorado School of Mines Quarterly* 65(4):73–87.

Dyni, J. R. 2003. Geology and resources of some world oil-shale deposits. *Oil Shale* 20(3):193–252.

Dyni, J. R., J. R. Donnell, W. D. Grundy, W. B. Cashion, L. A. Orlowski, and C. Williamson. 1991. Oil shale resources of the Mahogany Zone in eastern Uinta Basin, Uintah County, Utah. Denver, CO: U.S. Geological Survey Open-File Report 91-285.

Gualtieri, J. L. 1988. Geologic map of the Westwater 30' x 60' quadrangle, Grand and Uintah Counties, Utah, and Garfield and Mesa Counties, Colorado. Reston, VA: U.S. Geological Survey Miscellaneous Investigations I-1765, scale 1:100,000.

Haberlah, M., M. Owen, M., P. W. S. K. Botha, and P. Gottlieb. 2011. SEM-EDS based protocol for subsurface drilling mineral identification and petrological classification. In *Proceedings of the 10th International Congress for Applied Mineralogy (ICAM)*, ed. M. Broekmans. Berlin, Germany: Springer, pp. 265–273.

Keighin, C. W. 1977a. Preliminary geology map of the Burnt Timber Canyon quadrangle, Uintah County, Utah. Reston, VA: U.S. Geological Survey Miscellaneous Field Studies Map MF-875, scale 1:24,000.

Keighin, C. W. 1977b. Preliminary geologic map of the Cooper Canyon quadrangle, Uintah County, Utah. Reston, VA: U.S. Geological Survey Miscellaneous Field Studies Map MF-874, scale 1:24,000.

Keighin, C. W. 1977c. Preliminary geologic map of the Rainbow quadrangle, Uintah County, Utah. Reston, VA: U.S. Geological Survey Miscellaneous Field Studies Map MF-893, scale 1:24,000.

Pantea, M. P. 1987. Preliminary geologic map of the Davis Canyon quadrangle, Uintah County, Utah, and Garfield and Rio Blanco Counties, Colorado. Reston, VA: U.S. Geological Survey Miscellaneous Field Studies Map MF-1933, scale 1:24,000.

Pantea, M. P., and R. W. Scott. 1986. Preliminary geologic map of the Flat Rock Mesa quadrangle, Uintah County, Utah. Reston, VA: U.S. Geological Survey Miscellaneous Field Studies Map MF-1866, scale 1:24,000.

Pipiringos, G. N. 1978. Preliminary geologic map of the Bates Knolls quadrangle, Uintah County, Utah. Reston, VA: U.S. Geological Survey Miscellaneous Field Studies Map MF-1025, scale 1:24,000.

Pipiringos, G. N. 1979. Preliminary geologic map of the Agency Draw NE quadrangle, Uintah County, Utah. Reston, VA: U.S. Geological Survey Miscellaneous Field Studies Map MF-1078, scale 1:24,000.

Remy, R. R. 1992. Stratigraphy of the Eocene part of the Green River Formation in the south-central part of the Uinta Basin, Utah. Washington, DC: U. S. Geological Survey Bulletin 1787, Evolution of Sedimentary Basins - Uinta and Piceance Basins, Chapter BB.

Scott, R. W. and M. P. Pantea. 1985. Preliminary geologic map of the Dragon quadrangle, Uintah County, Utah, and Rio Blanco County, Colorado. Reston, VA: U.S. Geological Survey Miscellaneous Field Studies Map MF-1774, scale 1:24,000.

Scott, R. W., and M. P. Pantea. 1986. Preliminary geologic map of the Wolf Point quadrangle, Uintah County, Utah. Reston, VA: U.S. Geological Survey Miscellaneous Field Studies Map MF-1839, scale 1:24,000.

Smith, J. W., H. E. Thomas, and L. G. Trudell. 1968. Geologic factors affecting density logs in oil shale. Document ID SPWLA-1968-P, *Society of Petrophysicists and Well-Log Analysts Ninth Annual Logging Symposium*, New Orleans, LA.

Stanfield, K. E. and I. C. Frost. 1949. Method of assaying oil shale by a modified Fischer retort. Washington, DC: U.S. Bureau of Mines Report of Investigations RI-4477.

Tänavsuu-Milkeviciene, K. and F. J. Sarg. 2012. Evolution of an organic-rich lake basin—Stratigraphy, climate and tectonics: Piceance Creek basin, Eocene Green River Formation. *Sedimentology* 59(6):1735–1768.

Tixier, M. P. and M. R. Curtis. 1967. Oil shale yield predicted from well logs. Document ID WPC-12271, *Seventh World Petroleum Congress*, Mexico City, Mexico.

U.S. Bureau of Land Management (BLM). 2008. Oil shale and tar sands final Programmatic Environmental Impact Statement. http://ostseis.anl.gov/eis/guide/index.cfm.

U.S. Geological Survey Oil Shale Assessment Team. 2010. Oil shale resources of the Uinta Basin, Utah and Colorado. Reston, VA: U.S. Geological Survey Digital Data Series 69-BB, CD-ROM.

Vanden Berg, M. D. 2008. Basin-wide evaluation of the uppermost Green River Formation's oil-shale resource, Uinta Basin, Utah and Colorado. Salt Lake City, Utah: Utah Geological Survey Special Study 128, CD-ROM.

Vanden Berg, M. D., J. R. Dyni, and D. E. Tabet. 2006. Utah oil shale database. Utah Geological Survey Open-File Report 469, CD-ROM.

Weiss, M. P., I. J. Witkind, and W. B. Cashion. 1990. Geologic map of the Price 30' x 60' quadrangle, Carbon, Duchesne, Uintah, Utah, and Wasatch Counties, Utah. Reston, VA: U.S. Geological Survey Miscellaneous Investigations I-1981, scale 1:100,000.

Witkind, I. J. 1988. Geologic map of the Huntington 30' x 60' quadrangle, Carbon, Emery, Grand and Uintah Counties, Utah. Reston, VA: Utah Geological Survey Open-File Report 440DM, scale 1:100,000.

5

Chemical and Structural Characterization of Oil Shale from the Green River Formation

Thomas H. Fletcher, Ronald J. Pugmire, Mark S. Solum,
Charles L. Mayne, Anita M. Orendt, and Julio C. Facelli

CONTENTS

This chapter presents data on the macromolecular structure of the parent oil shale, demineralized kerogen, and bitumen from three sections of an oil shale core, Skyline 16, drilled in eastern Utah's Uinta Basin. The nuclear magnetic resonance (NMR) analysis presented here is the most advanced technique available for analyzing macromolecular structure. For the first time, lattice parameters have been calculated for the demineralized kerogen using data from the NMR analysis, providing further insight into its macromolecular structure. X-ray photoelectron spectroscopy (XPS) data on a similar demineralized Green River oil shale studied by Hillier and coworkers (Hillier 2011, Hillier et al. 2013) are also included. Finally, this chapter reviews the work of Orendt et al. (2013), who used a combination of ab initio and molecular mechanics calculations to develop a three-dimensional (3D) structural model of the Green River kerogen based on the two-dimensional (2D) structure proposed by Siskin et al. (1995).

5.1 Background

5.1.1 Oil Shales

Oil shales are fine-grained sedimentary rocks containing organic matter that, upon heating, can be converted to liquid shale oil. The shale oil can then be refined into a slate of products that are similar to those obtained from the refining of petroleum crude oil. There are two fractions of organic matter in oil shale: (1) bitumen, which is the fraction that is soluble in organic solvents, and (2) kerogen, which constitutes approximately 90% of the organic matter and is insoluble in common organic solvents (Miknis 1994). Source rock in the Green River Formation, one of the most extensive oil shale reserves in the world, contains hydrogen-rich algal kerogen (type I) with up to ~20 wt% organic matter in the form of amorphous kerogen solid integrated in a silicate- and carbonate-based mineral matrix (Maciel and Dennis 1981). The Green River Formation spans parts of eastern Utah, southern Wyoming, and western Colorado. It contains approximately 60% of the known world reserves of oil shale, with estimates as high as 4.3 trillion barrels of oil equivalent (Andrews 2006, Dyni 2006, Johnson et al. 2010).

Vandenbroucke has provided a general description of the major oil shale deposits found worldwide, including the Green River Formation from the Uinta Basin in Utah. "The Green River shale is an organic-rich formation that was deposited in two Paleocene/Eocene alkaline palaeolakes, Lake Gosiute in Wyoming, and Lake Uinta in Utah and Colorado..." (Vandenbroucke 2003). Although the resource in the Green River Formation is large, the oil produced from it cannot be refined by conventional technology due to its high paraffinic content. Vandenbroucke notes that the highly paraffinic oils and extracts created problems "because C_{25+} hydrocarbons reprecipitated during compound separation with usual solvents" (Vandenbroucke 2003). Additional information about Uinta Basin geology is detailed in Chapter 4.

5.1.2 Characterization by NMR

Solution-state proton (^{1}H) NMR has been used to characterize oil shale extracts since the early 1960s. As carbon-13 (^{13}C) NMR became routine in the 1970s, NMR has become a mainstay of research in this area. Netzel et al. (1981) provided an extensive study of shale oil distillates. Burnham and coworkers (Ward and Burnham 1984, Burnham 1991, Burnham et al. 1997) published a number of studies elucidating the extraction of shale oil from various oil-bearing shale formations. Using model compounds, Dalling et al. (1986) showed how assignments can be made in these complex mixtures even when multiple chiral centers complicate the spectra.

Likewise, solid-state NMR spectroscopy has been widely used to characterize oil-shale-related materials. Early experiments combining cross-polarization (CP) with magic angle spinning (MAS) provided a technique that gave higher sensitivity and greater resolution than static samples provided. This combined technique of CP/MAS has been used to study the structural characteristics of many types of carbonaceous materials including oil shale and kerogen concentrates derived from many oil shales, some of which are samples from the Green River Formation.

Bartuska et al. (1977) presented static CP spectra of a shale and kerogen and then demonstrated the improvement in resolution upon spinning. Resing et al. (1978) used CP/MAS to determine the aromaticity of a Green River shale and kerogen, but the kerogen appeared to be derived from a different shale sample source. Maciel et al. (2013) studied the correlation between the aliphatic resonances and oil yields in various shales and kerogen concentrates, including Green River samples. Miknis et al. (1979) provided CP/MAS spectra of 10 shale samples (including the Green River shale). They also analyzed several sets of shales and pyrolyzed shales (chars) using the CP/MAS technique, thus determining oil yields and aromaticities (Miknis et al. 1982). Maciel and Dennis (1981) used CP/MAS to study the aromatic structures of six shales and their kerogen concentrates including shales from the Green River Formation. They analyzed the spectra in terms of integrals of various ^{13}C chemical shift ranges corresponding to various functional groups and pointed out that great care must be taken in the phasing of the spectra. Hagaman et al. (1984) studied Green River shales from the Mahogany zone at different depths and related the integrals of the aliphatic carbon chemical shift range to the likely oil yield using an internal standard. Petsch et al. (2001) studied weathered kerogens, including a Green River sample, and found that polymethylenic, highly aliphatic kerogens were not altered, but aromatic and branched aliphatic kerogens were. Recently, Salmon et al. (2011) used liquid-state NMR to study extracts from a Green River kerogen using various solvents and high-resolution MAS on a deuterated-dimethyl-sulfoxide-swollen kerogen sample. These data displayed liquid-like resolution in 2D hydrogen–carbon (H–C) chemical shift correlation spectra of the swollen solid.

Solum et al. (1989) were successful in differentiating the basic structural features of all eight coals available from the Argonne Premium Coal Sample Bank. They defined 14 parameters that are useful in defining macromolecular structures as well as eight other lattice parameters that are applicable to fossil fuels. These NMR parameters can be used in conjunction with the elemental compositions to determine other chemical structure parameters such as the number of aromatic carbons per cluster and the molecular weight of clusters and side chains. Their procedure found wide application not only in coal structures (Solum et al. 1989, Van Niekerk et al. 2008) but also in modeling devolatilization of coal structures (Grant et al. 1989, Fletcher et al. 1990, 1992, Genetti et al. 1998), char structure (Fletcher et al. 1990, 1992), soot formation (Solum et al. 2001, Zhang et al. 2001, Jiang et al. 2002, Yan et al. 2005), and combustion deposits (Kelemen et al. 1998, Ulibarri et al. 2002). The definitions of these functional groups and lattice parameters are found in the appendix of Jiang et al. (2002) as well as in Table 5.1. This technique was previously applied to a demineralized Green River oil shale from Colorado and to the solid residue from pyrolysis of that demineralized kerogen (Hillier 2011, Hillier et al. 2013). All of these studies exploit the power of solid- and solution-state NMR combined with gas chromatography/mass spectrometry for increased understanding of oil shale as an important potential energy source.

5.1.3 Characterization by XPS

XPS is a surface technique that is useful in determining the forms and relative abundance of certain atoms. XPS is one of the few techniques that can determine the mole fraction ratios of the different forms of heteroatoms, such as whether an element exists in a single bond or double bond. Because of its unique capabilities, XPS was used by Kelemen and coworkers (Kelemen et al. 1999, 2002, 2007) to determine the types of nitrogen and oxygen bonds in different types of kerogens. Kelemen et al. (1994) first determined the shifts for different model compounds and compared those to the Argonne Premium Coals. A range of expected XPS shifts were reported for different types of nitrogen. These shifts were applied to various kerogens based on the data from the model compounds. Specifically, Kelemen et al. (2002) used the peaks that corresponded to the pyridinic, amine, pyrrolic, and quaternary types of nitrogen functionalities to fit the XPS spectra. The nitrogen curves were resolved using a full width at half maximum (FWHM) of ~1.7 eV and a mixed Gaussian–Lorentzian distribution curve.

5.1.4 Modeling Chemical Structure of Oil Shale

The work of Siskin et al. (1995) in the development of a 2D model of the Green River oil shale structure has been a major step forward in understanding its

TABLE 5.1

Definitions of the Structural and Lattice Parameters from the Standard One-Dimensional Analysis Procedure

Structural parameters

f_a = fraction of carbon atoms that are sp^2 hybridized (aromaticity)

f_a^C = fraction of carbon atoms that are in carboxyl or carbonyl groups

f_a^O = fraction of carbon atoms that are in a carbonyl group (aldehydes and ketones)

f_a^{OO} = fraction of carbon atoms that are in a carboxyl group (acids, esters, amides)

$f_{a'}$ = fraction of carbon atoms that are sp^2 hybridized excluding f_a^C (corrected aromaticity)

f_a^H = fraction of carbon atoms that are protonated aromatics

f_a^N = fraction of carbon atoms that are nonprotonated aromatics

f_a^P = fraction of carbon atoms that are aromatic with an oxygen atom attached

f_a^S = fraction of carbon atoms that are aromatic with a carbon chain attached (also includes biaryl carbons)

f_a^B = fraction of carbon atoms that are aromatic and a bridgehead carbon

f_{al} = fraction of carbon atoms that are sp^3 hybridized (aliphatic)

f_{al}^H = fraction of carbon atoms that are aliphatic but not methyls

f_{al}^* = fraction of carbons that are aliphatic and methyls

f_{al}^O = fraction of carbon atoms that are aliphatic and attached to an oxygen atom

Lattice parameters

χ_b = mole fraction of bridgehead carbon atoms

C = average aromatic cluster size

$\sigma + 1$ = average number of attachments on an aromatic cluster

P_0 = fraction of attachments that do not end in a side chain (methyl group)

B.L. = average number of attachments on an aromatic cluster that are bridges or loops (a loop is a bridge back to the same cluster)

S.C. = average number of side chains on an aromatic cluster

MW_{cl} = average molecular weight of an aromatic cluster including side chains and bridges

m_δ = average mass of a side chain or one-half of a bridge

chemical structure. The Siskin model was originally generated using nondestructive chemical derivatization and characterization via NMR and mass spectroscopy of materials isolated under mild conditions. The model was revisited in 2004 (Hobbs et al. 2003) and compared to results of NMR, XPS, and sulfur x-ray absorption near-edge structure (XANES) results. In 2007, a much larger (more than 10^4 core structures with approximately 10^6 atoms), more general 2D kerogen model was developed by Freund et al. (2007) using the data from various solid-state analyses to construct the cores.

However, a 3D model is necessary for modeling the structure, properties, and interactions of the kerogen. The 3D characteristics of kerogen define the manner in which the kerogen folds and interacts with both the extractable bitumen and the mineral matter. They also provide a new view of the structure, including information on which portions of the structure are exposed on the surface, which portions are accessible through channels, and/or which portions may be isolated in the interior of the structure. An understanding of where the various functional groups are located may serve as a useful guide for developing novel processing schemes for resource recovery. In addition, the surface exposure of polar functional groups provides new information on the interaction of the kerogen structure with the inorganic matrix that appears to bind tightly to the mineral matter (Vandegrift et al. 1980, Brons and Siskin 1988, Solomon et al. 1988, Niksa and Kerstein 1991, Fletcher et al. 2014). For the purposes of creating a 3D model, only the Siskin model is currently viable as the Freund model is much too large.

5.2 Sample Characterization

Oil shale samples were taken from a 1000 ft long, 4 in. diameter core drilled at the Uinta Skyline 16 location, as provided by the Utah Geological Survey and the Institute for Clean and Secure Energy at the University of Utah. Three 1 ft segments of this core were studied and are identified as GR1 (462–463 ft), GR2 (486–487 ft), and GR3 (548–549 ft). Samples were powdered with a Reutsch automated agate mortar and pestle in air until the resulting powder passed through a 100 mesh (149 μm) screen. Sized particles were stored under dry nitrogen (N_2) until used.

Moisture and ash analyses of the powdered oil shale samples were performed at Brigham Young University using American Society for Testing and Materials procedures with one exception: the ashing was performed for ~80 h in air at 505°C instead of 750°C to prevent carbon dioxide (CO_2) release from the carbonates in the shale. The analysis was performed twice, and the resulting data (moisture, ash, and organic content of the oil shales) were averaged as presented in Table 5.2. The GR1 shale was the

TABLE 5.2

Moisture and Ash Analyses of the GR1, GR2, and GR3 Samples

Wt% of Parent Shale	GR1	GR2	GR3	Average Standard Deviation
Moisture	0.415	0.265	0.38	0.066
Ash	73.64	85.44	79.11	0.038
Organic	25.95	15.80	20.51	0.71
Oil yield (gal/ton)[a]	60	28	22	—
Wt% oil from shale[b]	23%	11%	8%	—

[a] Results of Fischer oil assay reported by UGS.
[b] Based on a specific gravity of 0.9 for the organic material.

richest in organic content (almost 26 wt%), followed by GR3 (20.5 wt%), and then GR2 (~16 wt%). The moisture content of all samples was less than 0.5 wt%.

A portion of the shale core was demineralized at the University of Utah using a procedure similar to that of Vandegrift et al. (1980) except that 4:1 dichloromethane (DCM)/methanol (MeOH) was used instead of benzene for the extraction steps, and additional washings with boric acid and hydrochloric acid were performed. The nine-step demineralization procedure is outlined in Figure 5.1. The three demineralized kerogen samples are referred to as GR1.9, GR2.9, and GR3.9.

Ash tests were performed on the three demineralized kerogen samples to determine the residual mineral matter not removed by the procedure; ash contents of 5.0 wt%, 4.1 wt%, and 4.7 wt% were measured for the GR1.9, GR2.9, and GR3.9 samples, respectively. Scanning electron microscopy and energy-dispersive x-ray spectroscopy analysis was performed on the ash

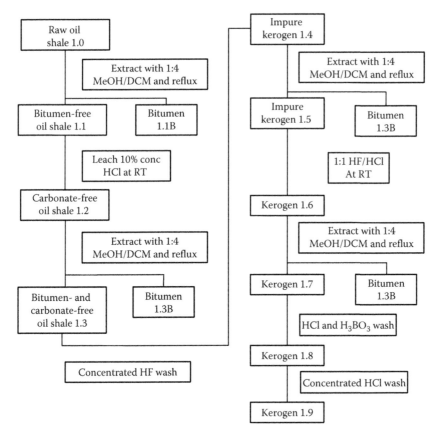

FIGURE 5.1
Outline of demineralization procedure to separate kerogen and bitumen from oil shale.

TABLE 5.3

Ultimate Analysis of GR1, GR2, and GR3 Samples
by Huffman Laboratories

	GR1.9	GR2.9	GR3.9
Moisture (wt% as received)	0.77	0.39	0.54
C (wt% daf[a])	77.4	77.5	76.2
H	9.8	10.0	9.5
N	2.8	2.6	2.5
O (difference)	8.2	8.0	8.1
S	2.0	2.0	3.7
Ash (wt% dry)	5.30	4.60	3.87

[a] Dry, ash-free.

from the fully burned GR2.9 and GR3.9 demineralized kerogens (Solum et al. 2014). The ash from both samples was high in iron, sulfur, and calcium. It seems likely that pyrite was not eliminated from the sample during the demineralization process.

The demineralized kerogen samples were also sent to an independent laboratory for ultimate analysis; see Table 5.3. All three kerogen samples showed a carbon (C) content of approximately 77 wt% on a dry, ash-free basis; hydrogen (H), nitrogen (N), and oxygen (O) contents were also similar. The GR3.9 sample had nearly twice the sulfur content as the other samples.

5.3 ^{13}C NMR Analysis of Bitumen Samples

The bitumen samples obtained from the kerogen isolation procedure for GR1, GR2, and GR3 were prepared at a concentration of 150 mg/mL in deuterated dicholoromethane (CD_2Cl_2); 0.7 mL of the solution was transferred to a 5 mm NMR tube with a screw cap and polytetrafluoroethylene-lined septum to prevent solvent loss during data acquisition. All high-resolution NMR data were acquired using an Agilent Technologies Direct Drive 500 MHz spectrometer with a 5 mm switchable broadband probe equipped with z-axis pulsed field gradient coil and sample temperature control. The sample temperature was controlled at 26°C using a stream of dry N_2 flowing at 10 liters per minute (L/min).

5.3.1 Quantitative High-Resolution ^{13}C Data

Two methods were used to acquire quantitative ^{13}C spectra. The first method, quantC, used the classical gated decoupling method to obtain decoupled spectra with the nuclear Overhauser effect (NOE) suppressed. This method retained both protonated and nonprotonated carbon resonances but sacrificed considerable sensitivity due to suppression of the NOE. Typical acquisition parameters were relaxation delay of 20 s, acquisition time of 2 s, 45-degree tip angle, 67,500 complex samples of the free induction decay (FID), and 22 h total acquisition time.

The second method, quantD, used the Agilent-supplied quantitative distortionless enhancement by polarization transfer (QDEPT) pulse sequence and quantdept macro (Crouch 2012). The technique involved a double array of the width of the final proton read pulse and the evolution time delay. The sum of all the members of the array yielded a quantitative carbon spectrum with nonprotonated carbons suppressed; however, it retained the proton-carbon magnetization transfer enhancement inherent in the distortionless enhancement by polarization transfer (DEPT) technique (Henderson 2004). Typical acquisition parameters were relaxation delay of 10 s, acquisition time of 4 s, 125,000 complex samples of the FID, and 22 h total acquisition time. These spectra were integrated using the standard Agilent VNMRJ software integration tools.

5.3.2 DEPT Spectra

DEPT spectra were obtained using the standard Agilent QDEPT pulse sequence to differentiate among nonprotonated methine (CH), methylene (CH$_2$), and methyl (CH$_3$) carbons. The QDEPT sequence incorporates shaped broadband inversion pulses to obtain more uniform inversion of the carbon magnetization across the entire carbon chemical shift range. This sequence also retains resonances from nonprotonated carbons, in contrast to the classical DEPT technique that suppresses nonprotonated carbon resonances.

5.3.3 NMR Analysis of GR1, GR2, and GR3 Bitumens

All three of the bitumen samples show essentially the same ^{13}C resonances with small differences in intensity, even though the GR1 and GR3 core segments are separated by 86 ft. These spectra are also substantially similar to spectra published previously on similar materials (Netzel et al. 1981, Ward and Burnham 1984, Dalling et al. 1986, Burnham 1991, Burnham et al. 1997). The numerous minor components of the bitumen consist mainly of branched hydrocarbons (i.e., isopranes) and, to a lesser extent, cyclic hydrocarbons.

Using the quantC method, the measured fraction of aromatic carbons is 11.2%, 9.3%, and 7.4% for bitumen isolated from core segments GR1 through GR3, respectively. The signal-to-noise (S/N) ratio in the aromatic region of the spectra is quite low, and these results should be interpreted to mean that the differences in aromatic content among the three bitumen samples are barely outside experimental error. However, the aromatic content of the bitumens is substantially lower than that of the corresponding kerogens (24%, 23%, and 24%) from these shale samples.

Although the S/N ratio in the aromatic region is low and the spectra are too complex to assign individual resonances to particular compounds, it is possible to separate protonated from nonprotonated carbons in the aromatic region by comparing the quantC (all aromatic carbons) and quantD (only protonated aromatic carbons) results, even though the ranges of the carbon chemical shifts overlap for these two types of aromatic carbons. Since nonprotonated carbons contribute a negligible amount to the integrals of the aliphatic region of the spectra, these integrals represent the same number of carbons in either the quantC or quantD spectra and are used to normalize the integrals of the aromatic region in each case. For the GR1, GR2, and GR3 bitumens, this treatment yields 30%, 46%, and 47% of the aromatic carbons as protonated carbons, respectively. Again, the errors in these estimates are large due to low S/N ratios, and the interpretation of these results is that a little less than half of the aromatic carbons are protonated in each of the three bitumen samples.

All three bitumen samples exhibit essentially the same results by NMR, with only 9% aromatic carbon (Solum et al. 2014). Figure 5.2 shows the aliphatic region of the spectrum from 10 to 45 ppm. The intense resonance near 30.1 ppm arises from the carbons located at least five carbons from the methyl end of a normal hydrocarbon chain. The high intensity of this resonance shows that the GR2 bitumen contains a significant amount of long-chain normal hydrocarbons. The corresponding methyl is located at 14.3 ppm with carbons 2, 3, and 4 located at 23.1, 32.3, and 29.8 ppm, respectively. Note that the peaks for carbons 1–4 have similar intensities and, in a long *n*-alkyl chain, each would represent two carbons (one from each end of the chain).

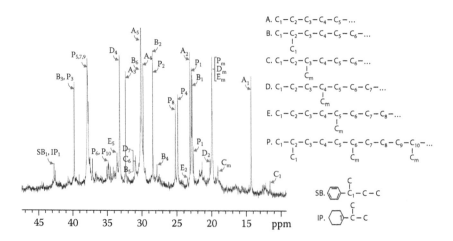

FIGURE 5.2
Expanded aliphatic region (with height cropped for the peak at 30.1 ppm) of the quantC spectrum of GR2 bitumen dissolved in CD_2Cl_2. Labels are suggested by Dalling et al. (1986).

Since the ratio of the intensity of the peak at 30.1 ppm to the average intensity of the first four peaks in the chain is about 8 (i.e., 16 carbons), one can assert that the average length of these normal chains is about 24 carbons. Of course, these n-alkyl chains can be attached to aromatic centers or to cyclic alkyl groups (Ward and Burnham 1984) or can have alkyl side-chains at either or both ends of the chain. When the n-alkyl chain is ten carbons or longer, the resonances from the carbons far from the substituent will contribute to the resonance at 30.1 ppm, and this effect can produce a systematic error, increasing the estimate of the chain length. However, the low abundance of these moieties would indicate that this error is small compared to other sources of error in the determination of peak areas.

The other spectral lines in Figure 5.2 indicate that the bitumen also contains branched alkanes that can be readily identified. If one carefully compares the chemical shifts of isoprenoid structures with the bitumen spectra, it is possible to identify chiral centers in isopranes in addition to the n-alkanes that are present, e.g., structures like pristane, phytane, farnesane, and squalane (Netzel et al. 1981, Dalling et al. 1986). The structures of these isopranes are superimposed on the spectrum, and the appropriate structural lines are identified for six isopranes. While these branched alkanes are not unique to this study, they represent candidate structural forms that are found in steranes/isoprenoids.

The complexity of the spectrum contains much information that is just above the noise level; thus, it is not possible to estimate the relative abundance of n-alkanes from the highly branched components. The chemical shift data were searched for any indication of the presence of cyclic alkyl (as well as aromatic) components that could serve as terminal structures for the alkyl chains/cross-links, but no such components were clearly identifiable.

Figure 5.3 illustrates the expanded aliphatic region of a set of DEPT spectra, showing separately the resonances of various carbons according to the number of attached protons. These spectra add additional evidence of isoprenoids due to the presence of peaks identified as CH and CH_2 groups. The vertical scale has been adjusted so that the intense CH_2 resonance at 30.1 ppm is off-scale to permit closer examination of the other resonances. Due to various instrumental limitations, cancelation of one carbon type from the spectrum of another carbon type is not perfect. Note, for example, that the CH_3 resonance at 14.5 ppm appears at diminished intensity in the CH_2 spectrum.

Although the DEPT spectra are not strictly quantitative, some sense of the relative amounts of the various carbon types can be gleaned by comparing the various spectra. Clearly, CH_2 is the dominant type, with CH_3 and CH at lower abundance. Since methines (CH) indicate a branch point in an aliphatic chain (i.e., an isoprenoid), it is evident that there are relatively few such branch points in this sample.

The CH_3 resonances upfield from 14.5 ppm arise from the terminal methyl of chains where the third carbon from the end is a methine. The spectrum representing quaternary carbons shows essentially no resonances; all the

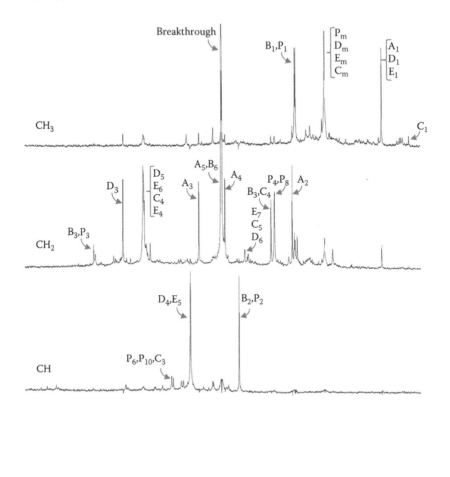

FIGURE 5.3
Aliphatic region of the DEPT spectrum of GR2 bitumen dissolved in CD_2Cl_2. Labels are the same as in Figure 5.2; see Dalling et al. (1986).

resonances seen in this particular linear combination of the raw data correspond to strong resonances of other types and are attributed to imperfect cancelation of these other types. Resonances from nonprotonated carbons are diminished in intensity by about a factor of four relative to those of protonated carbons, because nonprotonated carbons receive no transfer of proton magnetization. Nevertheless, it is clear that there are very few aliphatic, nonprotonated carbons in this sample.

The aromatic regions of the DEPT spectra show resonances over a wide range of chemical shift for both substituted and nonsubstituted aromatics but with no dominant resonances. These resonances indicate a wide variety of substituted aromatic molecules with differing substituents; no one molecular structure occurs in relatively high abundance. Note that the CH_2 trace shows no resonances as would be the case if chains terminated by an alkene moiety were present. This lack of terminal alkenes was observed previously (Netzel et al. 1981) for solvent extraction of bitumens as opposed to pyrolysis extraction of organic matter from oil shale where considerable amounts of terminal alkenes were observed (Hillier 2011, Hillier et al. 2013).

5.4 Solid-State NMR Analysis of Oil Shale and Demineralized Kerogen

5.4.1 ^{13}C Solid-State NMR Technique

Solid-state ^{13}C NMR analysis was performed on each of the shales and the demineralized kerogen samples. All of these experiments were run on a Varian Direct Drive oversampled spectrometer operating at a carbon frequency of 25.1562 MHz and a proton frequency of 100.02 MHz. The spectrometer utilizes a 7.5 mm pencil rotor probe with a ceramic housing to remove carbon background. The analysis included (1) a standard ^{13}C CP/MAS experiment with a 3 ms contact time, (2) a variable contact time experiment with 21 different contact times ranging from 5 μs to 25 ms, (3) a dipolar dephasing experiment using a 3 ms contact time, and (4) a single-pulse (SP) experiment. The details of these experimental protocols have been described by Solum and coworkers (Solum et al. 1989, 2001, Fletcher et al. 1990, 1992, 2001). Approximately 150 mg of each kerogen and 450 mg of each shale were required for this analysis. Applying the method of Solum and coworkers (Solum et al. 1989, 2001), the measured fraction of bridgehead carbons and the elemental analysis were used to determine average structural parameters, including the number of attachments per aromatic cluster, the molecular weight per aromatic cluster, and the average molecular weight of a side chain (see Figure 5.4). When applied to the unreacted kerogen samples, Solum's method for determining the fraction of attachments that were bridges (i.e., P_0) was subject to substantial errors due to the large number of CH_2 groups relative to CH_3 groups in the sample. The bleed-through from this larger band may have overestimated the aliphatic chain branching, resulting in more CH_3 groups being counted than existed. Other methods based on three-spin coherence might have given more information but were not possible with the available equipment (Mao and Schmidt-Rohr 2005, Mao et al. 2010).

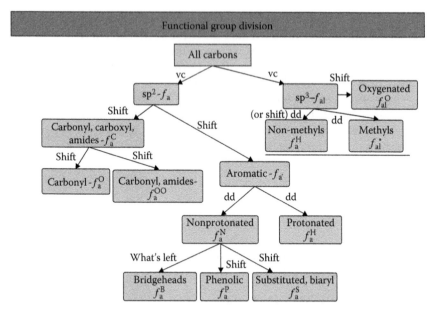

FIGURE 5.4

Definition and source of information for determining 14 structural and 8 lattice parameters used to define critical structural components of carbonaceous materials including oil shale kerogens. In this figure, "shift" is the fraction determined by integration of a selected chemical shift range, "vc" is the fraction determined by a *variable contact* time experiment and the magnetizations obtained by separately fitting the aromatic and aliphatic regions, and "dd" is the fraction determined from a *dipolar dephasing* experiment. The "dd" analysis works well for the aromatic region where the two time constants for the two components are quite different (12–40 µs and 300 µs–3 ms), but not for the aliphatic region with its molecular motion and similar time constants for each component (all <~130 µs).

5.4.2 Structural Parameters from Solid-State NMR Spectra of Shales and Kerogen Samples

Both SP/MAS spectra and CP/MAS spectra of the three oil shale samples (GR1, GR2, GR3) and their kerogen concentrates (GR1.9, GR2.9, GR3.9) were obtained (Fletcher and Pugmire 2014, Solum et al. 2014). The SP/MAS spectra contained both the inorganic (e.g., carbonates) and organic carbon, while the CP/MAS data primarily contained the organic carbon structural elements. Solum et al. (2014) showed that the inorganic carbon peak for the shale completely disappeared in the SP spectra of the kerogen concentrates. Within the small S/N ratio difference, the CP/MAS spectra are identical to the SP/MAS spectra for these samples. Therefore, the CP/MAS spectra of the kerogen concentrates could be analyzed without interference from inorganic carbon. From these spectra, the structural and lattice parameters given in Table 5.4 for the shales and their respective kerogen concentrates were obtained. Because of the large carbonate peak, aromaticities were not calculated for SP/MAS spectra of the shales.

TABLE 5.4

Structural and Lattice Parameters of the GR1, GR2, and GR3 Oil Shales and Their Kerogens

Structural Parameters

Compound	f_a	f_a^C	f_a^O	f_a^{OO}	f_a^*	f_a^H	f_a^N	f_a^P	f_a^S	f_a^B	f_{al}	f_{al}^H	f_{al}^*	f_{al}^O
GR1 (CP) cr	0.25	0.04	0.02	0.02	0.21	0.07	0.14	0.04	0.07	0.03	0.75	0.62	0.13	0.02
GR1.9 (CP)	0.24	0.04	0.01	0.03	0.20	0.06	0.14	0.03	0.07	0.04	0.76	0.65	0.11	0.00
$C_{100}H_{150}N_3O_8S_1$														
GR1.9 (SP)	0.25										0.75			
GR2 (CP) nc	0.22	0.04	0.02	0.02	0.18	0.06	0.12	0.03	0.06	0.03	0.78	0.65	0.13	0.00
GR2.9 (CP)	0.23	0.05	0.02	0.03	0.18	0.06	0.12	0.03	0.06	0.03	0.77	0.66	0.11	0.01
$C_{100}H_{153}N_3O_8S_1$														
GR2.9 (SP)	0.24										0.76			
GR3 (CP) cr	0.27	0.03	0.01	0.02	0.24	0.06	0.18	0.04	0.08	0.06	0.73	0.60	0.13	0.05
GR3.9 (CP)	0.24	0.04	0.01	0.03	0.20	0.05	0.15	0.03	0.07	0.05	0.76	0.63	0.13	0.00
$C_{100}H_{148}N_3O_8S_2$														
GR3.9 (SP)	0.25										0.75			

Lattice Parameters

Compound	χ_b	C	$\sigma+1$	P_0	B.L.	S.C.	MW_{cl}	m_δ
GR1 (CP)	0.143	8.4	4.4	−0.18	—	—		
GR1.9 (CP)	0.200	10.0	5.0	−0.10	—	—	776	131
GR2 (CP)	0.167	9.0	4.5	−0.44	—	—		
GR2.9 (CP)	0.167	9.0	4.5	−0.22	—	—	775	148
GR3 (CP)	0.250	12.0	6.0	−0.08	—	—		
GR3.9 (CP)	0.250	12.0	5.9	−0.30	—	—	946	135

1. cr—corrected for large aliphatic sidebands due to ferrimagnetic particles in raw shale.
2. nc—not corrected for very small aliphatic sidebands due to ferrimagnetic particles in raw shale.

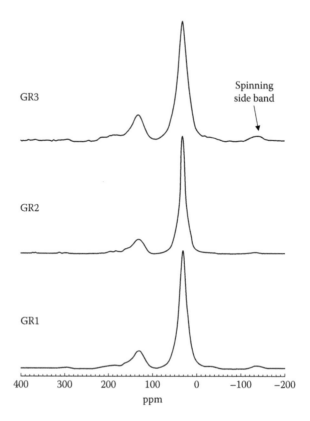

FIGURE 5.5
CP/MAS spectra of the three shales. The contact time was 3 ms and the pulse delay was 1 s.

A unique feature of the oil shale spectra, especially GR1 and GR3, is a sideband at approximately −133 ppm seen in the CP/MAS spectra of Figure 5.5. This feature is the spinning speed away from the aliphatic resonance centered at about 31 ppm. Due to the small chemical shift anisotropy of aliphatic carbons compared to aromatic carbons, one usually does not see spinning sidebands at this spinning speed. There is also a corresponding sideband under the carbonyl/carboxyl region between 165 and 240 ppm. This effect occurs when ferromagnetic or ferrimagnetic material is present in the sample while spinning. In these samples, this effect could come from ferrimagnetic iron(II,III) oxide (Fe_3O_4) or from other iron-containing minerals. It was difficult to accurately account for this sideband effect, particularly for the GR3 shale, in the calculation of structural parameters such as carbonyl/carboxyl, f_a^C, in Table 5.4. This aliphatic sideband is essentially missing in the high S/N CP/MAS spectra of the kerogen concentrates in Figure 5.6, showing that demineralization was effective in removing these minerals.

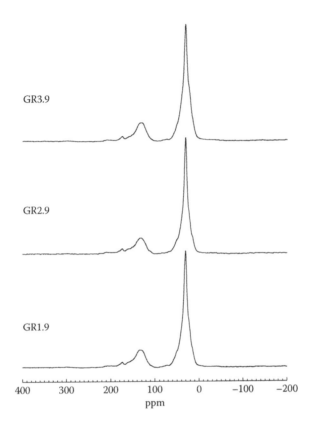

FIGURE 5.6
CP/MAS spectra of the three kerogen concentrates. The contact time was 3 ms and the pulse delay was 1 s.

In comparing the structural parameters in Table 5.4 for the three kerogens (GR1.9, GR2.9, GR3.9), there are no large differences among the three shales and their respective kerogen concentrates. There are much larger differences in the inorganic matter among the three cores (Fletcher and Pugmire 2014, Solum et al. 2014). The three kerogens have a total fraction of sp²-hybridized carbon of 0.23 or 0.24 with 0.04 or 0.05 of that being carbonyl/carboxyl groups. The SP/MAS aromaticity results are within 0.01 of the CP/MAS aromaticity for all samples, well within the experimental error. The carbonyl/carboxyl carbons (165–240 ppm) were subtracted from the total aromaticity, f_a, resulting in a corrected aromaticity, $f_{a'}$, consisting of only aromatic carbons. In preliminary NMR experiments on GR2.9, no evidence of alkene is seen either in 40,000 Hz spinning ^1H or 2D wide-line separation experiments at 800 MHz. Additionally, the high-resolution ^{13}C spectra of the unreacted kerogen concentrates do not appear to contain ethylene structures. However, the kerogen pyrolysis tars are known to contain alkane/alkene pairs (see Chapter 6).

5.4.3 Lattice Parameters from Solid-State NMR Spectra of Shales and Kerogen Samples

Using the CP/MAS spectra of the three shales (Figure 5.5), the bridge-head aromatic carbons were computed by subtracting the protonated (f_a^H), phenolic (f_a^P), and substituted (f_a^S) aromatic carbons from the corrected aromaticity. The bridgehead aromatic carbons were then used to estimate an "average" aromatic cluster size (Solum et al. 1989). For this estimate, biaryl linkages (zero mass bridges) were considered to be substituted aromatic carbons as their shift tensor and isotropic shift showed (Barich et al. 2001).

The data analysis indicates that the average aromatic cluster size (carbon atoms per cluster) is about 10 for the GR1.9 kerogen, 9.0 for the GR2.9 kerogen, and about 12.0 for the GR3.9 kerogen. The average cluster size is about the size of naphthalene, but nothing can be said about the distribution of cluster sizes. The number of attachments per cluster (bridges plus side chains), $\sigma + 1$, is on the order of 5 for the three kerogens, consistent with the observation that the oil yield in shale correlates neither with the aliphatic carbon nor with the aromatic carbon atoms that remain in the spent shale (Miknis et al. 1982, Walters et al. 2007, Fletcher et al. 2012, Lewis and Fletcher 2013).

The estimated average carbon chain length can be computed by dividing f_{al}, the fraction of carbon atoms that are sp^3 hybridized, by f_a^S, the fraction of carbon atoms that are aromatic. For the three kerogen concentrates, this chain length is about 11–13 carbons, which corresponds to half of a bridge between aromatic clusters or the size of an average side chain. This average carbon chain length is smaller than the size of extractable material analyzed by Bartuska et al. (1977) using Fourier transform infrared spectroscopy–mass spectrometry. In that work, the chains group were around 20–30 carbons. The average side chain mass based on cluster analysis (Solum et al. 1989) is 131, 148, and 135 for GR1.9, GR2.9, and GR3.9, respectively. The cluster analysis and the carbon chain length analysis therefore give consistent results.

The total cluster masses, including both aromatic centers and attachments, are 776, 775, and 946, respectively, for the three samples. These total cluster masses are much larger than those of the Argonne coals (Solum et al. 1989), which ranged from 269 to 408. The coals have much less bridge material and slightly larger aromatic clusters. In Table 5.4, the values for the fraction of intact bridges, P_0, are all negative. When this parameter is negative, the bridges and loops (bridge back on the same aromatic cluster) cannot be distinguished from side chains. This parameter should be between 0 and 1.0 if all chains terminate in a single methyl group, a reasonable approximation for coals (Solum et al. 1989). The negative values indicate significant branching in the aliphatic chains with multiple methyl groups on a single chain.

Dipolar dephased spectra of the kerogen concentrates using a dephasing time of 42 μs (Fletcher et al. 2014) show that the aliphatic region consists of three main peaks. The peak at 15 ppm is from the terminal CH_3 groups of aliphatic chains. The peak centered between 22 and 23 ppm is a composite peak from CH_3 on aromatic rings, and the first CH_2 group is from the terminal CH_3. The largest peak at about 30 ppm is from most of the CH_2 carbons in long chains. There must be a high degree of motion in some parts of these chains or this peak would be mostly suppressed at this dephasing time. A test molecule used in this laboratory, decylpyrene, had all nine of its CH_2 carbons suppressed at this dephasing time as is true for most static methylenes. Because of the dominance of the large CH_2 peak and the degree of motion in the sample, a better analysis would be to use spectral editing techniques based on the dipolar interaction. However, this type of analysis was not possible on the available spectrometer.

5.5 X-ray Photoelectron Spectroscopy

XPS analysis was used by Hillier and coworkers (Hillier 2011, Hillier et al. 2013) to determine chemical forms of nitrogen and oxygen in a demineralized Green River oil shale sample from Colorado. It is anticipated that these XPS data would be similar to those from Uinta Basin oil shales, including the Skyline 16 samples, because the NMR analyses of the shales from the two basins are similar.

5.5.1 Nitrogen Functionality

Using shifts similar to those determined by Kelemen and coworkers (2002), nitrogen forms and mole percentages were determined for the demineralized Colorado kerogen. Analysis was performed in two ways: (1) allowing the value of the FWHM to float (i.e., to be curve-fit) and (2) specifying the value of the FWHM. Two possible nitrogen distributions were used (Hillier 2011). As shown in Table 5.5, the resulting curve fit required four nitrogen species to resolve the curve and match more closely the results of Kelemen et al. (2002).

The kerogen studied by Hillier and coworkers had similar nitrogen percentages to the distributions determined by Kelemen and coworkers (2002) if the same shifts and FWHMs were used. If the FWHMs were not constrained, the distribution showed more pyridinic and no amino nitrogen functionality as seen in Table 5.5. The XPS results obtained by constraining both the peak locations and the FWHM for the kerogen are recommended.

5.5.2 Oxygen Functionality

Oxygen functionality can also be assigned through XPS data, but it must be done indirectly as described by Hillier (Hillier 2011, Hillier et al. 2013).

TABLE 5.5

Nitrogen Percentage Values Determined by Kelemen et al. (2002)
and by Hillier (2011) for a Floating FWHM or a Fixed FWHM

Type of Nitrogen	XPS Shift	Area	Percent of Total Nitrogen	N Atoms per 100 C Atoms
Keleman et al. (2002)				
Pyridinic	398.8	—	27%	—
Amino	399.4	—	10%	—
Pyrrolic	400.2	—	53%	—
Quaternary	401.4	—	10%	—
Floating FWHM				
Pyridinic	398.8	5720	39.7%	0.87
Amino	399.2	0	0.0%	0.00
Pyrrolic	400.2	5481	38.1%	0.83
Quaternary	401.4	3202	22.2%	0.48
Fixed FWHM				
Pyridinic	398.8	4583	30.1%	0.66
Amino	399.2	1033	6.8%	0.15
Pyrrolic	400.2	6581	43.2%	0.94
Quaternary	401.4	3021	19.9%	0.43

The carbon XPS data for the Colorado kerogen sample showed a peculiar hump on the low binding energy side (283 eV) corresponding to carbides, which were most likely artificially introduced in either the grinding process or the sample preparation. Ignoring the peak centered about 283 eV, this sample gave the distribution shown in Table 5.6. The principal peak corresponds to the peak centered at 284.7 eV with at least two carbon–oxygen bond types. These two shifts are indicative of carbonyl ($C=O$, 287.5 eV) and carboxyl ($O–C=O$, 289.0 eV) groups. There is a third shift corresponding to the ether/hydroxyl groups ($C–O$, 286.3 eV) that is difficult to resolve but can be determined by subtracting the two aforementioned groups from the total organic carbon. The mole percentages for the carbon–oxygen functionalities are listed in Table 5.6.

TABLE 5.6

Carbon–Oxygen Mole Percentages as Determined by
X-Ray Photoelectron Spectroscopy

	Shift (eV)	Area	Mole (%)
Aromatic and aliphatic carbon	284.7	29842	93.0
Carbonyl ($C=O$)	287.5	972	3.0
Carboxyl ($O—C=O$)	289	1289	4.0

5.6 Kerogen Structure Modeling

As mentioned in the introduction, the work of Siskin et al. (1995) in the development of a 2D model of the Green River oil shale kerogen structure, shown in Figure 5.7, was a major step forward in understanding the chemical structure of kerogen. However, in order to model the structure, properties, and interactions of the kerogen, a 3D chemical model was developed. A 3D model enables the exploration of the sensitivity of the experimental methods to the 3D structure and may validate and provide information on possible deficiencies with the 2D models.

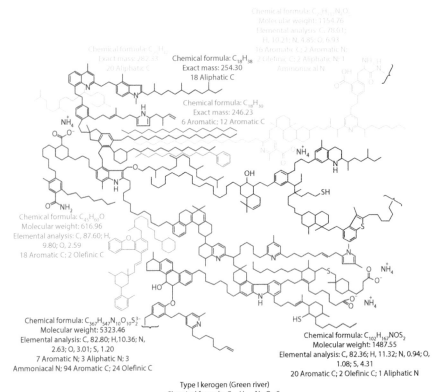

FIGURE 5.7
2D Siskin model of Green River kerogen from Siskin et al. (1995). (Reprinted with permission from Springer Science+Business Media B.V.: *Composition, Geochemistry and Conversion of Oil Shales, NATO ASI Series; Series C: Mathematical and Physics Sciences,* Detailed structural characterization of the organic material in Rundle Ramsay Crossing and Green River oil shales, Vol. 455, 1995, pp. 143–158, Siskin, M. et al.)

A 3D structure corresponding to Siskin's 2D kerogen model (Siskin et al. 1995) was built using HyperChem (Ref 1 (HyperChem)). A preliminary chemical structure was obtained via the molecular mechanics energy minimization routine in HyperChem using the MM+ force field (Ref 2 (Allinger)). This minimized structure was further optimized using the ab initio software package GAMESS (Ref 3 (Schmidt et al)) at the restricted Hartree–Fock (RHF) level of theory using the minimal STO-3G basis set (Ref 4 (Hehre et al)). A minimum energy structure identified by the aforementioned procedure was used to initiate a series of molecular mechanics calculations; simulated annealing (Ref 5 (Kirkpatrick et al.)) was used to generate several monomer conformations. The process was repeated until four monomer conformations were obtained from the parent. Each of these generated conformers was then locally optimized using GAMESS at the RHF/STO-3G level of theory in the same manner as the original 3D structure. The energies of these structures were compared, and the structure with the overall minimum energy (structure S4) was chosen as the "parent" for the next simulated annealing cycle.

From this annealing, five new structures, S4-1 through S4-5, shown in Figure 5.8, were generated and relaxed. Due to the size of these systems, it was not feasible to obtain a completely optimized structure. The lowest energy conformation obtained in the second annealing cycle (structure S4-5 in Figure 5.8) was used in the simulation of the pair distribution function (PDF) and the NMR spectra, which were compared to experimental spectra obtained on the GR1.9 demineralized kerogen sample discussed earlier in this chapter.

5.6.1 Simulated ^{13}C NMR Spectra

Simulated ^{13}C NMR spectra were calculated for structures S4-1 through S4-5, as shown in Figure 5.9. The simulated NMR spectra were obtained from density functional theory calculations with the PBE1PBE (Adamo and Barone 1999) exchange correlation functional and the 4-31G basis set (Ditchfield et al. 1971, Hehre et al. 1972) as implemented in the Gaussian09 suite of programs (Frisch et al. 2009). The calculated chemical shielding values were converted to chemical shifts on the tetramethylsilane (TMS) scale using the shielding calculation of methane at the same level of theory, 200.5 ppm, adjusted by −7 ppm, which is the chemical shift of dilute methane on the TMS scale (Jameson and Jameson 1987). Gaussian broadening of 2 ppm along with Lorentzian broadening of 1 ppm was applied on the aliphatic region, with 5 ppm Gaussian broadening used in the aromatic region to obtain the simulated solid-state NMR spectrum. All of the spectra are similar. However, they do show differences in the line shape in the aliphatic region.

The predicted and measured ^{13}C NMR spectra are shown in Figure 5.10. The agreement of the relative intensities is a reflection that the model and the experiment have the same ratios between aromatic and aliphatic

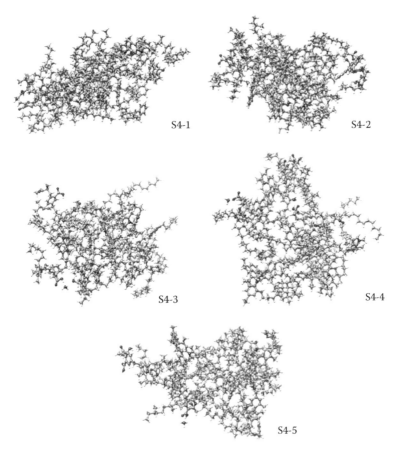

FIGURE 5.8
Local structures generated by subjecting kerogen structure S4 to the simulated annealing procedure. The atom colors are C, gray; O, red; N, blue; S, yellow; and H, white. The tubes represent
the molecule's backbone and the spheres represent the atoms.

components (28% aromatic/olefinic/carbonyl from the model and 24% from
the experimental NMR). Both the experiment and theory show the same tail
to higher chemical shifts due to the presence of the carbonyl carbons. This
tail is more pronounced in the experimental spectrum, possibly an indication of the higher oxygen content of the sample used in this work versus that
used in the construction of the 2D Siskin model.

5.6.2 Pair Distribution Function Plots

The PDFs were calculated using DISCUS and plotted using KUPLOT,
both part of the DIFFUSE (Proffen and Neder 1997) suite of packages.

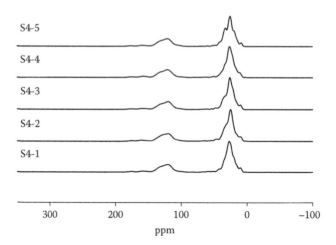

FIGURE 5.9
Simulated ¹³C nuclear magnetic resonance spectra for models S4-1 through S4-5.

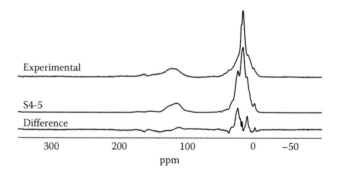

FIGURE 5.10
Comparison between simulated ¹³C NMR spectrum from model S4-5 and the experimental solid-state ¹³C NMR spectrum obtained on a Green River oil shale kerogen. The root mean square difference between S4-5 and experimental spectrum is 8 ppm.

Atomic coordinates of the model were used to calculate a PDF using the following equation:

$$G(r) = \frac{1}{r}\sum_{\nu}\sum_{\mu}\frac{f(0)_\nu\, f(0)_\mu}{f(0)^2}\,\delta(r - r_{\nu\mu}) - 4\pi r\rho_0 \tanh\left(S(R - r)\right) \qquad (5.1)$$

where
  r is the radius
  δ is the Dirac delta function
  ρ_0 is the average number density of the kerogen
  $f(0)_\nu$ and $f(0)_\mu$ are the x-ray atomic form factors for atoms ν and μ
  $\langle f(0)\rangle^2$ is the square of the average x-ray atomic form factors

The summation is made over all pairs of atoms ν and μ within the model separated by $r_{\nu\mu}$. The subtraction of $4\pi r\rho_0$ from the $G(r)$ in Equation 5.1 leads to the function being equal to zero at large radial distances. The shape term ($\tanh (S(R-r))$) corrects for the finite size and irregular boundaries of the kerogen model. This correction was derived from Equation 4 in the paper of Neder and Korsunskiy (2005), where S is related to the model shape and R to the model diameter.

The initial models consisting of a single kerogen unit were not sufficient to mimic the bulk kerogen. Therefore, a larger, 12-unit model was constructed in a manner that minimized the amount of "dead" spaces around the corners of the confining box as the calculation of the PDF is based on a rectangular box with no void spaces around the molecule. The larger molecule is shown in Figure 5.11.

The predicted PDF for the 12-unit molecule is shown in Figure 5.12. The first set of peaks in the PDF corresponds to C–H distances, whereas the second set corresponds to the geminal C–C distance; geminal denotes substituent atoms or groups, especially protons, attached to the same atom in a molecule. In the figure, the C–H distance is approximately 1.5 Å for aliphatic carbons and 1.4 Å for aromatic carbons. The geminal C–C distance is approximately 2.4 and 2.6 Å for aromatic and aliphatic carbons, respectively.

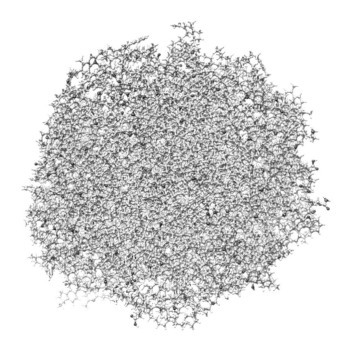

FIGURE 5.11
Three-dimensional structure of the 12-unit kerogen model. The atom colors and molecule description are the same as in Figure 5.8.

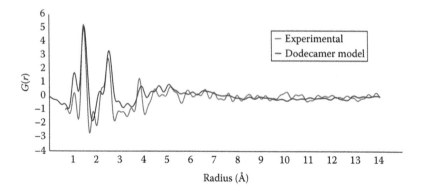

FIGURE 5.12
Comparison of experimentally determined for Green River kerogen and the 12-unit model. The dodecamer model was shape- and size-corrected using Equation 5.1 with $S = 0.05$ and $R = 19.3$ Å.

The peak at approximately 3 Å corresponds to the distance between carbons separated by four bonds in a cis configuration and the peak at approximately 3.8 Å to carbons in a trans configuration.

Overall, there is good agreement between the model and experimental PDFs, especially at shorter distances. However, the model is less accurate for distances between 4 and 6 Å. For distances above 6 Å, the PDF provides very poor resolution. A comparison of the peak intensities shows differences, especially in the intensity of the C–H peak. The fact that the C–H peak is smaller in the experiment versus the model-generated PDF is indicative of having fewer protonated carbons in the experimental sample and therefore more heteroatoms and aromatic carbons (which have a lower H to C ratio than aliphatic carbons). A greater proportion of heteroatoms (N, S, and O) in the sample compared to the model is consistent with the results of the elemental analysis of the kerogen samples presented in Table 5.3. While there is overall consistency between the model and experimental PDFs, this consistency does not provide any apparent structural information as the lack of features in the aforementioned model above approximately 5 Å makes it impossible to obtain longer-range structural information.

5.7 Conclusions

The purpose of this study was to examine various sections of a carefully prepared oil shale core taken from the Mahogany zone in the Green River Formation of eastern Utah (Uinta Skyline 16 location). Three 1 ft sections of the 4 in. diameter, 1000 ft long core were selected for the analysis. Samples from these 1 ft sections were separately crushed, sieved, and then demineralized. The organic content varied with each section (15.8, 20.51, and 25.95 wt%), but the organic structural components of the shale (the bitumen and kerogen isolates) were quite similar. Within the experimental error, the aromatic contents of the kerogens from GR1, GR2, and GR3 were essentially identical (24%, 23%, and 24%). The fraction of aromatic carbons in the extracted bitumen was 11%, 9%, and 7% for the three oil shale samples. The S/N ratio on the bitumen samples was very low, and the aromatic content was within experimental error for all three bitumen samples. The bitumen aliphatic structure was dominated by long chains with an average length of about 24 carbons. There was evidence that the amount of branching was small but not insignificant, which relates to the isoprenoids that have been observed in pyrolysis experiments.

The SP ^{13}C NMR experiments on the demineralized kerogen showed the absence of inorganic carbon structures (i.e., carbonates). The ^{13}C CP/MAS NMR experiments allowed the determination of kerogen lattice parameters for the first time, showing that the kerogen had approximately 10 aromatic carbons per cluster with an average side chain length of 11–13 carbons. The average molecular weight per cluster ranged from 775 to 946 for the three demineralized samples with an average molecular weight per side chain of 131–148. The number of attachments per cluster or coordination number (σ + 1) ranged from 4.5 to 5.9 for the three demineralized kerogen samples. The value of these chemical structure data and lattice parameters is in the interpretation of pyrolysis data and in the generation of chemical structure–based pyrolysis models, as described in Chapter 6.

Three-dimensional models for Green River kerogen were generated from Siskin's 2D model using an iterative molecular mechanics–based annealing followed by ab initio–based structure refinement. The models generated in this manner were used to simulate ^{13}C NMR spectra in order to explore the sensitivity of these spectra to the 3D structure. The simulated NMR spectra were shown to be in reasonable agreement with experimental spectra obtained for the kerogen extracted from the GR1, GR2, and GR3 samples. A larger (12-unit) kerogen molecule was required in order to obtain a simulated PDF spectrum that was similar to the experimental spectrum.

References

Adamo, C. and V. Barone. 1999. Toward reliable density functional methods without adjustable parameters: The PBE0 model. *J. Chem. Phys.* 110:6158–6169.

Allinger, N. L. 1977. Conformational analysis 130. MM2. A hydrocarbon force field utilizing V1 and V2 torsional terms. *J. Am. Chem. Soc.*, 99:8127–8134.

Andrews, A. 2006. Oil shale: History, incentives, and policy. CRS report for Congress, Washington, DC.

Barich, D. H., R. J. Pugmire, D. M. Grant, and R. J. Iuliucci. 2001. Investigation of the structural conformation of biphenyl by solid state ^{13}C NMR and quantum chemical NMR shift calculations. *J. Phys. Chem. A* 105(28):6780–6784.

Bartuska, V. J., G. E. Maciel, J. Schaefer, and E. O. Stejskal. 1977. Prospects for carbon-13 nuclear magnetic resonance analysis of solid fossil-fuel materials. *Fuel* 56:354–358.

Brons, G. and M. Siskin. 1988. Particle size reduction studies on Green River oil shale. *Energy Fuels* 2(5):628–633.

Burnham, A. K. 1991. Oil evolution from a self-purging reactor: Kinetics and composition at 2°C/min and 2°C/h. *Energy Fuels* 5(1):205–214.

Burnham, A. K., J. E. Clarkson, M. F. Singleton, C. M. Wong, and R. W. Crawford. 1982. Biological markers from Green River kerogen decomposition. *Geochim. Cosmochim. Acta* 46(7):1243–1251.

Burnham, A. K., H. R. Gregg, R. L. Ward, K. J. Knauss, S. A. Copenhaver, and J. G. Reynolds. 1997. Decomposition kinetics and mechanism of *n*-hexadecane-1,2-^{13}C2 and dodec-1-ene-1,2-^{13}C2 doped in petroleum and *n*-hexadecane. *Geochim. Cosmochim. Acta* 61(17):3725–3737.

Crouch, R. 2012. Acquiring quantitative small-molecule NMR data with QDEPT. Agilent Technologies Application Note. www.agilent.com. Accessed on January 30, 2012.

Dalling, D. K., R. J. Pugmire, D. M. Grant, and W. E. Hull. 1986. The use of high-field carbon-13 NMR spectroscopy to characterize chiral centers in isopranes. *Magn. Resonan. Chem.* 24(3):191–198.

Ditchfield, R., W. J. Hehre, and J. A. Pople. 1971. Self-consistent molecular orbital methods. 9. Extended Gaussian-type basis for molecular-orbital studies of organic molecules. *J. Chem. Phys.* 54:724.

Dyni, J. R. 2006. Geology and resources of some world oil-shale deposits. Scientific Investigations Report 2005-5294. Reston, VA: U.S. Geological Survey.

Fletcher, T. H., D. C. Barfuss, and R. J. Pugmire. 2014. Modeling oil shale pyrolysis using the Chemical Percolation Devolatilization model. Paper presented at the *34th Oil Shale Symposium*, Golden, CO.

Fletcher, T. H., R. Gillis, J. Adams et al. 2014. Characterization of macromolecular structure elements from a Green River oil shale, II. Characterization of pyrolysis products by ^{13}C NMR, GC/MS, and FTIR. *Energy Fuels* 28:2959–2970.

Fletcher, T. H., A. R. Kerstein, R. J. Pugmire, and D. M. Grant. 1990. Chemical percolation model for devolatilization. 2. Temperature and heating rate effects on product yields. *Energy Fuels* 4(1):54–60.

Fletcher, T. H., A. R. Kerstein, R. J. Pugmire, M. S. Solum, and D. M. Grant. 1992. Chemical percolation model for devolatilization. 3. Direct use of carbon-13 NMR data to predict effects of coal type. *Energy Fuels* 6(4):414–431.

Fletcher, T. H., H. R. Pond, J. Webster, J. Wooters, and L. L. Baxter. 2012. Prediction of tar and light gas during pyrolysis of black liquor and biomass. *Energy Fuels* 26:3381–3387.

Fletcher, T. H. and R. J. Pugmire. 2014. *Rates and Mechanisms of Oil Shale Pyrolysis: A Chemical Structure Approach.* Salt Lake City, UT: University of Utah Institute for Clean and Secure Energy.

Fletcher, T. H., M. S. Solum, D. M. Grant, S. Critchfield, and R. J. Pugmire. 1990. Solid-state ¹³C and ¹H NMR studies of the evolution of the chemical structure of coal char and tar during devolatilization. *23rd Symposium (International) on Combustion*, Pittsburgh, PA, pp. 1231–1237.

Fletcher, T. H., M. S. Solum, D. M. Grant, and R. J. Pugmire. 1992. Chemical structure of char in the transition from devolatilization to combustion. *Energy Fuels* 6(5):643–650.

Freund, H., S. R. Kelemen, C. C. Walters et al. 2007. Predicting oil and gas compositional yields via chemical structure-chemical yield modeling (CS-CYM): Part 1—Concepts and implementation. *Organ. Geochem.* 38(2):288–305.

Frisch, M. J., G. W. Trucks, H. B. Schlegel et al. 2009. Gaussian 09, Revision B.01. Wallingford CT: Gaussian, Inc.

Genetti, D., T. H. Fletcher, and R. J. Pugmire. 1998. Development and application of a correlation of ¹³C NMR chemical structural analyses of coal based on elemental composition and volatile matter content. *Energy Fuels* 13(1):60–68.

Grant, D. M., R. J. Pugmire, T. H. Fletcher, and A. R. Kerstein. 1989. Chemical model of coal devolatilization using percolation lattice statistics. *Energy Fuels* 3(2):175–186.

Hagaman, E. W., F. M. Schell, and D. C. Cronauer. 1984. Oil-shale analysis by CP/MAS-¹³C NMR spectroscopy. *Fuel* 63(7):915–919.

Hehre, W. J., R. Ditchfield, and J. A. Pople. 1972. Self-consistent molecular orbital methods. 12. Further extensions of Gaussian-type basis sets for use in molecular-orbital studies of organic molecules. *J. Chem. Phys.* 56:2257.

Hehre, W. J., R. F. Stewart, and J. A. Pople. 1969. Self-consistent molecular orbital methods I. Use of Gaussian expansions of Slater type atomic orbitals. *J. Chem. Phys.* 51:2657.

Henderson, T. J. 2004. Sensitivity-enhanced quantitative ¹³C NMR spectroscopy via cancellation of 1JCH dependence in DEPT polarization transfers. *J. Am. Chem. Soc.* 126(12):3682–3683.

Hillier, J. L. 2011. Pyrolysis kinetics and chemical structure considerations of a Green River oil shale and its derivatives. Ph.D. dissertation, Brigham Young University, Provo, UT.

Hillier, J. L., T. H. Fletcher, M. S. Solum, and R. J. Pugmire. 2013. Characterization of macromolecular structure of pyrolysis products from a Colorado Green River oil shale. *Ind. Eng. Chem. Res.* 52(44): 15522–15532.

Hobbs, M. L., K. L. Erickson, T. Y. Chu et al. 2003. CPUF—A chemical-structure-based polyurethane foam decomposition and foam response model. Sandia Report SAND2003-2282. Albuquerque, NM and Livermore, CA: Sandia National Laboratories.

HyperChem(TM) Professional 7, Hypercube, Inc., 1115 NW 4th Street, Gainesville, Florida 32601, USA. Accessed March 28, 2016.

Jameson, A. K. and C. J. Jameson. 1987. Gas-phase ¹³C chemical shifts in the zero-pressure limit: Refinements to the absolute shielding scale for ¹³C. *Chem. Phys. Lett.* 134:461.

Jiang, Y. J., M. S. Solum, R. J. Pugmire, D. M. Grant, H. H. Schobert, and P. J. Pappano. 2002. A new method for measuring the graphite content of anthracite coals and soots. *Energy Fuels* 16(5):1296–1300.

Johnson, R. C., T. J. Mercier, M. E. Brownfield, M. P. Pantea, and J. G. Self. 2010. An assessment of in-place oil shale resources in the Green River Formation, Piceance Basin, Colorado. In *Oil Shale and Nahcolite Resources of the Piceance Basin, Colorado, U.S. Geological Survey Digital Data Series 69–Y, Chapter 1*. Reston, VA: U.S. Geological Survey.

Kelemen, S. R., M. Afeworki, M. L. Gorbaty et al. 2002. XPS and 15N NMR study of nitrogen forms in carbonaceous solids. *Energy Fuels* 16(6):1507–1515.

Kelemen, S. R., M. Afeworki, M. L. Gorbaty et al. 2007. Direct characterization of kerogen by X-ray and solid-state ^{13}C nuclear magnetic resonance methods. *Energy Fuels* 21(3):1548–1561.

Kelemen, S. R., H. Freund, M. L. Gorbaty, and P. J. Kwiatek. 1999. Thermal chemistry of nitrogen in kerogen and low-rank coal. *Energy Fuels* 13(2):529–538.

Kelemen, S. R., M. L. Gorbaty, and P. J. Kwiatek. 1994. Quantification of nitrogen forms in Argonne premium coals. *Energy Fuels* 8(4):896–906.

Kelemen, S. R., M. Siskin, H. S. Homan, R. J. Pugmire, and M. S. Solum. 1998. Fuel, lubricant and additive effects on combustion chamber deposits. SAE technical paper 982715.

Kirkpatrick, S., C. D. Gelatt, and M. P. Vecch. 1983. Optimization by simulated annealing. Science 220:671–680.

Lewis, A. D. and T. H. Fletcher. 2013. Prediction of sawdust pyrolysis yields from a flat-flame burner using the CPD model. *Energy Fuels* 27:942–953.

Maciel, G. E. and L. W. Dennis. 1981. Comparison of oil shales and kerogen concentrates by ^{13}C nuclear magnetic resonance. *Organ. Geochem.* 3(4):105–109.

Mao, J., X. Fang, Y. Lan et al. 2010. Chemical and nanometer-scale structure of kerogen and its change during thermal maturation investigated by advanced solid-state ^{13}C NMR spectroscopy. *Geochim. Cosmochim. Acta* 74(7):2110–2127.

Mao, J. D. and K. Schmidt-Rohr. 2005. Methylene spectral editing in solid-state ^{13}C NMR by three-spin coherence selection. *J. Magn. Resonan.* 176(1):1–6.

Miknis, F. P. 1994. Introduction to mass spectrometric techniques for fossil fuel analysis. In *Composition, Chemistry and Conversion of Oil Shales, NATO ASI Series; Series C: Mathematical and Physics Sciences*, Vol. 455, ed. C. Snape, pp. 69–91. Dordrecht, the Netherlands: Kluwer Academic Publishers.

Miknis, F. P., G. E. Maciel, and V. J. Bartuska. 1979. Cross polarization magic-angle spinning ^{13}C NMR spectra of oil shales. *Organ. Geochem.* 1(3):169–176.

Miknis, F. P., N. M. Szeverenyi, and G. E. Maciel. 1982. Characterization of the residual carbon in retorted oil shale by solid-state ^{13}C NMR. *Fuel* 61(4):341–345.

Neder, R. B. and V. I. Korsunskiy. 2005. Structure of nanoparticles from powder diffraction data using the pair distribution function. *J. Phys. Condens. Matter* 17:S125.

Netzel, D. A., D. R. McKay, R. A. Heppner et al. 1981. ^{1}H- and ^{13}C-NMR studies on naphtha and light distillate saturate hydrocarbon fractions obtained from in-situ shale oil. *Fuel* 60(4):307–320.

Niksa, S. and A. R. Kerstein. 1991. FLASHCHAIN theory for rapid coal devolatilization kinetics. 1. Formulation. *Energy Fuels* 5(5):647–665.

Orendt, A. M., I. S. O. Pimienta, S. R. Badu et al. 2013. Three-dimensional structure of the Siskin green river oil shale kerogen model: A comparison between calculated and observed properties. *Energy Fuels* 27:702–710.

Petsch, S. T., R. J. Smernik, T. I. Eglinton, and J. M. Oades. 2001. A solid state ^{13}C-NMR study of kerogen degradation during black shale weathering. *Geochim. Cosmochim. Acta* 65(12):1867–1882.

Proffen, T. and R. B. Neder. 1997. DISCUS: A program for diffuse scattering and defect-structure simulation. *J. Appl. Crystallogr.* 30:171.

Resing, H. A., A. N. Garroway, and R. N. Hazlett. 1978. Determination of aromatic hydrocarbon fraction in oil shale by ^{13}C N.M.R. with magic-angle spinning. *Fuel* 57(8):450–454.

Salmon, E., F. Behar, and P. G. Hatcher. 2011. Molecular characterization of Type I kerogen from the Green River Formation using advanced NMR techniques in combination with electrospray ionization/ultrahigh resolution mass spectrometry. *Organ. Geochem.* 42(4):301–315.

Schmidt, M. W., K. K. Baldridge, and J. A. Boatz et al. 1993. General atomic and molecular electronic structure system. *J. Comput. Chem.* 14:1347–1363.

Siskin, M., C. G. Scouten, K. D. Rose et al. 1995. Detailed structural characterization of the organic material in Rundle Ramsay Crossing and Green River oil shales. In *Composition, Geochemistry and Conversion of Oil Shales, NATO ASI Series; Series C: Mathematical and Physics Sciences*, Vol. 455, ed. C. Snape, pp. 143–158. Dordrecht, the Netherlands: Kluwer Academic Publishers.

Solomon, P. R., D. G. Hamblen, R. M. Carangelo, M. A. Serio, and G. V. Deshpande. 1988. General model of coal devolatilization. *Energy Fuels* 2(4):405–422.

Solum, M. S., C. L. Mayne, A. M. Orendt, R. J. Pugmire, J. Adams, and T. H. Fletcher. 2014. Characterization of macromolecular structure elements from a Green River oil shale, I. Extracts. *Energy Fuels* 28:453–465.

Solum, M. S., R. J. Pugmire, and D. M. Grant. 1989. Carbon-13 solid-state NMR of Argonne premium coals. *Energy Fuels* 3(2):187–193.

Solum, M. S., A. F. Sarofim, R. J. Pugmire, T. H. Fletcher, and H. Zhang. 2001. ^{13}C NMR analysis of soot produced from model compounds and a coal. *Energy Fuels* 15(4):961–971.

Ulibarri, T. A., D. K. Derzon, K. L. Erickson et al. 2002. Preliminary investigation of the thermal decomposition of Ablefoam and EF-AR20 foam (Ablefoam replacement). Sandia Report SAND2002-0183. Albuquerque, NM and Livermore, CA: Sandia National Laboratories.

Van Niekerk, D., R. J. Pugmire, M. S. Solum, P. C. Painter, and J. P. Mathews. 2008. Structural characterization of vitrinite-rich and inertinite-rich Permian-aged South African bituminous coals. *Int. J. Coal Geol.* 76(4):290–300.

Vandegrift, G. F., R. E. Winans, R. G. Scott, and E. P. Horwitz. 1980. Quantitative study of the carboxylic acids in Green River oil shale bitumen. *Fuel* 59(9):627–633.

Vandenbroucke, M. 2003. Kerogen: From types to models of chemical structure. *Oil Gas Sci. Technol.* 58(2):243–269.

Walters, C. C., H. Freund, S. R. Kelemen, P. Peczak, and D. J. Curry. 2007. Predicting oil and gas compositional yields via chemical structure–chemical yield modeling (CS–CYM): Part 2 – Application under laboratory and geologic conditions. *Energy Fuels* 38:306–322.

Ward, R. L. and A. K. Burnham. 1984. Identification by ^{13}C NMR of carbon types in shale oil and their relation to pyrolysis conditions. *Fuel* 63(7):909–914.

Yan, S., Y.-J. Jiang, N. D. Marsh, E. G. Eddings, A. F. Sarofim, and R. J. Pugmire. 2005. Study of the evolution of soot from various fuels. *Energy Fuels* 19(5):1804–1811.

Zhang, H., T. H. Fletcher, S. T. Perry, M. S. Solum, and R. J. Pugmire. 2001. Soot formation during coal pyrolysis. Paper presented at the *11th International Conference on Coal Science*, San Francisco, CA.

6

Oil Shale Pyrolysis Rates and Mechanisms

Thomas H. Fletcher, Ronald J. Pugmire,
Mark S. Solum, and Charles L. Mayne

CONTENTS

This chapter discusses rates of oil shale pyrolysis, yields of pyrolysis products, characterization of pyrolysis products, mechanistic pyrolysis modeling, and changes in char aromaticity during pyrolysis. The samples used for these experiments and analyses are the same Green River oil shale and demineralized kerogen samples discussed in Chapter 5. Mass release rates of these samples during pyrolysis were measured in a thermogravimetric analyzer (TGA), and corresponding kinetic rates were determined. Product analyses were then performed by a combination of carbon-13 nuclear magnetic resonance (^{13}C NMR) spectroscopy, gas chromatography/mass spectrometry (GC/MS), and Fourier transform infrared (FTIR) spectroscopy techniques, including solid-state ^{13}C NMR analysis of the pyrolyzed kerogen char samples obtained at different temperatures and liquid-state ^{13}C NMR analysis of the associated tars collected during pyrolysis.

6.1 Background

The pyrolysis of kerogen yields light gas, tar (oil), and char (coke). Tar is defined as any volatile matter that will later condense at room temperature and is thus the oil fraction of the pyrolysis products. Char is the organic material left behind after pyrolysis. Reviews of pyrolysis rates and chemical analyses of volatile products from oil shale were recently published (Burnham 2010, Hillier 2011, Hillier and Fletcher 2011, Tiwari 2012, Tiwari and Deo 2012a,b, Hillier et al. 2013). Light gases have been studied by several research groups (Campbell et al. 1978, Huss and Burnham 1982, Meuzelaar et al. 1986, Braun et al. 1992, Burnham 2010). Huizinga et al. (1988) used pyrolysis MS to view the nature and distribution of products of kerogen pyrolysis of an immature Green River sample from the Red Point mine in Piceance, Colorado. The experiment included pressures of 100 psi (6.8 atm) of helium (He) over a mixture of the sample and water. Huizinga found that the pyrolysis products included *n*-alkane/alkene pairs with a preference for odd carbon numbers. Reynolds et al. (1991) performed on-line MS to analyze the gas evolution and species composition during oil shale pyrolysis. Tiwari and Deo (Tiwari 2012, Tiwari and Deo 2012a,b) used TGA coupled with MS to study the evolution of species in the 300 amu range during pyrolysis of a Utah oil shale at heating rates of 0.5–10 K/min. Lighter species evolved slightly earlier on the temperature scale and significant amounts of alkenes were reported.

Hagaman et al. (1984) used ^{13}C NMR spectroscopy to estimate potential oil yields from shales by utilizing the aliphatic content. They found that throughout the Mahogany zone (the zone of highest oil yield per ton of rock) of a Green River oil shale, the aliphatic carbon fraction was nearly constant. Boucher et al. (1991) utilized NMR and a complementary ruthenium oxide (RuO_2) mild oxidation technique on a type II kerogen, reporting a carbon aromaticity of 25.8% with predominantly methylene (CH_2) groups in rings or chains or attached to oxygen.

Miknis and coworkers (Miknis et al. 1982, 1987, Netzel and Miknis 1982) performed NMR on several oil samples and residual carbons from oil shale retorts, showing differences between shales of different geological formations. Maciel et al. (1978) reported that oil yields from oil shale correlated with aliphatic carbon content as measured by ^{13}C NMR spectroscopy, and a more general correlation was developed by Miknis and Conn (1986). Solomon and Miknis (1980) developed a similar correlation of oil yield using aliphatic carbon content as measured by FTIR spectroscopy on crushed oil shale pressed into potassium bromide pellets.

6.2 Oil Shale Pyrolysis Rates

6.2.1 Thermogravimetric Analysis Data

TGA data were obtained for three samples of crushed and sieved oil shale from the Green River Formation in Utah (GR1, GR2, and GR3 samples from the Skyline 16 core described in Chapter 5). Mean particle diameters were 60 μm. Approximately 10 mg of each sample was heated to 850°C at three heating rates (1, 5, and 10 K/min) and two pressures (1 and 40 bar) in the TGA. The sample hold time at 850°C was 5 min. Three separate experiments were performed at each condition. Flow rates of the He were kept constant for all runs in order to minimize randomizing factors.

Characteristic TGA data are shown in Figure 6.1. The actual data are shown superimposed on the model data. Two separate reactions were observed in the TGA mass versus time curves: kerogen and calcium carbonate pyrolysis. At atmospheric pressure, the kerogen pyrolyzes at approximately 450°C and the calcium carbonate at 650°C. It has been shown that the temperature where pyrolysis occurs increases slightly with increases in heating rate (Hillier and Fletcher 2011, Tiwari and Deo 2012a).

6.2.2 Rate Determination

Once the data were reduced to eliminate noise and the effects of buoyancy, the data were fit with a first-order reaction rate expression, Equation 6.1, or a first-order distributed activation energy model (DAEM), Equation 6.2, following the method of Hillier (Hillier et al. 2010, 2013, Hillier and Fletcher 2011).

$$\frac{dm}{dt} = -A e^{-E/RT} \, m \tag{6.1}$$

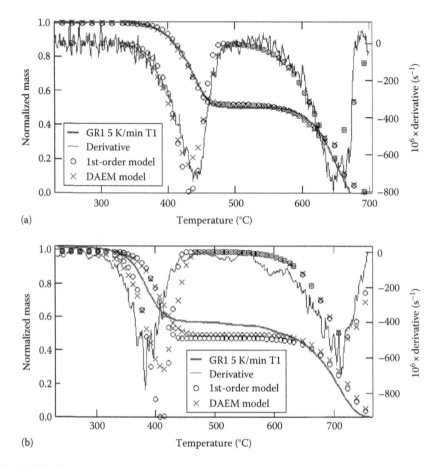

(a)

(b)

FIGURE 6.1

TGA pyrolysis data and best-fit model calculations for GR1 oil shale, crushed and sieved to 60 μm mass mean diameter and heated at 5 K/min. (a) Atmospheric pressure, (b) 40 bar.

$$\frac{dm}{dt} = -A e^{-(E \pm \sigma)/RT} \, m \qquad (6.2)$$

where
  m is the sample mass
  A and E are Arrhenius kinetic coefficients
  R is the gas constant
  T is the temperature

The standard deviation (σ) of the activation energy (E) is a fitting parameter in the first-order DAEM. An optimization program was used to determine the coefficients A, E, and σ. A sequential form of the DAEM, where the activation energy becomes a function of the extent of conversion, was used,

TABLE 6.1

Kinetic Coefficients Determined from TGA Pyrolysis of Green River Oil Shale Samples

Sample		First Order		DAEM	
		1 atm	40 bar	1 atm	40 bar
GR1	A (1/s)	8.90×10^{13}	2.80×10^{14}	9.20×10^{13}	1.00×10^{14}
	E (kJ/mol)	221	219	223	215
	σ (kJ)	—	—	4	2.6
GR2	A (1/s)	4.50×10^{13}	8.00×10^{13}	2.60×10^{14}	3.00×10^{14}
	E (kJ/mol)	216.9	210	228.1	219.4
	σ (kJ)	—	—	2.6	6.7
GR3	A (1/s)	9.50×10^{13}	1.50×10^{14}	9.40×10^{13}	3.50×10^{14}
	E (kJ/mol)	220	217	222	225
	σ (kJ)	—	—	4.6	5.3

and only one activation energy was used at each time step (Fletcher et al. 1992). This method fits the raw mass versus time data as well as the derivative of the mass versus time curve for several different heating rates and for repeat experiments simultaneously without any approximate solutions. This method is therefore thought to be the most accurate method of extracting rate coefficients from TGA data.

Separate kinetic coefficients were determined for each of the three oil shale samples (GR1, GR2, and GR3) at each pressure for both the first-order model and the DAEM model (Fletcher and Pugmire 2014). Figure 6.1 shows TGA data with the corresponding curve fits for a sample heated at a rate of 5 K/min at both atmospheric pressure and 40 bar. The kinetic coefficients were determined by fitting data from all three heating rates (1, 5, and 10 K/min). Table 6.1 shows the resulting kinetic coefficients for all three GR samples.

6.2.3 CO_2 Formation Rate

The CO_2 formation from the oil shale pyrolysis at temperatures above 575°C was also fit with mass release kinetic models. This analysis can be found in Fletcher and Pugmire (2014) and Hillier and Fletcher (2011).

6.3 Yields and Characterization of Pyrolysis Products

6.3.1 Experimental Methodology

6.3.1.1 Kerogen Retort

A pyrolysis reactor, referred to as a kerogen retort, was built to pyrolyze the demineralized oil shale (i.e., kerogen) samples described in Chapter 5 and to

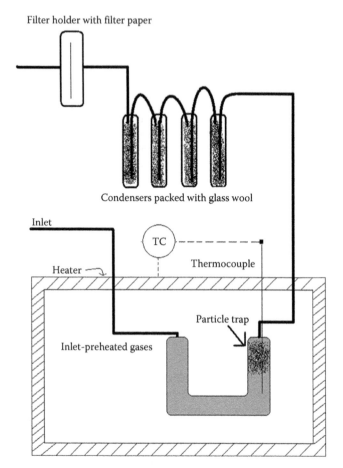

FIGURE 6.2
Schematic of the kerogen retort used for generating chars and tars.

collect char and tar samples in sufficient quantity for further analysis (see Figure 6.2). This retort, described in detail by Hillier (Hillier 2011, Hillier and Fletcher 2011), was designed to mimic heating rate conditions (10 K/min) found in previous TGA experiments (Hillier et al. 2013).

The retort was made from ¾-in. stainless steel tubing bent into a U-like shape as seen in Figure 6.2. Inlet tubing inside the heater served to preheat the gases. The heater outlet, made of ¼-in. stainless steel tubing, was designed to allow for thermocouple access. A glass wool plug was included in the outlet region of the reactor to prevent particles from becoming entrained and traveling into the condensers while allowing the vaporized tars and other gases to freely exit the furnace. Four condensers were immersed in a cooling bath (isopropanol and dry ice) to aid tar condensation; the tars were collected on the glass wool and were easily extracted via a dichloromethane

(DCM) wash. A filter holder with filter paper was included downstream of the condensers to verify that tars did not travel downstream. The filter turned brown in experiments that lacked proper or insufficient packing of the glass wool. Outlet gases were directed into a collection bag for gas analysis.

Approximately 1 g of sample, spread out inside the reactor tube, was used for each experiment. The nitrogen (N_2) purge gas flowed at approximately 1 liter per minute (L/min). An additional experiment was conducted with He and the results did not differ from the N_2 experiments. The heating rate used in these experiments was 10 K/min. The retort was physically removed from the furnace once the desired temperature was achieved for each experiment, thus cooling the retort and samples at a rate of approximately 100 K/min. The final temperature was kept below 525°C to prevent CO_2 release from the carbonates in the shale.

6.3.1.2 Elemental Analysis

The elemental analyses of the chars from the kerogen retort are reported in Table 6.2. Plots of carbon (C), hydrogen (H), and H to C ratio are shown in Figures 6.3 and 6.4. The hydrogen was released preferentially to carbon above a temperature of 450°C. The final carbon content of the char was 81%–85% on a dry, ash-free (daf) basis.

TABLE 6.2

Elemental Analysis of the Chars from the Kerogen Retort

	T (°C)	C (%)	H (%)	N (%)	O by Diff	S (%)
			Wt% Dry, Ash-Free			
GR1.9	300	78.85	9.58	2.90	6.56	2.11
	375	81.25	9.96	3.09	3.73	1.98
	410	80.38	9.67	3.13	4.91	1.93
	445	81.57	9.65	3.42	3.51	1.85
	495	85.16	6.14	5.42	−1.28	4.56
GR2.9	425	80.44	9.60	3.05	4.38	2.52
	445	82.65	9.30	3.42	2.58	2.04
	475	83.40	8.40	4.04	1.21	2.95
	525	86.80	3.98	5.89	−3.76	7.09
GR3.9	400	79.77	9.45	2.91	5.19	2.68
	434	80.84	9.22	3.12	2.93	3.89
	450	81.22	8.61	3.46	3.44	3.27
	460	82.64	8.32	3.78	1.78	3.48
	470	82.31	7.89	3.96	1.84	3.99
	490	82.50	4.93	5.22	0.86	6.49
	525	82.64	3.77	5.52	1.25	6.81

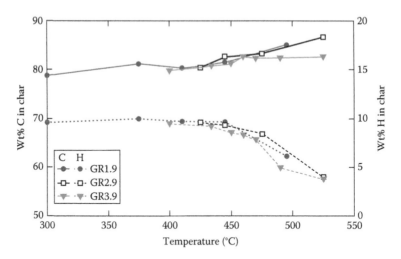

FIGURE 6.3
Carbon and hydrogen contents of the chars from the kerogen retort collected at different temperatures. The heating rate was 10 K/min.

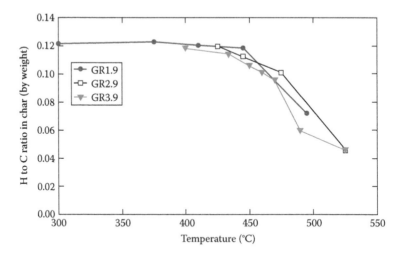

FIGURE 6.4
Hydrogen to carbon ratios of the chars from the kerogen retort collected at different temperatures. The heating rate was 10 K/min.

6.3.1.3 ^{13}C NMR Analysis

Solid-state ^{13}C NMR analysis was performed on the demineralized kerogen samples (GR1.9, GR2.9, and GR3.9) and on the char remaining after pyrolysis in the kerogen retort; approximately 120–250 mg of char was required. This analysis included careful integration of the characteristic chemical shift ranges in both the normal cross-polarization with magic angle spinning (CP/MAS) data and the dipolar dephased data, as described by Solum et al. (1989).

The method of Solum and coworkers (Solum et al. 1989, 2001) was used to derive 14 structural elements plus 8 lattice parameters derived from the structural elements. Lattice parameters derived for the pyrolyzed char samples include the number of attachments per aromatic cluster ($\sigma + 1$), the number of aromatic carbons per aromatic cluster (C), and the number of aliphatic carbons in a side chain. Solum's method to determine the fraction of attachments that are bridges (i.e., p_0) for the unreacted kerogen samples was subject to large errors due to the large number of CH_3 and CH_2 groups in the sample. This error in p_0 may have been due to aliphatic chain branching resulting in more methyl terminating groups than could be accounted for without chain branching.

The tar samples from pyrolysis of the GR1.9, GR2.9, and GR3.9 kerogen concentrates at various temperatures were collected on glass wool and then extracted into deuterated dicholoromethane (CD_2Cl_2). The resulting samples had varying total volumes of solvent plus tar. Each sample was transferred to a 5 mm NMR tube with a screw cap and polytetrafluoro-ethylene-lined septum to prevent solvent loss during data acquisition. The total volume was then adjusted to 0.7 mL either by adding CD_2Cl_2 or by evaporating some of the solvent using a stream of dry N_2. All solution-state ^{13}C NMR data were acquired using an Agilent Technologies Direct Drive 500 MHz spectrometer with a 5 mm switchable broadband probe equipped with z-axis pulsed field gradient coil and sample temperature control. The sample temperature was controlled at 26°C using a stream of dry N_2 flowing at 10 L/min. The two methods (quantC and quantD) described in Chapter 5 were used to acquire quantitative ^{13}C spectra. Distortionless enhancement by polarization transfer (DEPT) spectra was obtained using the standard Agilent quantitative DEPT (QDEPT) pulse sequence to differentiate among nonprotonated CH, CH_2, and CH_3 carbons, as described in Chapter 5.

6.3.1.4 GC/MS Analysis

A TSQ 7000 triple quadrupole Finnigan mass spectrometer was used to analyze the composition of the tar samples. The inlet for most experiments was a Varian 3400 CX gas chromatograph operating in splitless injection mode (Fletcher and Pugmire 2014, Fletcher et al. 2014). The DB-1 column (J&W Scientific) was 20 m long with a 0.1 mm I.D. microbore, and a 0.4 μm film. The temperature program for the GC method started at 50°C for 1 min. Next, the column was heated at 8 K/min to 300°C and held at that temperature for 15 min. The solvent used to bring the pyrolyzate into solution was DCM. Additional details of the GC/MS system and technique are found in Hillier (2011).

6.3.1.5 FTIR Analysis

FTIR experiments were performed to determine the nature of the species in the gas phase. The species of interest were the light hydrocarbons, carbon monoxide (CO), and methanol. The experiments were carried out by collecting all the effluent gas from the kerogen retort at three specific temperature ranges (200°C–295°C, 300°C–395°C, and 400°C–495°C) designed to span the early, middle, and late gas generation phases during kerogen pyrolysis. The gases were collected in a 100L Tedlar® gas sampling bag. The FTIR was prepared by purging the sample cell with N_2, and the contents were pumped through the gas cell for a period of time to ensure a representative sample was present. More details are reported elsewhere (Hillier et al. 2013, Fletcher and Pugmire 2014, Fletcher et al. 2014).

6.3.2 Analysis of Pyrolysis Products from Kerogen Retort

6.3.2.1 Pyrolysis Product Yields

Yields of tar, char, and light gas from the pyrolysis of the demineralized kerogen samples in the kerogen retort at a heating rate of 10 K/min are shown in Figure 6.5 with tabulated values in Table 6.3. The experiments at 575°C likely were influenced by the decomposition of the carbonates in the shale minerals. Based on the data at 525°C, the total volatiles yields on a daf basis were 75.2%, 78.7%, and 83.8% for the GR1.9, 2.9, and 3.9 samples. Tar yields for the same samples were 59.2%, 57.5%, and 69%, respectively. The average uncertainty based on the difference from the mean for duplicate samples was 1 wt% for the total volatiles yield and 4 wt% for the tar yield. It is interesting that the daf yields of both tar and total volatiles were highest for the GR3.9 demineralized kerogen since that kerogen was derived from the oil shale with the lowest Fischer oil assay and with a midrange value for total organic content (see Chapter 5, Table 5.2). However, the yields of bitumen and kerogen from the extraction process were not well known, so it is difficult to compare the yields in the kerogen retort to the Fischer assay yields.

6.3.2.2 NMR Analysis of Char

Representative solid-state ^{13}C NMR chemical structure data and derived lattice parameters for the GR3.9 kerogen concentrate sample and resulting chars produced in the kerogen retort are presented in Table 6.4. Data for the other samples were reported by Fletcher and coworkers (Fletcher and Pugmire 2014, Fletcher et al. 2014). The ^{13}C CP/MAS spectra for the chars from the GR3.9 sample are shown in Figure 6.6. As reported in Chapter 5, the carbon aromaticity ($f_{a'}$) was about 20% in all three kerogen concentrates, with 9–12 aromatic carbons per cluster (C) and 11–12 aliphatic carbons per side chain, a value that was lower than expected from the mass spectrometer analysis

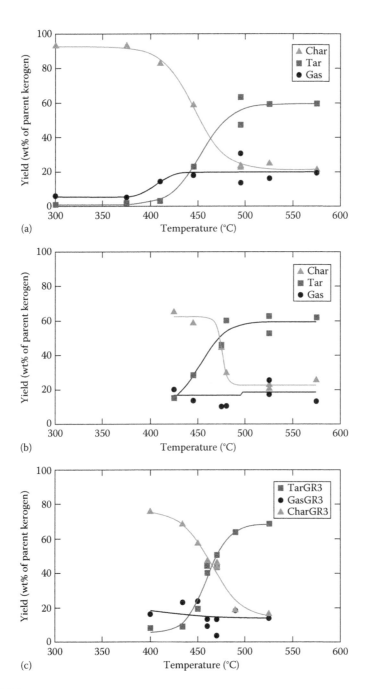

FIGURE 6.5
Yields of char, tar, and light gas from pyrolysis of (a) GR1.9, (b) GR2.9, and (c) GR3.9 in the kerogen retort in N_2 at a heating rate of 10 K/min. Multiple data points at a given temperature represent repeat experiments. Lines are sigmoidal fits to the data and are given for readability only.

TABLE 6.3

Yields of Char, Tar, and Light Gas from the Kerogen
Retort (wt% of Parent Kerogen)

Temperature (°C)	Char	Tar	Light Gas
GR1.9			
300	92.9	1.0	6.1
375	92.9	2.0	5.1
410	82.8	3.0	14.1
445	59.0	23.0	18.0
495	22.4	47.0	30.6
495	23.2	63.3	13.5
525	24.8	59.2	16.1
575	21.1	59.7	19.2
GR2.9			
425	65.1	15.0	19.9
445	58.5	28.0	13.5
475	44.1	45.9	10.0
480	29.7	60.0	10.3
525	20.2	62.6	17.2
525	22.5	52.4	25.1
575	25.3	61.6	13.1
GR3.9			
400	75.7	8.1	16.2
434	68.5	8.7	22.8
450	57.2	19.2	23.6
460	46.8	44.1	9.1
460	46.9	40.0	13.1
470	43.7	43.3	13.0
470	45.6	50.5	3.8
490	18.4	63.5	18.1
525	16.7	69.5	13.9
525	15.7	68.6	15.7

but still within a reasonable error. The fraction of aliphatic carbon (f_{al}) in
the three kerogen chars decreased from approximately 76% to as low as 8%
during pyrolysis, as shown in Figure 6.7 and Table 6.4, with a corresponding
increase in the carbon aromaticity. The decrease in aliphatic carbon was very
similar to the decrease in mass due to devolatilization, as shown by the m/m_0
curve in Figure 6.7. Changes in aromaticity and aliphatic carbon with temperature were similar for all three kerogen chars (GR1.9, GR2.9, and GR3.9);
the bulk of the changes occurred in a narrow temperature window from
425°C to 525°C with the most rapid change at 475°C. The high aromaticity

TABLE 6.4

Carbon Structural Parameters Determined from ^{13}C NMR Spectroscopy for Demineralized Kerogen (GR3.9) and Resulting Chars Collected at Different Temperatures with a Heating Rate of 10 K/min

Structural Parameter	GR3.9	400°C	430°C	434°C	450°C	460°C	470°C	490°C	525°C
Aromatic carbon, $f_a = f_{a'} + f_a^C$	0.24	0.32	0.38	0.39	0.42	0.47	0.51	0.80	0.87
Carbonyl, f_a^C	0.04	0.03	0.04	0.03	0.04	0.04	0.03	0.07	0.06
Aldehydes and ketones, f_a^O	0.01	0.01	0.02	0.01	0.02	0.02	0.01	0.03	0.02
Acids and esters, f_a^{OO}	0.03	0.02	0.02	0.02	0.02	0.02	0.02	0.04	0.04
Aromatic carbon, carbonyl subtracted, $f_{a'}$	0.20	0.29	0.34	0.36	0.38	0.43	0.48	0.73	0.81
Protonated aromatic carbon, f_a^H	0.05	0.09	0.13	0.14	0.11	0.13	0.12	0.15	0.16
Nonprotonated aromatic carbon, $f_a^N = f_a^P + f_a^S + f_a^B$	0.15	0.20	0.21	0.22	0.27	0.30	0.36	0.58	0.65
Aromatic carbon with oxygen attachment (phenolic), f_a^P	0.03	0.04	0.04	0.04	0.04	0.04	0.05	0.08	0.09
Aromatic carbon with alkyl attachment, f_a^S	0.07	0.09	0.11	0.10	0.11	0.12	0.14	0.18	0.22
Aromatic bridgehead and inner carbon f_a^B	0.05	0.07	0.06	0.08	0.12	0.14	0.17	0.32	0.34
Aliphatic carbon, f_{al}	0.76	0.68	0.62	0.61	0.58	0.53	0.49	0.2	0.13
Aliphatic CH and CH$_2$, f_{al}^H	0.63	0.57	0.53	0.51	0.46	0.41	0.37	0.12	0.08
Aliphatic CH$_3$ and nonprotonated carbon, f_{al}^*	0.13	0.11	0.09	0.10	0.12	0.12	0.12	0.08	0.05
Aliphatic with oxygen attachment, f_{al}^O	0.0	0.01	0.0	0.01	0.01	0.01	0.01	0.02	0.03

(Continued)

TABLE 6.4 (Continued)

Carbon Structural Parameters Determined from ^{13}C NMR Spectroscopy for Demineralized Kerogen (GR3.9) and Resulting Chars Collected at Different Temperatures with a Heating Rate of 10 K/min

Structural Parameter	GR3.9	400°C	430°C	434°C	450°C	460°C	470°C	490°C	525°C
Aromatic bridgehead carbons, X_b	0.25	0.241	0.176	0.222	0.316	0.326	0.354	0.438	0.42
Average number of carbons per cluster, C	12	11.6	9.3	10.8	15.6	16.1	17.6	21.4	20.6
Total attachments per cluster, $\sigma + 1$	5.9	5.2	4.1	4.2	6.3	5.9	7	7.6	7.9
Fraction of intact bridges per cluster, p_0	−0.3	0.15	0.4	0.29	0.2	0.25	0.37	0.69	0.84
Bridges and loops per cluster, B.L.	—	0.8	1.6	1.2	1.3	1.5	2.6	5.2	6.6
Side chains per cluster, S.C.	—	4.4	2.5	3	5	4.4	4.4	2.4	1.3
Molecular weight per cluster, MW_d	946	602-		446	607	544	535	427	370
Molecular weight per side chain, m_6	135	88		74	66	59	46	22	15
Ratio f_{al}/f_a^S	10.86	7.56	5.64	6.10	5.27	4.42	3.50	1.11	0.59
Mass release (wt% of parent kerogen)		24.3%	n.a.	31.5%	42.8%	53.2%	55.3%	81.6%	83.8%

Note: Italicized symbols are the same as used by Solum and coworkers (Solum et al. 1989, 2001).

* p_0 for the kerogen has a large error due to high quantities of CH_2.

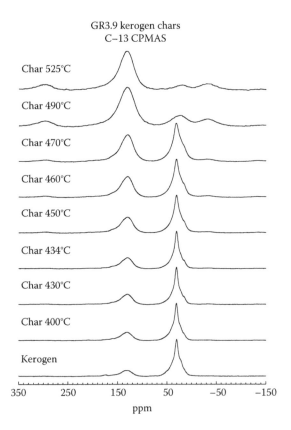

FIGURE 6.6
CP/MAS spectra of the GR3.9 kerogen concentrate and chars made from the concentrate. The spectra were taken with a 3 ms contact time and 1 s pulse delay.

in kerogen chars above 500°C is similar to the high aromaticity in spent oil shales observed by Miknis et al. (1987).

The increase in aromaticity seen in Figure 6.7 is not merely due to the effect of losing aliphatic carbon. The average number of aromatic carbons per cluster (C) increases from roughly 10 to about 20, as illustrated in Figure 6.8, and corresponds to an average growth in the number of aromatic rings per cluster from 2 to 5. The number of attachments per cluster also increases from about 5 to 8, meaning that some cross-linking has occurred. The average molecular weight per cluster is 830 and is determined in conjunction with the elemental analysis of the chars. The approximate average length of the aliphatic side chains for these three samples, computed by dividing f_{al} by f_a^S (Kelemen et al. 1998), decreases from about 11 at room temperature to less than 1 at the highest temperatures. It should be noted that the parameter f_a^S includes biaryl zero mass bridges that become more important at higher

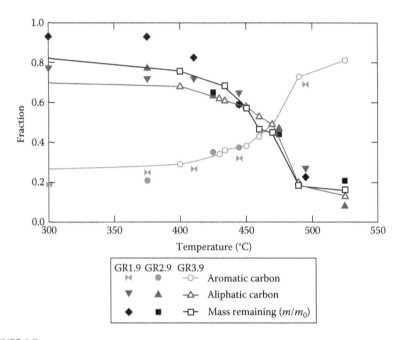

FIGURE 6.7
Aromatic and aliphatic carbon fractions for the kerogen chars formed at 10 K/min in the kerogen retort. Solid lines are drawn through the GR3.9 data marked by open symbols. Data below 300°C are reported elsewhere. (From Fletcher, T.H. et al., *Energy Fuels*, 28, 2959, 2014).

temperatures or for higher-rank materials and that the quantity of biaryl bridges is not well understood.

6.3.2.3 Solution-state ^{13}C NMR Analysis of Tars

The quantC and quantD carbon spectra of the various tar samples condensed from the effluent of the kerogen pyrolysis experiments are qualitatively very similar but differ somewhat in the relative amounts of the various major components. An example of the GR2.9 pyrolysis tar data dissolved in DCM is shown in Figure 6.9 for the kerogen pyrolyzed at 445°C. The spectra are similar using the quantC and quantD experimental protocols (Fletcher and Pugmire 2014, Fletcher et al. 2014). At 19%, the aromatic carbon content of this tar is similar to that of the unreacted kerogen concentrate. Within experimental error, the tars from all temperatures and kerogen concentrates have approximately the same aromatic carbon content. In addition, the two carbon resonances observed at 114 and 140 ppm are consistent with the presence of terminal alkene functional groups in the pyrolysis tar. These two functional groups are also consistent with the alkane/alkene pairs found in the GC/MS data. These peaks were not observed in the bitumen spectra (Solum et al. 2014).

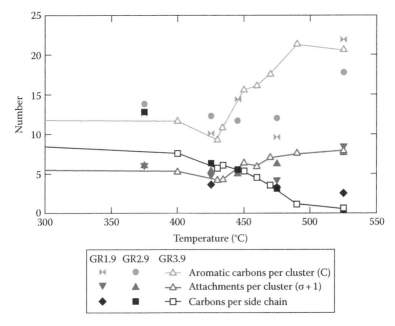

FIGURE 6.8
Aromatic carbons per cluster (C), attachments per cluster (σ + 1), and estimated carbons per side chain in kerogen chars. Solid lines are drawn through the GR3.9 data marked with open symbols. Data below 300°C are reported elsewhere. (From Fletcher, T.H. et al., *Energy Fuels*, 28, 2959, 2014).

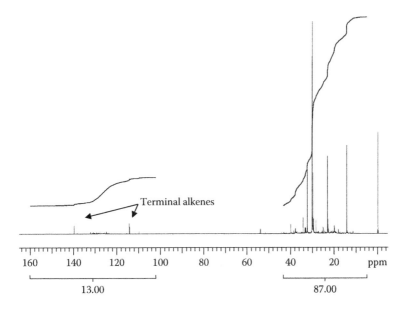

FIGURE 6.9
^{13}C quantD spectra of tar from the pyrolysis of GR2.9 kerogen at 445°C dissolved in DCM.

Figure 6.10 presents the aliphatic region of a set of tar DEPT spectra from GR2.9 containing resonances of various carbons according to the number of attached protons; the scale has been adjusted so that the intense methylene (CH_2) resonance at 30.1 ppm is off-scale to permit closer examination of the other resonances. The letters and subnumbers in Figure 6.10 are explained in Figure 6.11. For instance, A_n represents the *n*th carbon in a straight-chain normal paraffin. The letters B, C, D, E, and P represent different types of

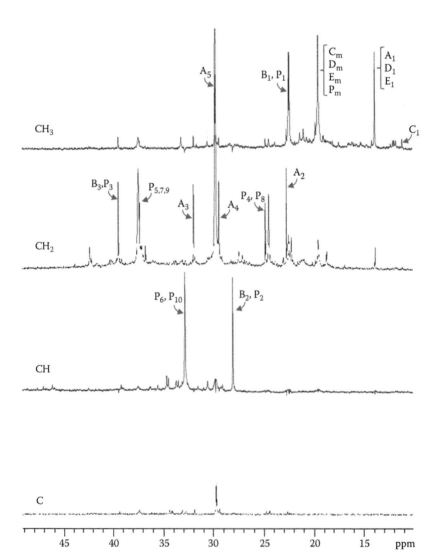

FIGURE 6.10
Aliphatic region of the DEPT spectrum of GR2.9 tar dissolved in DCM. Peak labels are explained in Figure 6.11.

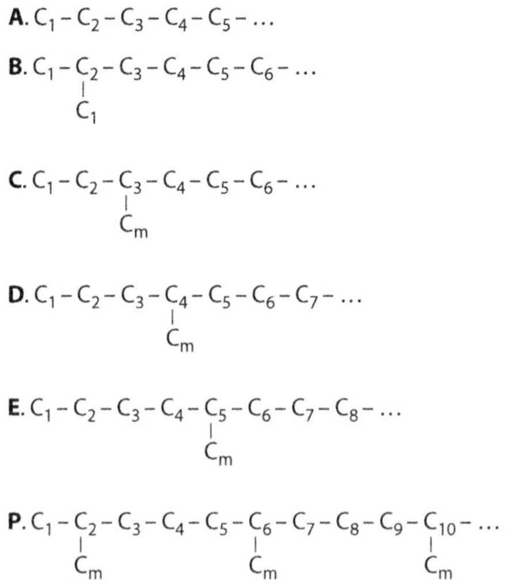

A. $C_1 - C_2 - C_3 - C_4 - C_5 - \dots$

B. $C_1 - C_2 - C_3 - C_4 - C_5 - C_6 - \dots$
 $\quad\quad\quad |$
 $\quad\quad\, C_1$

C. $C_1 - C_2 - C_3 - C_4 - C_5 - C_6 - \dots$
 $\quad\quad\quad\quad\quad |$
 $\quad\quad\quad\quad C_m$

D. $C_1 - C_2 - C_3 - C_4 - C_5 - C_6 - C_7 - \dots$
 $\quad\quad\quad\quad\quad\quad |$
 $\quad\quad\quad\quad\quad C_m$

E. $C_1 - C_2 - C_3 - C_4 - C_5 - C_6 - C_7 - C_8 - \dots$
 $\quad\quad\quad\quad\quad\quad\quad |$
 $\quad\quad\quad\quad\quad\quad C_m$

P. $C_1 - C_2 - C_3 - C_4 - C_5 - C_6 - C_7 - C_8 - C_9 - C_{10} - \dots$
 $\quad\quad |\quad\quad\quad\quad\quad\quad |\quad\quad\quad\quad\quad |$
 $\quad\, C_m\quad\quad\quad\quad\quad C_m\quad\quad\quad\quad C_m$

FIGURE 6.11
Explanation of peak labels in Figure 6.10 as suggested by Dalling et al. (1986).

carbon branching found in isopranes, etc. The subterm C_m is representative of branching points of various lengths. Due to various instrumental limitations, cancellation of one carbon type from the spectrum of another carbon type is not perfect. Note, for example, that the methyl (CH_3) resonance at 14.5 ppm appears at diminished intensity in the CH_2 spectrum.

As noted in Chapter 5, while the DEPT spectra are not strictly quantitative, the relative amounts of the various carbon types can be gleaned by comparing the spectra. Similar to the parent kerogens, CH_2 is the dominant type, with CH_3 and methines (CH) at lower abundance. Since CH indicates a branch point in an aliphatic chain, there are relatively few such branch points in these tars. The weak CH_3 resonances upfield from 14.5 ppm arise from the terminal methyl of chains where the third carbon from the end is a CH. As confirmed by the integrals in Figure 6.9, these terminal methyls are much less abundant than those of the long-chain *n*-alkyl moieties.

The spectrum representing quaternary carbons (labeled "C" in Figure 6.10) shows essentially no resonances. All of the resonances seen in this particular linear combination of the raw data correspond to strong resonances of other types and are attributed to imperfect cancelation of these other types. Resonances from nonprotonated carbons are diminished in intensity by a factor of about four relative to those of protonated carbons because nonprotonated carbons receive no transfer of proton magnetization. Nevertheless, it is clear that there are very few aliphatic nonprotonated carbons in this sample. Hence, one can compare the quantC (all aromatic carbons) and quantD (only protonated carbons) integrals such as in Figure 6.9. Assuming that the integrals of the aliphatic region represent the same number of carbons in both cases, the integrals of the aromatic region are normalized as shown in Equation 6.3 to compute the percentage (64%) of aromatic carbons that are protonated.

$$\frac{13.00*81.08}{18.92*87.00} = 64.03\% \tag{6.3}$$

While the signal to noise (S/N) in the aromatic region is low, making the experimental error rather high, one can confidently say that over half of the aromatic carbons are protonated.

Figure 6.12 shows the aliphatic region of the quantD spectra of the GR3.9 kerogen tars. The spectra are all dominated by *n*-alkyl chains. The intense resonances near 30.1 ppm arise from all the carbons located at five or more carbons from the methyl end of each *n*-alkyl chain. The high intensity of this resonance compared to all others (e.g., Figure 6.10) shows that all of these tars contain a significant amount of long-chain normal hydrocarbons. The corresponding CH_3 is located at 14.3 ppm with carbons 2, 3, and 4 located at 23.1, 32.3, and 29.8 ppm, respectively. Note that the peaks for carbons 1–4 have similar intensities and, in a long *n*-alkyl chain, each would represent two carbons (one from each end of the molecule). Since the ratio of the intensity of the peak at 30.1 ppm to the average intensity of the first four peaks in the chain is about four (i.e., 8 carbons), the average length of these normal chains is about 16 carbons. This result is consistent with the GC/MS results discussed in the next section.

There is a source of systematic error in the determination of the chain length that cannot be compensated for. While *n*-alkyl chains can be attached to aromatic centers or have alkyl side-chains, as long as the *n*-alkyl part of the molecule is 10 carbons or longer, the resonances from carbons at least five carbons removed from the aromatic ring or branch point will contribute to the resonances discussed earlier.

Although the same resonances appear in both the bitumen (see Chapter 5) and the kerogen pyrolysis spectra (see Figure 6.12), intensities for the condensate resonances corresponding to carbons other than the *n*-alkyl carbons are much lower than for corresponding resonances in the bitumens. This is true for all the pyrolysis temperatures studied, but there is a definite trend toward a higher proportion of *n*-alkyl carbons with increasing maximum pyrolysis temperature as shown in Figure 6.12.

The DEPT spectrum of the pyrolyzed kerogen shows CH_2 at 114 ppm and CH at 140 ppm in all of the pyrolysis tars but peaks are not seen in the corresponding bitumen spectra (see Chapter 5). These two resonances arise from the first and second carbons (C_1 and C_2), respectively, of *n*-alkenes with chain length of at least 7. These functional groups are consistent with alkane/alkene pairs that are found in the GC/MS data discussed in the following section. It is presumed that they arise from thermal cleavage of

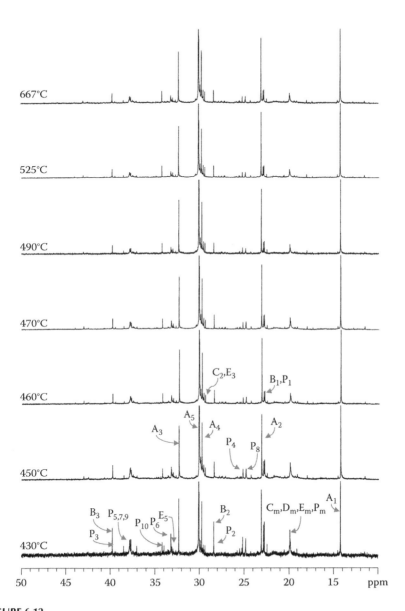

FIGURE 6.12

^{13}C quantD spectra of tar from the pyrolysis of GR3.9 kerogen at various maximum temperatures. Each spectrum is scaled so that the CH_3 resonance at 14.3 ppm is near full-scale. The intense CH_2 resonance at 30.1 ppm is off-scale.

n-alkyl groups from aromatic rings during pyrolysis. These two resonances are present in every pyrolysis experiment and seem to have little dependence on maximum pyrolysis temperature. Long-chain dialkenes (e.g., 1,7-octadiene) would also give rise to these two characteristic resonances, but molecules of this type were not detected by GC/MS.

6.3.2.4 GC/MS Analysis of Tars

Gas chromatograms of the tars generated at different temperatures in the kerogen retort from the GR3.9 kerogen sample are presented in Figure 6.13; chromatograms of the other samples show similar results. Prominent peaks from the gas chromatograms were identified with the mass spectrometer using fragmentation patterns and were reported by Fletcher et al. (2014). The lightest identified molecule was 1-heptene, which was eluted from the GC column only shortly after the solvent. The heaviest molecule identified was tetracosane ($C_{24}H_{50}$), which was detected nearly 50 min after injection. The remaining molecules were primarily alkane/1-alkene pairs between 7 and 24 carbons in length with the largest peaks observed for chains with 15–17 carbons. A notable exception to the alkane/alkene pairs was aromatic pyrene, which was emitted about halfway through the process. Pyrene is a common species found in residues from incomplete combustion processes.

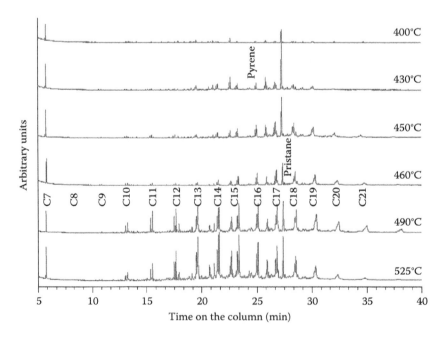

FIGURE 6.13
Gas chromatogram of components of GR3.9 tar collected at various temperatures during pyrolysis (10 K/min).

These tars were collected as a cumulative sample, so any compounds generated at lower temperatures would also be present at higher temperatures in addition to the compounds generated at higher temperatures. For example, any compounds present in the 375°C sample should also be present in the 525°C sample, but some compounds present in the 525°C may not be present in the 375°C sample. The main features of tars collected at all temperatures were similar although more small, irresolvable peaks were present at higher temperatures. No significant differences in the composition of the tars from the three demineralized kerogen samples were observed in the GC/MS data.

6.3.2.5 FTIR Analysis of Light Gases

Figure 6.14 shows FTIR spectra of light gases collected from the kerogen retort during pyrolysis of GR1.9 kerogen. Spectra for the GR2.9 and GR3.9 tars showed very similar results. The temperatures over which the gases were collected are listed in Figure 6.15. The peaks for CO_2, alkanes, and H_2O were observed at all temperature ranges for both samples. The peaks for methane (CH_4) and hydrocarbon compounds with two or more carbons (C_2+)

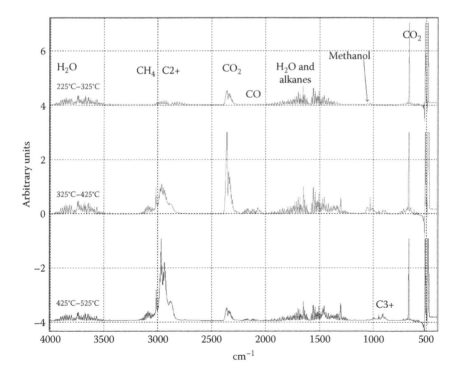

FIGURE 6.14
FTIR spectra of light gas collected from pyrolysis of the GR1.9 kerogen sample in the kerogen retort at the listed temperatures.

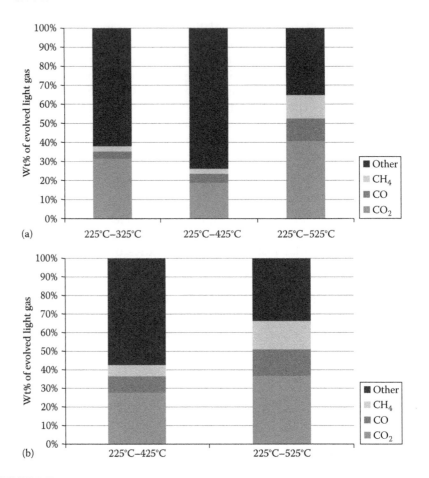

FIGURE 6.15
Composition of light gases evolved over different temperature ranges from (a) GR1.9 kerogen and (b) GR3.9 kerogen determined by quantitative FTIR analysis. Corresponding light gas yields are given in Figure 6.5.

increased dramatically at higher temperatures. The CO peak was small but increased at higher temperatures.

Quantitative analysis of the FTIR spectra was performed using calibrated gas standards, allowing the determination of the concentration of CH_4, CO, and CO_2 in the sample gas at each of the three temperatures. Using a concentration coupled with a known factor of dilution, the amount of each component as a percentage of the original sample or of the amount of light gas evolved was determined; results are shown in Figure 6.15. No significant differences are seen between the gases from the two kerogen samples (GR1.9 and GR3.9). The fraction of light gases that are CH_4, CO, and CO_2 increases to about 65 wt% for the gas samples collected at 525°C. The "other" category in this figure is comprised of H_2O and light hydrocarbons (presumably alkanes).

6.3.2.6 Discussion

The results of the pyrolysis experiments and analyses are very similar to data for an oxidized Colorado Green River oil shale of unknown origin reported by Hillier and coworkers (Hillier 2011, Hillier et al. 2013). In the present study, the tar yield in the 10 K/min pyrolysis experiments ranged from 60 to 69 wt% of the parent kerogen at the highest temperatures in the kerogen retort (525°C and 575°C) with total volatile yields ranging from 755 to 84 wt% of the demineralized kerogen (see Table 6.3). The slight differences in the volatile and tar yields from the demineralized kerogens do not correlate with depth or with the amount of organic material in the parent shale.

The solid-state NMR data for the char residue from pyrolysis of the demineralized kerogen showed 80% carbon aromaticity at the highest temperatures, with increases in the number of aromatic carbons per cluster and the number of attachments per cluster. These data indicate the manner in which the char residue is cross-linking in addition to losing aliphatic material. Therefore, a generalized oil shale pyrolysis model must treat both the cross-linking and the cleavage of aliphatic bonds. The measured time-dependent changes in the kerogen structural and lattice parameters shown in this chapter provide both insight and quantitative comparisons for validation of generalized oil shale pyrolysis models. Only by modeling the correct chemical mechanisms can one be confident in predictions of yield and quality at a range of possible heating rate and pressure conditions in future planned kerogen recovery processes.

The chemical structural data for the pyrolysis products from these pristine Utah oil shale cores matched the data reported by Hillier et al. (2013), implying that the NMR technique is consistent and that the degree of oxidation in the sample used by Hillier was low. Therefore, the data reported by Hillier et al. can also be used to develop better generalized oil shale pyrolysis models without questioning the effects of oxidation on the data obtained.

The condensable pyrolysis gases (termed tar in this chapter) consist mainly of alkane/1-alkene pairs with average chain lengths of 15–17 carbons. The composition data from liquid-state NMR and GC/MS analysis agree for these tars. Relatively few changes are seen in the tar composition as a function of the maximum temperature achieved in the kerogen retort or of the depth of the parent oil shale. The average chain lengths are similar to those obtained from demineralized kerogen by Boucher et al. (1991) using mild oxidation by ruthenium(IV) oxide. Boucher reported that the most common straight-chain hydrocarbon was C_{11}, with C_{10} through C_{13} being the dominant species.

Two possible reactions for the formation of alkane/1-alkene pairs are discussed by Hillier and coworkers (Hillier 2011, Hillier et al. 2013). Breakage of the bond between the aromatic ring and an α-carbon would form an alkene; subsequent hydrogenation would form an *n*-alkane. Breakage of the bond between the α-carbon and the β-carbon with hydrogen abstraction would produce the alkanes as well. Very little evidence was seen for dialkenes.

The light gas analyses presented here are similar to those for the Colorado oil shale reported by Hillier. At the highest temperatures, 40% of the light gases were not fully characterized by FTIR but seemed to be a combination of H_2O and light hydrocarbons (likely alkane/alkene pairs that did not condense).

6.4 Modeling Kerogen Pyrolysis with the CPD Model

As knowledge of the characterization of oil shale increases due to data derived from modern solid-state NMR, MS, and GC, a description of oil shale pyrolysis products based on chemical structure is possible. Many current models empirically fit the mass released during pyrolysis (e.g., Campbell et al. 1978) and may also empirically fit curves of individual species or tar. Reviews of simple empirical models of oil shale pyrolysis were recently performed (Burnham 2010, Hillier and Fletcher 2011, Tiwari and Deo 2012a). Mechanistic models describe the actual chemical processes involved and may be able to describe oil shale pyrolysis over a broader range of heating conditions (heating rate, temperature, and pressure). A very detailed mechanistic model (CS-CYM) based on the chemical structure of various kerogens, including oil shale, was reported by Freund et al. (2007) and Walters et al. (2007). Core molecular structures were developed based on elemental analyses, solid-state ^{13}C NMR, x-ray photoelectron spectroscopy (XPS), x-ray absorption near-edge structure, and pyrolysis-GC. The core structures were expanded stochastically to describe large macromolecules, which were then used to model pyrolysis reactions at a range of conditions, including geological conditions.

The Chemical Percolation Devolatilization (CPD) model is a mechanistic pyrolysis model originally developed by Fletcher et al. (Grant et al. 1989, Fletcher et al. 1992) to describe coal pyrolysis. Although not as sophisticated as the CS-CYM model, the CPD model is intended to be an engineering model to predict tar, char, and light gas yields of pyrolysis at different temperatures, heating rates, and pressures based on chemical structure. Coal is modeled as a system of aromatic clusters connected by aliphatic bridges and short, nonbridging attachments to the aromatic structures. During pyrolysis, aliphatic bridges are either (1) cleaved, leaving two side chains, or (2) transformed into a stable char bridge by releasing their aliphatic portion. A Bethe lattice is used to describe the parent chemical structure with features such as cluster molecular weight, side chain molecular weight, and the number of attachments per cluster determined by solid-state ^{13}C NMR measurements. Percolation lattice statistics are used to relate the temperature-dependent rate of aliphatic bridge breaking to the number of clusters that are detached from the original infinite lattice. Detached clusters form a liquid pool that can evaporate as tar, depending on the vapor pressure of each cluster. Detached clusters that remain too long with the solid will cross-link to the char. In addition to coal, the CPD model has been used to model pyrolysis of black liquor (Fletcher et al. 2012), biomass (Fletcher et al. 2012, Lewis and Fletcher 2013), and polyurethane foam (Clayton 2002, Ulibarri et al. 2002, Hobbs et al. 2003).

6.4.1 Chemical Structure Data

The chemical structure of oil shale was recently measured in great detail using solid-state ^{13}C NMR, FTIR, and GC/MS (Hillier et al. 2013, Fletcher et al. 2014, Solum et al. 2014), as described in Chapter 5 and in this chapter. The NMR techniques of Solum et al. (1989) were used to describe 14 structural parameters and then combined with the elemental analysis to calculate 8 lattice parameters for Green River oil shales and the solid char remaining after pyrolysis (Hillier et al. 2013, Fletcher et al. 2014, Solum et al. 2014). These lattice parameters have been particularly useful in developing models of hydrocarbon pyrolysis, including CPD (Grant et al. 1989, Fletcher et al. 1992), FG/DVC (Solomon et al. 1988), and Flashchain (Niksa and Kerstein 1991). The lattice is assumed to be composed of sites connected by labile (i.e., breakable) bridges. A generic model of a hydrocarbon molecule is shown in Figure 6.16 that depicts aromatic clusters as sites connected by aliphatic bridges. This figure also depicts side chains, which are hydrocarbon chains that are attached to the aromatic cluster but not to an adjacent cluster.

Average values of chemical structure parameters are calculated from the solid-state ^{13}C NMR data, including the average number of aromatic carbons per cluster, the average molecular weight of a side chain (assuming that a side chain is one-half the molecular weight of a labile bridge), the average number

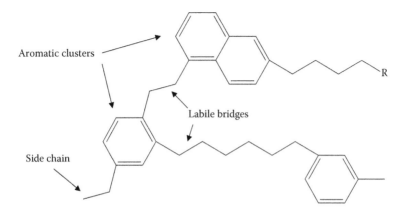

FIGURE 6.16
Representative hydrocarbon molecule showing aromatic clusters, labile bridges, and side chains in a lattice.

TABLE 6.5

Chemical Structure Parameters Measured for the GR1.9, GR2.9, and GR3.9 Kerogen Samples

	GR1.9	GR2.9	GR3.9
Molecular weight per cluster (MW_{cl})	776	775	946
Molecular weight per side chain ($m\delta$)	131	148	135
Attachments per cluster ($\sigma + 1$)	5.0	4.5	5.9
Fraction of attachments that are bridges (p_0)[a]	0.5	0.5	0.5

Source: Solum, M.S. et al., *Energy Fuels*, 28, 453, 2014.

[a] p_0 could not be measured by the NMR technique for oil shale, but was assumed to be 0.5.

of attachments (i.e., bridges plus side chains) per cluster, and the fraction of attachments that are bridges (as opposed to side chains). Lattice structure parameters reported by Solum et al. (2014) from the solid-state analysis of the GR1.9, GR2.9, and GR3.9 kerogen samples are shown in Table 6.5. The GR3.9 parameters are used in the CPD model analysis.

6.4.2 Model Description

During pyrolysis, labile (i.e., breakable) bridges are cleaved as the temperature increases. Depending on the lattice structure, cleaving one bridge does not necessarily release pyrolysis products. Therefore, a treatment of the relationship between the lattice structure and the probability of creating liberated fragments is necessary. The liberated fragments may contain one or more aromatic clusters and hence will contain a molecular weight distribution. The liberated fragments with high molecular weight may not vaporize, depending on the temperature and pressure, so vapor pressures of the different fragments must be used to determine phase behavior. Fragments that do not vaporize may cross-link with remaining solids to form char. Side chains are likely released at a different rate than the rate of cleavage of labile bridges, since the side chains have a different chain length.

The bridge-breaking scheme in the CPD model is shown in Figure 6.17. An aliphatic "labile" bridge ($\pounds$) is activated ($\pounds^*$) and then is either cleaved to form two side chains (δ) with rate k_δ or transformed into a stable bridge (c) (such as a biaryl bridge) with rate k_c while releasing the aliphatic material

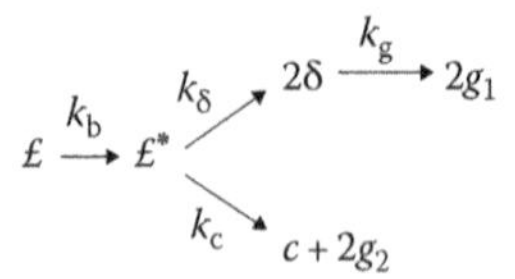

$$\pounds \xrightarrow{k_b} \pounds^* \begin{cases} \xrightarrow{k_\delta} 2\delta \xrightarrow{k_g} 2g_1 \\ \xrightarrow{k_c} c + 2g_2 \end{cases}$$

FIGURE 6.17
Bridge-breaking scheme used in the CPD model.

as gases. The side chains will eventually degrade to form gases as well with rate k_g. Experience has shown that the ratio of k_g/k_c is relatively constant. The gas that is produced by the side chains that break off is referred to as g_1, whereas the gas that is produced by the bridge transformation to a char bridge is referred to as g_2.

6.4.3 Results

Predictions were made with the CPD model using the chemical structure parameters in Table 6.5 with the rate parameters for bridge breaking obtained from coal pyrolysis. These predictions did not compare well with TGA data (Hillier and Fletcher 2011) nor with tar and total yield data obtained in a kerogen retort (Hillier et al. 2013, Fletcher et al. 2014). The chemical structure of the kerogen in oil shale is much different than the chemical structure in coal, so this result was not surprising.

Next, rate coefficients were determined for a first-order distributed DAEM using TGA pyrolysis data on the three samples referenced in Table 6.5. These rate coefficients (see Table 6.1) were used in the CPD model for both k_δ and k_c. Predicted and measured yields of char and tar by the CPD model using the rate coefficients from Table 6.1 resulted in a tar yield of only 5%, which is much lower than measured. The CPD model was formulated assuming that the tar consists only of material containing an aromatic cluster, a good assumption for coal. Any material released from breaking a labile bridge or a side chain is assumed to become a light gas (i.e., non-condensable). This assumption works well for coal pyrolysis because of the low molecular weights of side chains and bridges in coals. However, it is apparent that this approach does not work well for pyrolysis of kerogen from oil shale.

The side chains in the Green River oil shale are much larger than those in coal, with molecular weights reported to be 128–148 amu (Hillier et al. 2013, Fletcher et al. 2014). By estimating that 80% of the aliphatic side chain and bridge material was of high enough molecular weight to be condensable and counted as tar, the resulting calculations for GR3.9 kerogen matched the measured tar yield as shown in Figure 6.18. However, the onset of light gas was predicted to occur at a higher temperature than indicated by the data. This discrepancy was also seen in the predictions for the GR1.9 and GR2.9 samples. Therefore, a method was sought to give a better prediction of the light gas as well as to reduce the empiricism of the tar prediction.

It is not possible to determine the distribution of molecular weights of the side chains in the unreacted kerogen from the NMR data. The only data on molecular weight distributions are for species after pyrolysis. Therefore, the tar yields obtained from the kerogen retort data of Fletcher et al. (2014) and described in Section 6.3 are as good as any data available for determining the split between condensable (tars/oils) and noncondensable gases.

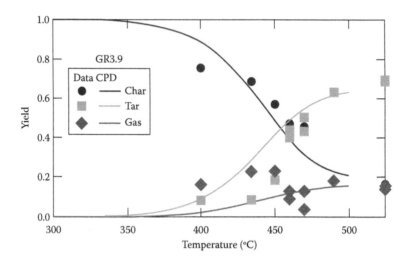

FIGURE 6.18

CPD model calculations of tar and char yields for GR3.9 kerogen pyrolysis at 10 K/min using rate coefficients from Table 6.1 and counting 80% of the "light gas" as tar. (Data are from Fletcher, T. H. et al., *Energy Fuels*, 28, 2959, 2014.)

It was postulated that there might be a difference in the two pathways for "light gas" formation shown in Figure 6.18. Since a labile bridge would have to be small to collapse into a char bridge, g_2 was assigned as light gas. The pathway for breaking a bridge was thought to be easier for longer chains, so g_1 was assigned as heavy, noncondensable gas. The molecular weight of the heavy gas (MW_{hg}) was calculated from Equation 6.4 given the molecular weight of the light gas (MW_{lg}), set to an average of 20 amu, and the molecular weight of the combined gas, calculated from the NMR parameters by the CPD model (Fletcher et al. 1992):

$$y_{lg}MW_{lg} + y_{hg}MW_{hg} = MW_{all\ gases} \tag{6.4}$$

where

$y_{lg} = g_2/(g_1 + g_2)$
$y_{hg} = g_1/(g_1 + g_2)$
y_{lg} and y_{hg} were set to 0.2 and 0.8, respectively

The reaction rate coefficients were then adjusted slightly to obtain a best fit with the kerogen retort data of Fletcher et al. (2014). The new rate coefficients are shown in Table 6.6.

CPD model predictions of char and tar yields with these new rate coefficients are shown in Figure 6.19 for the GR3.9 sample. Similar results were obtained for the GR1.9 and GR2.9 samples using this method. No empirical factor was used to count part of the light gas as tar, other than

TABLE 6.6

Rate Coefficients Determined for Bridge Breaking in the CPD Model for the GR1.9, GR2.9, and GR3.9 Kerogen Samples to Fit the Kerogen Retort Data of Fletcher et al. (2014)

	GR1.9	GR2.9	GR3.9
A_b (s^{-1})	9.8×10^{12}	9.8×10^{12}	1.8×10^{15}
E_b/R (T^{-1})	23,900	23,900	25,918
σ_{Eb}/R (T^{-1})	0	0	0
A_g (s^{-1})	1.58×10^{10}	2.58×10^{10}	1.58×10^{14}
E_g/R (T^{-1})	21,000	21,000	27,600
σ_{Eg}/R (T^{-1})	300	300	300
k_δ/k_c	0.9	0.9	1.8

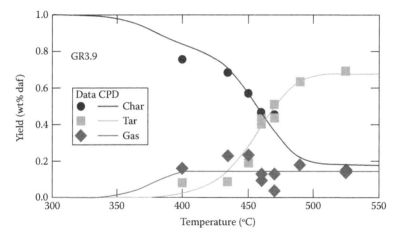

FIGURE 6.19

CPD model calculations of tar and char yields for GR3.9 kerogen pyrolysis at 10 K/min using rate coefficients from Table 6.6 and assigning g_2 as the "light gas" and g_1 as the condensable "heavy gas." (Data are from Fletcher, T.H. et al., *Energy Fuels*, 28, 2959, 2014.)

assigning g_1 as tar. This distinction between light gas and heavy gas allowed the model to predict the early light gases that came off before the tar. The need for two different reaction constants is apparent in Figure 6.19; much of the light gas has left the kerogen before tar is released. The overall agreement between experimental tar and char yields and results from the CPD model using this approach is very good.

The calculated bridge variables for the prediction in Figure 6.19 are shown in Figure 6.20. The mass remaining (m/m_0) is shown for reference. The labile bridges (£) start to react at about 350°C in this calculation, with a corresponding increase in side chains (δ) as some of the labile bridges break apart. The labile bridges have all reacted by 400°C. Char bridges (represented by $g_2/2$) are also shown, corresponding to the release of light gas as a bridge

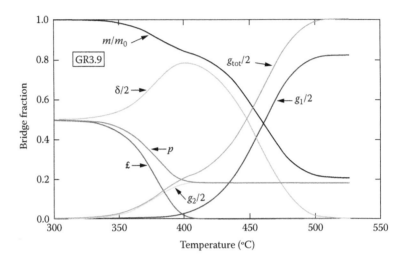

FIGURE 6.20
CPD model calculations of bridge variables for GR3.9 kerogen pyrolysis at 10 K/min using rate coefficients from Table 6.6 and assigning g_2 as the "light gas" and g_1 as the condensable "heavy gas."

collapses into a stable char bridge. The char bridges reach a maximum at the same temperature that the labile bridges stop reacting, and the total intact bridges ($p = £ + c$) become constant after 400°C. There is a delay in the formation of "heavy" gas, even though the activation energy (E_g) for formation of g_1 from δ is lower than the activation energy for labile bridge scission (E_b), since the pre-exponential factor A_g is much lower than A_b. The ratio of k_b/k_g is one to two orders of magnitude at temperatures of 350°C–500°C. The formation of g_2 preferentially to g_1 is what causes the light gas to be released earlier than the tar in Figure 6.19.

In the GR1.9 calculations, when 99% of the labile bridges were broken (i.e., $£ = .005$), 56% of the original mass of kerogen still remained, corresponding to the temperature at which 70% of the volatiles were released. Hence, approximately 30% of the volatiles were released after the labile bridge reactions were completed, corresponding to the region where side chains (δ) react to form g_1, which in this formulation is composed of high molecular weight alkyl chains. These high molecular weight gases are condensable and counted as tar, and their presence gives rise to the high wax content of "unconventional" oil from Green River oil shale. Similar behavior was calculated for the GR2.9 and GR3.9 samples. However, due to different best-fit kinetic coefficients, the labile bridges were cleaved at a later temperature in the calculations for the GR3.9 sample. In addition, the GR3.9 calculations showed a larger peak in the side chains than the calculations for the other two samples (Fletcher et al. 2015).

The good agreement with data is partially a result of the curve fit of the kinetic parameters for each kerogen and partially a result of the model formulation, which takes into account the direct measurement of chemical structure features such as molecular weight per aromatic cluster, molecular weight per side chain, and attachments per cluster. The resulting model has the potential to describe the pyrolysis of other oil shales around the world if the parent chemical structure is measured.

6.5 Analysis of Changes in Char Aromaticity

In order to understand the pyrolysis of oil shale, the aromaticity of the char (the organic material left behind after pyrolysis) was investigated (Fletcher and Pugmire 2014). Throughout pyrolysis, the aromaticity of the carbons in the char increased from 0.2 to 0.8 (Fletcher et al. 2014). Assuming that no new aromatic carbons were produced during pyrolysis, a mass balance of the aromatic carbons was used to predict the final amount in the char. This approach helps determine if any aliphatic carbons became aromatic in the experiment. It requires the elemental analyses of the unreacted kerogen, tar, and light gas from the kerogen retort, which are shown in Table 6.7. Note that the tar composition comes from the literature for a similar sample.

The carbon in the parent shale (C_{shale}) was divided into the mass fraction of carbon in the tar (C_{tar}), gas (C_{gas}), and char (C_{char}), as shown by the carbon balance in the following Equation:

$$C_{shale} = C_{tar} * f_{tar} + C_{gas} * f_{gas} + C_{char} * f_{char} \qquad (6.5)$$

where f is obtained from the experimental yields. Everything in Equation 6.5 is known except for C_{char}. The calculated value of C_{char} was 0.818, which is in the range of the measured carbon contents in the char; see Table 6.2.

TABLE 6.7

Elemental Compositions of Oil Shale and Pyrolysis Products

| | Weight Fractions | | | | | Carbon |
	Carbon	Hydrogen	Oxygen	Sulfur	Nitrogen	Aromaticity
Extracted Kerogen[a]	0.762	0.095	0.081	0.037	0.025	0.2
Tar[b]	0.851	0.114	0.013	0.007	0.018	0.19
Gas[c]	0.273	0.0478	0.651	0.0	0.0	0.0

[a] The composition for the extracted kerogen comes from GR3.9 (Solum et al. 2014).
[b] The composition for the tar comes from Netzel and Miknis (1982).
[c] The composition of the gas comes from Fletcher et al. (2014). It is assumed that the "other" portion of the light gases can be averaged as water.

A similar balance on aromatic carbon is shown in Equation 6.6, again assuming that aromatic carbons were neither created nor destroyed in the retort. The only unknown in the following equation is the char aromaticity, $f_{a'char}$:

$$C_{shale} * f_{a'shale} = C_{tar} * f_{a'tar} * f_{tar} + C_{char} * f_{a'char} * f_{char} \qquad (6.6)$$

where

   $f_{a'shale}$ is the aromaticity of the shale
   $f_{a'tar}$ is the aromaticity of the tar (Table 6.4)

The calculated value of $f_{a'char}$ was 0.281, but the measured value was 0.81 (Fletcher and Pugmire 2014, Fletcher et al. 2014). This difference between the actual aromaticity and the calculated aromaticity shows that there were carbons that became aromatic as the reaction progressed. Possible mechanisms were suggested by Fletcher and Pugmire (2014). These findings highlight the need to consider changes in the char structure that pyrolysis produces. Pyrolysis is not just the simple bridge-breaking mechanism that was previously assumed. Rather, pyrolysis contains many reactions that link the char and may contribute to the type of products in the tar and the gas. The calculated amount of cross-linking in the CPD model was negligible and would not add aromatic carbons anyway. In order to predict the possible products of pyrolysis and the chemical structure changes in the char, further work is needed to elucidate the exact mechanism of the aromatic production and to model aromaticity.

6.6 Summary and Conclusions

A large set of experiments was performed on a single set of oil shale samples as described in this chapter and in Chapter 5. The source for all samples was a well-characterized core from Utah's Green River oil shale, which is a type I kerogen. Demineralized samples were obtained from three different depths in the core. The combination of analytical techniques used in this research is unique and adds important information to the existing body of knowledge regarding the chemistry of oil shale.

This chapter specifically focused on thermal decomposition of the samples. Crushed oil shale samples were pyrolyzed at several heating rates (0.5–10 K/min) in a TGA at ambient pressure and at 40 bar, and rate coefficients were regressed for both the first-order model and the DAEM. The effect of pressure on the pyrolysis rate was shown to be small. Pyrolysis at higher heating rates ($\geq$ 60 K/min) was subject to mass and heat transfer effects and was thus not considered in this study.

Pyrolysis experiments on the demineralized kerogen were performed in a novel kerogen retort at a heating rate of 10 K/min to generate light gas, tar, and char for further analysis. The char, tar, and gas samples were analyzed by several methods, including ^{13}C NMR, GC/MS, and FTIR. The tar, defined as the products that condensed after cooling to room temperature, corresponded to the oil produced. The volatiles yield was about 80% on a daf basis, with a 60 wt% daf tar yield. The tar yield did not correlate with depth or organic content of the original sample, but the total volatile yield increased slightly with depth of the parent shale.

NMR analysis of the demineralized kerogens showed a carbon aromaticity of 20% with an average number of aromatic carbons per cluster of 10.3 but an average cluster molecular weight of 830. The carbon aromaticity of the post-pyrolysis char increased to approximately 80% with large decreases in the side chain molecular weight (from about 11 carbons to less than 1 carbon). These chemical structural data are consistent with the idea that the aromatic portion of the kerogen largely remains in the char, and the aliphatic chains break off and vaporize.

The liquid-phase NMR analysis of the tars from the pyrolysis of the demineralized kerogen showed more detail than previously reported in the literature with peaks for carbons present as *n*-alkanes as well as branched alkanes. The NMR spectra were basically the same for all tars collected, regardless of the temperature achieved during pyrolysis or the depth of the parent shale. The tar was approximately 20% aromatic with long-chain alkane structures prevalent and terminal alkene groups present that were not observed in previous analyses of the bitumen from these shales. It was estimated that more than half of the aromatic carbons in the tar were protonated.

The pyrolysis products were analyzed by GC/MS and showed large amounts of alkane/1-alkene pairs with an average chain length of 15–17 carbons. The analysis of tars collected cumulatively at different maximum temperatures showed similar chemical structures dominated by the *n*-alkane/1-alkene pairs. The FTIR analysis showed that 60% of the light gases at the highest temperature (525°C) consisted of CH_4, CO, and CO_2 with the balance likely a combination of H_2O and alkane/alkene pairs less than C_7.

The detailed chemical structure analysis of the demineralized kerogen from well-characterized oil shale samples discussed in Chapter 5 along with the analysis of pyrolysis products discussed in this chapter are important data for the development of improved pyrolysis models based on chemical structure rather than mere empiricism. The quantitative, time-dependent changes in chemical structure reported here serve as a good source for validating generalized models of oil shale pyrolysis that are applicable to wide ranges of heating rate and pressure.

A mechanistic model of oil shale kerogen pyrolysis based on the CPD model was developed to calculate tar (oil) yields. The chemical structure inputs to the CPD model were taken from detailed ^{13}C NMR measurements of demineralized kerogen from three Green River oil shale samples (GR1, GR2, GR3). Rate coefficients were modified in the CPD model to fit tar and char yields from kerogen retort experiments at 10 K/min. It was also necessary to model a light gas and a heavy gas; light gases were assumed to come only from char bridge formation and heavy gases were assumed to come only from aliphatic bridge cleavage. The heavy gas became part of the tar because of its high molecular weight. This assignment of light and heavy gases resulted in very good agreement between calculated and measured tar and total volatile yields for the three kerogen samples studied at atmospheric pressure and a heating rate of 10 K/min. The mechanism that best fits the tar and light gas yields showed that the labile bridges between clusters were cleaved first, with corresponding generation of side chains and char bridges. The side chains were assumed to be the long-chain aliphatics seen in the NMR data and were released subsequent to the cleavage of all labile bridges.

A balance was performed on the aromatic carbon in the char, presuming that no new aromatic carbon was formed during pyrolysis. This aromaticity balance did not match the change in aromaticity in the char, meaning that there are mechanisms present that form substantial amounts of aromatic carbon in the late stages of pyrolysis.

References

Boucher, R. J., G. Standen, and G. Eglinton. 1991. Molecular characterization of kerogens by mild selective chemical degradation—Ruthenium tetroxide oxidation. *Fuel* 70(6):695–702.

Braun, R. L., A. K. Burnham, and J. G. Reynolds. 1992. Oil and gas evolution kinetics for oil shale and petroleum source rocks determined from pyrolysis—TQMS data at two heating rates. *Energy & Fuels* 6(4):468474.

Burnham, A. K. 2010. Chemistry and kinetics of oil shale retorting. In *Oil shale: A solution to the Liquid Fuel Dilemma*, eds. O. I. Ogunsola, A. M. Hartstein, and O. Ogunsola, pp. 115–134. Washington, DC: American Chemical Society.

Campbell, J. H., G. H. Koskinas, and N. D. Stout. 1978. Kinetics of oil generation from Colorado oil shale. *Fuel* 57(6):372–376.

Clayton, D. J. 2002. Modeling flow effects during polymer decomposition using percolation statistics. Ph.D. dissertation, Provo, UT: Brigham Young University.

Dalling, D. K., R. J. Pugmire, D. M. Grant, and W. E. Hull. 1986. The use of high-field carbon-13 NMR spectroscopy to characterize chiral centers in isopranes. *Magnetic Resonance in Chemistry* 24(3):191–198.

Fletcher, T. H., D. Barfuss, and R. J. Pugmire. 2015. Modeling light gas and tar yields from pyrolysis of Green River oil shale demineralized kerogen using the CPD Model. Energy & Fuels, 29:4921–4926. DOI: 10.1021/acs.energyfuels.5b01146.

Fletcher, T. H., R. Gillis, J. Adams et al. 2014. Characterization of macromolecular structure elements from a Green River oil shale, II. Characterization of pyrolysis products by ^{13}C NMR, GC/MS, and FTIR. *Energy and Fuels* 28:2959–2970.

Fletcher, T. H., A. R. Kerstein, R. J. Pugmire, M. S. Solum, and D. M. Grant. 1992. Chemical percolation model for devolatilization. 3. Direct use of ^{13}C NMR data for predict effects of coal type. *Energy & Fuels* 6(4):414–431.

Fletcher, T. H., H. R. Pond, J. Webster, J. Wooters, and L. L. Baxter. 2012. Prediction of tar and light gas during pyrolysis of black liquor and biomass. *Energy & Fuels* 26:3381–3387.

Fletcher, T. H. and R. J. Pugmire. 2014. Rates and mechanisms of oil shale pyrolysis: A chemical structure approach. Topical report submitted to the U.S. Department of Energy, National Energy Technology Laboratory, DOE Award No. DE-FE0001243. Salt Lake City, UT: University of Utah, Institute for Clean and Secure Energy.

Freund, H., C. C. Walters, S. R. Kelemen et al. 2007. Predicting oil and gas compositional yields via chemical structure-chemical yield modeling (CS-CYM): Part 1—Concepts and implementation. *Organic Geochemistry* 38(2):288305.

Grant, D. M., R. J. Pugmire, T. H. Fletcher, and A. R. Kerstein. 1989. Chemical model of coal devolatilization using percolation lattice statistics. *Energy & Fuels* 3(2):175186.

Hagaman, E. W., F. M. Schell, and D. C. Cronauer. 1984. Oil-shale analysis by CP/MAS-13C NMR spectroscopy. *Fuel* 63(7):915–919.

Hillier, J., T. Bezzant, and T. H. Fletcher. 2010. Improved method for the determination of kinetic parameters from non-isothermal thermogravimetric analysis (TGA) data. *Energy & Fuels* 24(5):2841–2847.

Hillier, J. L. 2011. Pyrolysis kinetics and chemical structure considerations of a Green River oil shale and its derivatives. Ph.D. dissertation, Provo, UT: Brigham Young University.

Hillier, J. L. and T. H. Fletcher. 2011. Pyrolysis kinetics of a Green River oil shale using a pressurized TGA. *Energy & Fuels* 25:232–239.

Hillier, J. L., T. H. Fletcher, M. S. Solum, and R. J. Pugmire. 2013. Characterization of pyrolysis products from a Colorado Green River oil shale by ^{13}C NMR, XPS, GC/MS, and FTIR. *Industrial and Engineering Chemistry Research* 52(44):1552215532.

Hobbs, M. L., K. L. Erickson, T. Y. Chu et al. 2003. CPUF—A chemical-structure-based polyurethane foam decomposition and foam response model. Sandia Report SAND2003-2282. Albuquerque, NM and Livermore, CA: Sandia National Laboratories.

Huizinga, B. J., Z. A. Aizenshtat, and K. E. Peters. 1988. Programmed pyrolysis-gas chromatography of artificially matured Green River kerogen. *Energy & Fuels* 2(1):74–81.

Huss, E. B. and A. K. Burnham. 1982. Gas evolution during pyrolysis of various Colorado oil shales. *Fuel* 61(12):1188–1196.

Kelemen, S. R., M. Siskin, H. S. Homan, R. J. Pugmire, and M. S. Solum. 1998. Fuel, lubricant and additive effects on combustion chamber deposits. SAE technical paper 982715.

Lewis, A. D. and T. H. Fletcher. 2013. Prediction of sawdust pyrolysis yields from a flat-flame burner using the CPD model. *Energy & Fuels* 27:942–953.

Maciel, G. E., V. J. Bartuska, and F. P. Miknis. 1978. Correlation between oil yields of oil shales and ^{13}C nuclear magnetic resonance spectra. *Fuel* 57(8):505–506.

Meuzelaar, H. L. C., W. Windig, J. H. Futrell, A. M. Harper, and S. R. Larter. 1986. Pyrolysis mass spectrometry and multivariable analysis of several key world oil shale kerogens and some recent alginites. In *Mass Spectrometric Characterization of Shale Oils, ASTM STP 902*, pp. 81–105. Philadelphia, PA: American Society for Testing and Materials.

Miknis, F. P. and P. J. Conn. 1986. A common relation for correlating pyrolysis yields of coals and oil shales. *Fuel* 65(2):248–250.

Miknis, F. P., N. M. Szeverenyi, and G. E. Maciel. 1982. Characterization of the residual carbon in retorted oil shale by solid-state ^{13}C NMR. *Fuel* 61(4):341–345.

Miknis, F. P., T. F. Turner, G. L. Berdan, and P. J. Conn. 1987. Formation of soluble products from thermal decomposition of Colorado and Kentucky oil shales. *Energy & Fuels* 1(6):477–483.

Netzel, D. A. and F. P. Miknis. 1982. NMR study of US eastern and western shale oils produced by pyrolysis and hydropyrolysis. *Fuel* 61(11):1101–1109.

Niksa, S. and A. R. Kerstein. 1991. FLASHCHAIN theory for rapid coal devolatilization kinetics. 1. Formulation. *Energy & Fuels* 5(5):647–665.

Reynolds, J. G., R. W. Crawford, and A. K. Burnham. 1991. Analysis of oil shale and petroleum source rock pyrolysis by triple quadrapole mass spectrometry: Comparisons of gas evolution at the heating rate 10 degree C/min. *Energy & Fuels* 5(3):507–523.

Solomon, P. R., D. G. Hamblen, R. M. Carangelo, M. A. Serio, and G. V. Deshpande. 1988. General model of coal devolatilization. *Energy & Fuels* 2(4):405422.

Solomon, P. R. and F. P. Miknis. 1980. Use of Fourier Transform infrared spectroscopy for determining oil shale properties. *Fuel* 59(12):893–896.

Solum, M. S., C. L. Mayne, A. M. Orendt, R. J. Pugmire, J. Adams, and T. H. Fletcher. 2014. Characterization of macromolecular structure elements from a Green River oil shale, I. Extracts. *Energy and Fuels* 28:453–465.

Solum, M. S., R. J. Pugmire, and D. M. Grant. 1989. Carbon-13 solid-state NMR of Argonne premium coals. *Energy & Fuels* 3(2):187–193.

Solum, M. S., A. F. Sarofim, R. J. Pugmire, T. H. Fletcher, and H. Zhang. 2001. ^{13}C NMR analysis of soot produced from model compounds and a coal. *Enery & Fuels* 15:961–971.

Tiwari, P. 2012. Oil shale pyrolysis: Benchscale experimental studies and modeling. Ph.D. dissertation, Salt Lake City, UT: University of Utah.

Tiwari, P. and M. Deo. 2012a. Detailed kinetic analysis of oil shale pyrolysis TGA data. *AIChE Journal* 58(2):505–515.

Tiwari, P. and M. Deo. 2012b. Compositional and kinetic analysis of oil shale pyrolysis using TGA–MS. *Fuel* 94:333–341.

Ulibarri, T. A., D. K. Derzon, K. L. Erickson et al. 2002. Preliminary investigation of the thermal decomposition of Ablefoam and EF-AR20 foam (Ablefoam replacement). Sandia Report SAND2002-0183. Albuquerque, NM and Livermore, CA: Sandia National Laboratories.

Walters, C. C., H. Freund, S. R. Kelemen, P. Peczak, and D. J. Curry. 2007. Predicting oil and gas compositional yields via chemical structure–chemical yield modeling (CS-CYM): Part 2—Application under laboratory and geologic conditions. *Energy & Fuels* 38:306–322.

7

Core-Scale Oil Shale Pyrolysis

Pankaj Tiwari, Josh Staten, and Milind Deo

CONTENTS

Previous chapters discussed the large set of experiments that were performed on a single set of oil shale samples (GR1, GR2, and GR3) and their pyrolysis products from the Skyline 16 oil shale core, drilled in the Green River Formation of eastern Utah. This chapter focuses on heat and mass transport effects in larger core samples from the Skyline 16 core and from other cores elsewhere in the Mahogany zone of the Green River Formation in Utah. It describes experiments for determining the magnitude of the effects of four factors (core size, heating regime, pressure, and temperature) on products from oil shale pyrolysis. It is shown that each of the four factors contributes to oil yield and to product distribution. These factors may be used in enhancing the techniques that exist for the production of shale oil from oil shale pyrolysis.

7.1 Background

Different in situ and ex situ technologies have been proposed for commercial oil shale development. Royal Dutch Shell built a pilot-scale facility that employed the in situ conversion process (ICP) in which the oil shale is gradually heated (Wellington et al. 2005, Ryan et al. 2010) with electric heaters to a temperature of about 350°C. ExxonMobil's ElectroFrac™ process fractures the oil shale using a hydraulic fluid then fills the fractures with a conductive material, creating a planar heater (Symington et al. 2006). For ex situ processing, the Petrobras Petrosix gas combustion retort is a conventional oil shale process, pyrolyzing 7800 tons of crushed oil shale per day (Lee 1990). Red Leaf's EcoShale® in-capsule process technology was operated successfully in a 2009 field pilot in which a capsule filled with oil shale was heated by a gas-fired, closed-loop piping system (Red Leaf Resources 2015). Each method varies in implementation, but the underlying physical restraints of kinetics, thermodynamics, heat transfer, and mass transport must be overcome at a low enough cost to make the technologies economically feasible for commercial oil production from oil shale.

Pyrolysis of oil shale occurs at approximately 300°C, the temperature at which the endothermic energy barrier is overcome; the decomposition rate then accelerates as the temperature increases (Franks and Goodier 1922). There are four physical phenomena that need to be addressed to optimize shale oil production via oil shale pyrolysis. One, heat transfer through the oil shale is the key to unlocking the oil from the oil shale kerogen in an economical manner. Two, reaction kinetics shows the optimum heat regime for maturing the kerogen into oil. Three, mass transfer moves products of the matured kerogen through the rock to a recoverable area. Four, thermodynamics affects each of the other phenomena; as the products are formed, their intrinsic properties change. Each of the four phenomena is affected by the nature of the rock and of the kerogen, by the type of heat source used, by the pressure of the system, and by the time the oil shale is held at a given set of conditions.

A good model of oil shale pyrolysis will bridge the gap between in situ and ex situ retorting, enabling experts in the field to compare the costs and benefits of different technologies. For example, the kinetic model of kerogen decomposition will be identical in both processes. Heat and mass transfer models will be dictated by the process and by the particle characteristics. The accuracy of predictions from in situ models will be governed by the heterogeneity of the formation, whereas predictions from the more tightly controlled ex situ processes will have narrower uncertainty bands. To predict oil yields for in situ pyrolysis, any model must be validated against available experimental data. As described in the next section, Tiwari (2012) performed the initial steps in the validation process by obtaining thermogravimetric analysis (TGA) data for determining kinetic rates, high-pressure TGA (HPTGA) data for determining pressure effects on kinetic rates, TGA with mass spectrometry (TGA-MS) data for determining stoichiometric composition, and data from the multiscale pyrolysis of different sizes of core samples for determining scaling factors due to size.

7.2 Oil Shale Pyrolysis Experiments

This section describes the experimental procedures that were performed and presents selected results from the characterization, pyrolysis, and pyrolysis product analysis of three sets of oil shale samples. The first two sets of samples were provided by the Utah Geological Survey and were from the Mahogany zone of the Green River Formation in Utah. Sample #1 (S1) was a powdered oil shale sample, while Sample #2 (S2) was a set of four samples, including three cores (0.75 in., 1 in., and 2.5 in. in diameter) and one powdered sample. The third set of samples was from the 4 in. diameter Skyline 16 core discussed in Chapters 4, 5 and 6. This set included three fresh, organic-rich (Mahogany zone) samples: GR1 (461.1–462.1 ft in depth), GR2 (485.9–486.9 ft), and GR3 (548.1–549.1 ft). More detailed results can be found in Tiwari (2012).

7.2.1 Oil Shale Characterization

7.2.1.1 S1 and S2

The prepyrolysis (e.g. raw) S1 and S2 oil shale samples were tested for inherent moisture and characterized using elemental analysis and x-ray diffraction (XRD). To test for inherent moisture, small quantities of powdered S1 and S2 samples were dried at 100°C for 4 h. No significant weight loss was observed, so the samples were used as received. Elemental analysis was done to characterize the common elements found in the organic portion of the oil shale. A LECO CHNS-932 analyzer was used for carbon (C), hydrogen (H), nitrogen (N), and sulfur (S) and a VTF-900 analyzer for oxygen (O). Using the O/C and H/C ratios, both S1 and S2 were located on a van Krevelen chart as type I kerogens. The results of the elemental analysis are shown in Table 7.1.

The raw oil shale samples were ground to 325 mesh in a micronizing mill and then characterized using XRD. The mineral composition is important for determining the likelihood that the TGA, and multiscale pyrolysis results are affected by mineral interactions. The results of the XRD characterization are given in Table 7.2. While there are many constituents, dolomite (33.5 wt% and

TABLE 7.1

Elemental Analysis (CHNSO) Results for the Raw Oil Shale Samples

Element	Sample #1		Sample #2 (Powder)	
	Weight %	Std Dev	Weight %	Std Dev
Carbon	17.45	0.26	22.09	1.00
Hydrogen	1.60	0.08	2.14	0.12
Nitrogen	0.53	0.06	0.65	0.06
Sulfur	0.18	0.04	0.11	0.02
Oxygen	15.69	0.79	16.54	0.97
H/C (molar)	1.10	—	1.17	—
O/C (molar)	0.67	—	0.56	—

Note: Molar H/C and O/C ratios are calculated for kerogen typing.

TABLE 7.2

Results from XRD of Oil Shale Rock

Mineral	Weight %		Chemical Formula
	Sample #1	Sample #2	
Quartz	7.7	7.7	SiO_2
Plagioclase	19.5	7.60	$CaAl_2Si_2O_8$
Calcite	6.9	3.95	$CaCO_3$
Illite	5.8	2.84	$(K,H_3O)(Al,Mg,Fe)_2(Si,Al)_4O_{10}[(OH)_2,(H_2O)]$
Dolomite	33.5	62.93	$Ca\,Mg\,(CO_3)_2$
Orthoclase	12.4	10.88	$KAlSi_3O_8$
Aragonite	11.7	–	$CaCO_3$
Analcime	2.4	4.13	$NaAlSi_2O_6 \cdot H_2O$

62.93 wt% in S1 and S2, respectively) is the predominant mineral in both samples. Illite and analcime, also found in both samples, may add to the complexity of modeling the pyrolytic reactions because water is released from these minerals at a relatively low temperature. Another modeling complexity is the release of carbon dioxide (CO_2), oxygen (O_2), and other mineral emissions as the inorganic portion of the oil shale decomposes. These potential interactions and decompositions are not considered in the model developed here.

7.2.1.2 GR1, GR2, and GR3

Elemental (CHNS) and TGA analyses were conducted on uniformly mixed, powdered (100 mesh) shale samples of GR1, GR2, and GR3. TGA experiments were also performed on the isolated kerogens from these samples (GR1.9, GR2.9, and GR3.9). The kerogens were isolated from the homogenous powdered samples of GR1, GR2, and GR3 shales using a series of strong acids (demineralization process) as described in Chapter 5. The elemental analysis was repeated three times for each sample, and the average values with

TABLE 7.3

Elemental Analysis (CHNS) of Skyline 16 (GR) Samples

Sample ID	C (wt%)	H (wt%)	N (wt%)	S (wt%)
GR1	33.93 ± 5.76	3.21 ± 0.21	1.17 ± 0.27	0.56 ± 0.68
GR2	19.80 ± 5.23	1.40 ± 0.64	0.47 ± 0.19	0.13 ± 0.17
GR3	20.44 ± 1.00	1.84 ± 0.05	0.71 ± 0.11	0.18 ± 0.15

TABLE 7.4

TGA Analysis (Weight Loss) of Skyline 16 (GR) Samples and Their Isolated Kerogens (GR X.9) at Heating Rates of 10°C/min

	Samples ID	Sample (mg)	Organic (%)	Mineral (%)	Coke (%)
Oil shale	GR1	18.16	21.13	17.86	1.63
	GR2	17.00	7.20	29.85	0.0
	GR3	23.11	11.16	20.43	0.34
Isolated kerogen	GR1.9	4.32	70.58	—	15.57
	GR2.9	2.3	80.81	—	11.98
	GR3.9	6.77	67.58	—	10.92

standard deviation are reported in Table 7.3. The TGA results from the original shale samples and from the kerogen isolates are summarized in Table 7.4. Details of the TGA experiments are found in Section 7.2.2.2.

The samples analyzed reflect considerable variation in the organic versus mineral composition as well as in the elemental composition. GR1 had the highest organic content of the three samples at 21.13 wt%, while GR2 had only 7.2% organic content. GR2 had the highest content of minerals that decompose at high temperature (29 wt%). A similar trend was observed in the elemental analyses. The weight percents of C, H, N, and S were higher in the organic-rich GR1 sample than in the organic-lean GR2 sample.

7.2.2 Experimental Procedures

Four different sets of experimental procedures, summarized in Table 7.5, were employed to study the phenomena associated with production of shale oil via oil shale pyrolysis. These procedures were carried out on the powdered and core samples discussed above (or in the previous section). S1 was used in all four sets of experimental procedures, S2 was used in three sets (excluding TGA-MS), and the Skyline 16 samples were used in two sets.

7.2.2.1 TGA Pyrolysis Experiments on S1 and S2 Samples

TGA was performed on S1 and S2 (powder) to obtain intrinsic reaction rates of the oil shale. This procedure was similar to that performed on the Skyline

TABLE 7.5

Experimental Procedures Performed on Powdered and/or Oil Shale Core Samples

Oil Shale Sample	Size	Experimental Procedure
Sample #1	Powder	TGA
		HPTGA
		TGA-MS
		Multiscale pyrolysis (1 in. Swagelok reactor)
Sample #2	Powder	TGA
		HPTGA
	0.75 in. diameter	Multiscale pyrolysis (1.25 in. Swagelok reactor)
	1 in. diameter	Multiscale pyrolysis (1.25 in. flange reactor)
	2.5 in. diameter	Multiscale pyrolysis (3.0 in. flange reactor)
GR1	Powder	TGA
	1 in. diameter	Multiscale pyrolysis (1 in. Swagelok reactor)
GR2	Powder	TGA
	1 in. diameter	Multiscale pyrolysis (1 in. Swagelok reactor)
GR3	Powder	TGA
	1 in. diameter	Multiscale pyrolysis (1 in. Swagelok reactor)

Notes: All samples are from the Mahogany zone of the Green River Formation in Utah. Multiscale pyrolysis was performed at various heating rates, final temperatures, pressures, and operating modes (batch, semibatch, and continuous).

16 oil shale samples in Chapter 5. TGA measures the change in weight of a sample undergoing pyrolysis to indirectly observe the progress of reactions. TA Instrument's Q500 was the TGA equipment used in these experiments. It controlled the temperature of the reaction chamber up to a maximum of 1000°C using electrical heating, accommodated a total gas flow rate of 100 mL/min and provided heating rates from 0.1°C/min to 100°C/min.

The TGA was purged with nitrogen (N_2) for 15 min prior to initiating the experiment. Approximately 20 mg of sample was placed in a platinum basket attached to a microbalance with a small wire. Mass or heat transport effects were assumed to be negligible because of the size of the particles. The reactor was then heated to the desired temperature at a set heating rate. For the TGA and TGA-MS experiments, the balance gas (N_2) was maintained at flow rates of 40 and 10 mL/min and the purge gas at 60 and 90 mL/min, respectively. For the isothermal tests, the samples were rapidly heated (heating rate of 100°C/min) to a predetermined final temperature between 200°C and 550°C. For the non-isothermal tests, the heating rate was varied between 0.5°C/min and 50°C/min.

The HPTGA experiments were performed on S1 and powdered S2 samples to determine pressure effects on intrinsic reaction rates and on product distribution. The Cahn TherMax 500 HPTGA from Thermo Fischer was used.

TGA-MS experiments were performed on S1 to determine at which temperature the targeted components evolved. To conduct the experiments, the TGA described earlier was coupled with a Thermostat model GSD 301 TC from Pfeiffer Vacuum MS.

7.2.2.2 TGA Pyrolysis Experiments on GR Samples

TGA pyrolysis/combustion experiments were performed on powdered samples of GR1, GR2, and GR3 at a heating rate of 10°C/min to measure the organic and mineral content of the samples and to estimate the coke formed during pyrolysis. The final temperature for the pyrolysis experiments was 1000°C and for the combustion experiments was 600°C. The final hold times were 10 min for all experiments. The pyrolysis was divided into two stages: the first peak (organic), which appeared from ambient to 600°C (depending on heating rate), and the second peak (mineral decomposition), which occurred from 600°C to 1000°C. After pyrolysis, the sample was cooled to 400°C in an N_2 environment without opening the furnace. Then, air was supplied to combust the sample, and the sample was heated to 600°C. Finally, the sample was held isothermally at 600°C for 10 min to ensure complete combustion of the any coke that had formed. This scheme can be seen in Figure 7.1.

Similar TGA pyrolysis/combustion experiments were also performed on the isolated kerogens from these samples (GR1.9, GR2.9, and GR3.9). The pyrolysis experiments were performed at three heating rates, 5°C/min, 10°C/min, and 20°C/min. The second peak (mineral decomposition) does not appear in these kerogen samples.

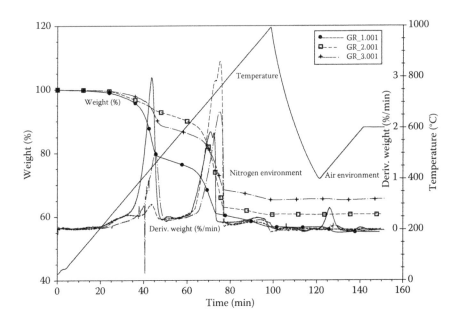

FIGURE 7.1
Scheme of different stages of TGA analysis to estimate the organic, mineral and coke content of the material (raw or spent shales). Example of GR powdered samples.

7.2.2.3 *S1 and S2 Multiscale Pyrolysis Experiments*

The multiscale pyrolysis experiments were run in batch, semibatch, and continuous flow conditions at temperatures of 350°C, 425°C, and 500°C and at pressures of psig (ambient pressure) and 500 psig. A specified heating rate was applied until the desired temperature was reached. The samples were then tested at 6, 12, 18, 24, and 48 h. In batch mode, the system was closed and no products escaped the reaction chamber. This mode simulates an in situ condition where the pyrolysis products do not reach a pressure that fractures the rock. In semibatch mode, products were allowed to escape out of the top of the reactor, but no sweep gas was used. This mode simulates an in situ condition where the products escape the system by their generated pressure and by the given conditions. In continuous mode, the reactor was swept with gas; any product that was exuded was swept out of the reaction chamber and the secondary reactions were quenched. This condition simulates ideal conditions in a reservoir with high flow capacity. However, immediate removal and quenching of pyrolysis products from the heated zone is impractical at an industrial scale with current in situ technology.

Pyrolysis experiments were performed on S1 under batch, semibatch, and continuous reactor conditions to determine how the primary reaction products may continue to break down with differing residence times at reaction conditions. Then, pyrolysis experiments were performed on S2 (0.75–2.5 in. diameter cores) under continuous reactor conditions to evaluate transport resistances (mass and heat) and to study the effects of pressure (0 and 500 psig), core size (0.75–2.5 in. in diameter), heating rate (nonisothermal, 1°C/min–10°C/min; isothermal, 100°C/min), and final temperature (300°C–500°C) on product yield and distribution.

Four different cylindrical reactors were constructed for the analysis: 1 in. × 6 in. (diameter by length), 1.25 in. × 8 in., 1.25 in. × 12 in., and 3 in. × 10 in. These reactor sizes were chosen to house the powdered and core samples with minimal dead volume. All reactors were constructed of 316 stainless steel rated to 4000 psi at 600°C. Heating tape heated the three smaller reactors and a ceramic heater band heated the 3 in. × 10 in. reactor. The reactors and fittings were insulated using glass wool and high-temperature silicon tape. High-pressure Swagelok fittings were used for the 1 in. × 6 in. and 1.25 in. × 12 in. reactors. These reactors were used to test the powdered core samples (S1 and S2) and the 0.75 in. core (S2), respectively. Graphite flanges were used to seal both ends of the 1.25 in. × 8 in. and 3 in. × 10 in. reactors. The flange reactors were used to test the S2 1 in. and 2.5 in. cores, respectively.

Type K thermocouples were used to find the temperature profile through the core for each case. The thermocouples were inserted 0.6 in. into the core through 0.128 in. diameter holes drilled through the reactor and the core sample as illustrated in Figure 7.2. The 1.25 in. flange reactor had holes designed to monitor the temperature at the center of the core (TC-1), the surface of the core (TC-4), and the surface of the reactor (TC-5). The two Swagelok

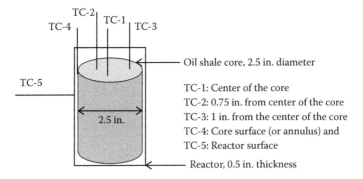

FIGURE 7.2
Schematic of thermocouple placement for the 2.5 in. diameter core sample. All thermocouples used in the tests were of type K.

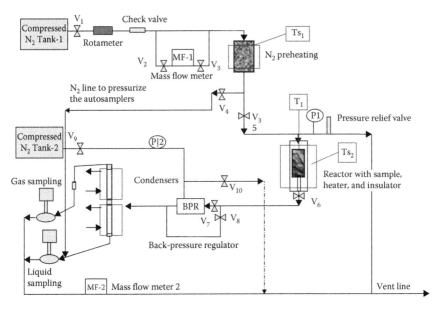

FIGURE 7.3
Experimental setup used to study the effect of operational parameters on yield and quality of the pyrolysis product distribution.

reactors were designed to monitor TC-1 and TC-5. The 3 in. flange reactor was designed to monitor all five of the thermocouples in Figure 7.2.

A schematic for the continuous flow experiments is shown in Figure 7.3. The system allowed for the immediate collection of the products formed from the pyrolysis reactions. The products were then condensed, sampled, and analyzed. The walls of the chamber were heated, and the reactor temperature was controlled using SPECVIEW and either TC-1 or TC-5, depending on whether the experiment was isothermal or nonisothermal. The N_2 was preheated before entering the reaction chamber. Temperature and gas flow rates were recorded using LabVIEW. Pressure was controlled using a Swagelok back-pressure regulator. A heater kept the line from the reactor to the back-pressure regulator at a constant temperature of 200°C. The condensers were cooled using a Brookfield TC501 bath with controller. Samples were taken using a 12 port autosampler with a VCOM interface.

7.2.2.4 Skyline 16 Multiscale Pyrolysis Experiments

Each core (GR1, GR2, GR3) was dissected into three sections (Figure 7.4). These sections were then tested in continuous flow isothermal experiments at 350°C, 425°C, and 500°C with hot N_2 flow (~100 mL/min) for 24 h. The temperature of the reactor surface (TC-5) was used to control the pyrolysis temperature. Temperature profiles were recorded at three points: reactor surface (TC-5), core surface (TC-4), and at the center of the core (TC-1). The steady-state temperature difference between the reactor surface and the center of the core was about 50°C while that between the core surface and center of the core was 20°C.

7.2.2.5 Pyrolysis Product Analysis

Oil and gas samples were collected for compositional analyses after pyrolysis. The fluid collected in the condensers after pyrolysis was analyzed using gas chromatography (GC) and GC/MS (only selected samples) to determine composition. The GC chromatograms were converted to carbon number distribution by using simulated distillation procedures. Produced oil was analyzed using FTIR to find the wax appearance temperature of the oil, densitometry to find the density of the oil, and rheometry to find the viscosity of the oil.

Pyrolyzed samples (e.g., spent shales) from the multiscale pyrolysis reactor (either powders or cores) were subjected to TGA analysis to estimate the amount of unreacted organic material and of coke/char left in the rock. To perform these tests, the solid material collected after the reactor pyrolysis was

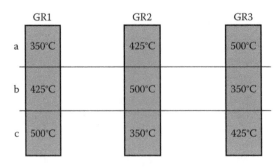

FIGURE 7.4
GR core sections subjected to isothermal pyrolysis under different temperatures.

crushed/ground to a powder (if the sample was a core) or used "as-is" (if the sample was a powder). Then, the TGA analysis discussed in Section 7.2.2.2 was performed. For these samples, the first pyrolysis peak (organic) was the unreacted organic material remaining in the sample after the pyrolysis reactor experiment. The weight loss during the combustion stage corresponded to the coke that formed during pyrolysis.

7.3 Pyrolysis Experimental Results and Discussion

Results of the nonisothermal TGA experiments and of the isothermal and nonisothermal, multiscale pyrolysis experiments are presented in this section; isothermal TGA results were not used in the kinetic model and are not presented in this chapter. For full results and details, including results from the HPTGA and TGA-MS experiments, see Tiwari (2012).

7.3.1 Determination of Oil Shale Pyrolysis Rates Using TGA Data

Results from the nonisothermal TGA of S1 are seen in Figure 7.5. The different symbols connected by lines represent the mass loss at heating rates

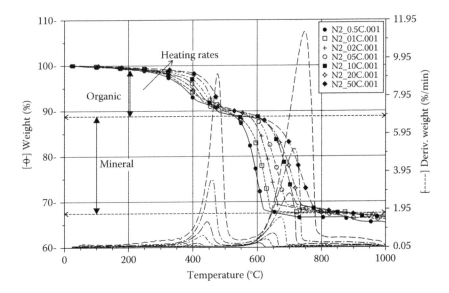

FIGURE 7.5
Nonisothermal thermogravimetric analysis pyrolysis thermograms of S1 at heating rates from 0.5°C/min to 50°C/min in an N_2 environment. The solid lines are weight curves and the dashed lines are the corresponding derivatives. The first derivative peak represents organic weight loss and the second peak represents mineral weight loss.

ranging from 0.5°C/min to 50°C/min, while the lines represent their corre-sponding time derivatives. Based on these TGA results and on results from other researchers (Franks and Goodier 1922), it is inferred that pyrolysis begins when the oil shale reaches approximately 300°C.

From the nonisothermal TGA data of S1, kinetic parameters were calcu-lated using the advanced isoconversional method, described briefly here and more extensively in (Vyazovkin and White 2000). In this method, the TGA weight loss data are used to calculate the activation energy as a function of α, the normalized weight loss, or sample conversion (Equation 7.1), using Equation 7.2:

$$\alpha = \frac{W_0 - W_t}{W_0 - W_\infty} \tag{7.1}$$

$$\frac{d\alpha}{dt} = f(T) \cdot f(\alpha) = A \cdot e^{\frac{-E}{RT(t)}} \cdot f(\alpha) \tag{7.2}$$

where
  W_0 is the initial sample weight
  W_t is the weight at any given time (t)
  W_∞ is the final weight
  A is the pre-exponential factor
  E is the activation energy
  R is the universal gas constant
  $T(t)$ is the temperature at time t.

Figure 7.6 shows the activation energies for the kerogen in the oil shale range from 93 to 245 kJ/mol and the values of $A \cdot f(\alpha)$ vary from 1.42E6 to 4.46E16 min^{-1}.

Figure 7.7 shows the fit of the model to the S1 experimental data for nor-malized conversion of kerogen to products (initial and final conversion set to zero and one, respectively) versus temperature for heating rates ranging from 0.5°C/min to 50°C/min. The lower heating rates show the decompo-sition reaction starting just above 250°C, but there is a thermal induction period for higher heating rates. Inferring from Figure 7.7, slower heating rates require lower temperatures but longer process time to reach full con-version (~830 min at 0.5°C/min versus ~10 min at 50°C/min).

7.3.2 TGA Results for GR Samples and Their Kerogen Isolates

The TGA results from the original GR1, GR2, and GR3 powdered samples and from their powdered kerogen isolates were presented in Table 7.4. The onset points of GR1.9 kerogen decomposition coincide with those of organic

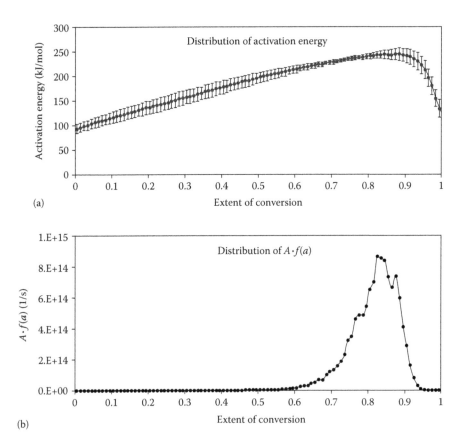

FIGURE 7.6
Kinetic parameters obtained using the TGA data shown in Figure 7.5. Distribution of activation energies and pre-exponential factors as a function of conversion are shown.

matter in the original GR1 shale under identical conditions as shown in Figure 7.8. The raw shale and isolated kerogen have similar decomposition patterns, which suggests that the kerogen decomposition is not affected by the inorganic material present. Also, the results from the kerogen isolates (GR1.9, GR2.9, and GR3.9) show that the onset points (start and end) in the pyrolysis zone of all three kerogens are identical (Tiwari 2012).

For the kerogen isolates, the coke formed during pyrolysis varies in the range of 10–15 wt% of initial kerogen weight and is highest for GR1.9; see Table 7.4. This trend is also observed during the TGA pyrolysis of the raw oil shale samples, where the amount of coke formed during pyrolysis correlates to the amount of the organic matter in the original sample. Coke formation is a maximum in the organic-rich GR1 sample, while there is no coke formation during the GR2 pyrolysis.

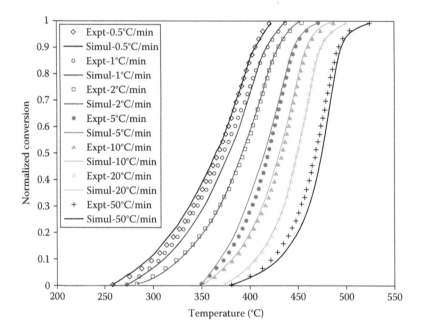

FIGURE 7.7
Experimental and simulated conversion profiles at different heating rates using the advanced isoconversional method.

7.3.3 Multiscale Pyrolysis Results

Pyrolysis was performed at four scales (powder and 0.75 in., 1 in., and 2.5 in. diameter cores) using the apparatus shown in Figure 7.3. The results and findings are presented in this section.

7.3.3.1 Pyrolysis of Powder Sample (S1)

S1 was pyrolyzed in the 1 in. Swagelok reactor in batch, semibatch, and continuous flow modes at various temperatures (350°C, 425°C, and 500°C) and pressures (ambient pressure and 500 psig).

Selected results for the batch experiments can be seen in Figure 7.9; batch experiments were considered to be isothermal with heating rates of 100°C/min. As noted in Section 7.2.2.5, the weight percent of unreacted organic material and that of coke after pyrolysis were determined using TGA (in inert and then oxidizing environments). Hold time and pressure appear to affect coke formation as the only significant coke that formed (0.28 and 1.20 wt%) in the batch experiments came after 6 and 18 h of heating at 500°C and an initial pressure of 500 psig. There was negligible coke formed at 500°C and ambient initial pressure. Also at ambient pressure, no appreciable products formed until the temperature reached 425°C. There is a distinct trend of more product at higher pressure.

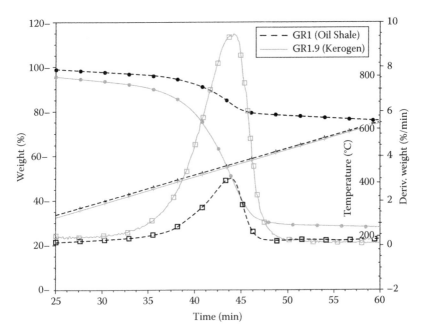

FIGURE 7.8
Comparison of the onset points for organic matter decomposition during the pyrolysis (10°C/min in N_2 environment) of isolated kerogen (GR1.9) and of the raw oil shale (GR1). The straight lines correspond to the temperature.

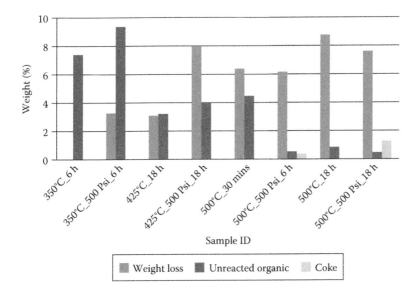

FIGURE 7.9
Weight loss, unreacted organic content, and coke results from the batch pyrolysis of S1 followed by the TGA analysis of spent shale.

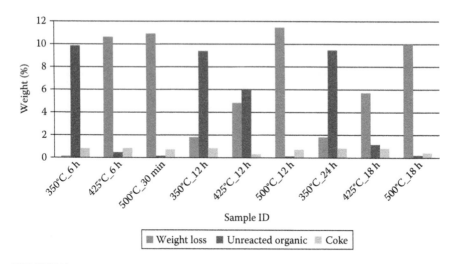

FIGURE 7.10

Weight loss, unreacted organic content, and coke results the semibatch pyrolysis of S1 followed by the TGA analysis of spent shale.

Selected results for the semibatch mode can be seen in Figure 7.10. All semibatch experiments were considered to be isothermal with heating rates of 100°C/min. Hold time had a negligible effect on the higher-temperature experiments and a small effect on the experiments at 350°C (2% weight loss after 24 h compared to negligible weight loss after 6 h). In contrast to the batch experiments, coke formation was observed at all temperatures, although it never exceeded 1% by weight.

Under continuous flow conditions, no coke was found for heating rates below 5°C/min, but there was 0.89 wt% coke found in the sample heated to 500°C at a rate of 10°C/min. The oil yield reflected the heating rate, i.e., higher heating rates yielded more oil.

These powdered oil shale experiments provide insight into the challenges faced with in situ retorting of oil shale to produce shale oil. The results from the batch and semibatch pyrolysis of S1 show that the product composition is dependent on temperature, pressure, and residence time; higher values for these parameters create more coke. The continuous flow experiments show that maximum oil production can be achieved by optimizing the heating rate and the oil extraction rate; oil production is maximized at high heating rates and high oil extraction rates (instantaneous product extraction and quenching). However, for in situ processes, this situation is not possible.

7.3.3.2 Pyrolysis of 0.75 in. Diameter S2 Core

Table 7.6 lists the elemental analysis (CHNSO) of three 0.75 in. diameter S2 oil shale core samples after pyrolysis (spent shale) and of the pyrolysis products

TABLE 7.6

Elemental Analysis of Three 0.75 in. Diameter S2 Oil Shale Cores and Their Pyrolysis Products (shale oil)

Samples	Pyrolysis Temperature	C (wt%)	H (wt%)	N (wt%)	S (wt%)	O (wt%)	Total	H/C (molar)	O/C (molar)
Oil shale (powdered)	—	22.09	2.14	0.65	0.11	16.54	41.53	1.17	0.56
Spent shale	300°C	14.12	0.44	0.26	0.01	25.42	40.24	0.38	1.35
	350°C	14.10	0.82	0.47	0.02	20.87	36.28	0.70	1.11
	400°C	13.06	0.21	0.27	0.01	27.99	41.54	0.19	1.61
Shale oil	300°C	79.72	10.72	2.34	0.65	2.36	95.79	1.61	0.02
	350°C	79.91	10.91	2.34	0.62	1.93	95.71	1.64	0.02
	400°C	80.89	11.10	2.05	0.65	2.13	96.82	1.65	0.02

Note: Oil shale refers to the prepyrolysis samples, while spent shale refers to the post-pyrolysis sample.

at three temperatures (300°C, 350°C, and 400°C). For comparison, the analysis of the powdered S2 oil shale sample prior to pyrolysis is also included. This analysis gives material constraints for the products formed during pyrolysis; in particular, the H/C ratio gives the best idea of what type of oil or quantity of coke will result from pyrolysis.

Results from the pyrolysis of the 0.75 in. diameter cores are shown in Table 7.7. For the ambient pressure experiments, the test at 400°C had the highest oil yield and weight loss. For the high-pressure experiments (500 psig), the test at 500°C had the highest oil yield and weight loss. It can be inferred from Table 7.7 that high pressure results in a higher gas yield, while the oil yield does not change significantly.

While the kinetic model for oil shale pyrolysis described in Section 7.3.1 predicts that higher pressure results in higher coke formation, in the 0.75 in. core pyrolysis, the high-pressure experiment at 400°C results in lower coke yields (0.22 wt%) than the 400°C experiment at ambient pressure (5.78 wt%) (Table 7.7). These results illustrate that pressure is not the only factor influencing coke formation and that core heterogeneity is a large source of uncertainty (the S2 cores were not taken from the exact same location/depth).

7.3.3.3 Pyrolysis of 1 in. Diameter S2 Core

Results from the pyrolysis of 1 in. diameter S2 core samples at 500°C and ambient pressure for three heating rates (1°C/min, 5°C/min, 10°C/min) can be seen in Table 7.8. The results indicate that the lower heating rates yield more oil at a cost of process time. This increase in oil yield at low heating rates may be due to the oil shale spending more time in a temperature range that limits secondary reactions such as coking and cracking while the oil is expelled from the core. Another advantage of the lower heating rate is that a higher grade of oil (e.g., lighter compounds) is produced as seen in the combined naphtha and middle distillate condenser fractions in Figures 7.11 and 7.12 for heating rates of 5°C/min and 10°C/min, respectively.

7.3.3.4 Pyrolysis of 2.5 in. Diameter S2 Core

The 2.5 in. diameter core sample had the greatest heat and mass transport effects of the sizes tested. The temperature distributions for a series of isothermal experiments are shown in Figure 7.13; please refer to Figure 7.2 for the locations of the thermocouples. The anomalies in the temperature readings, listed below, are most likely from the heterogeneity of the core, the deviance from ideal conditions (e.g., symmetry, insulation), and/or the secondary reactions:

TABLE 7.7

Results for the Isothermal Pyrolysis of the 0.75 in. Diameter S2 Oil Shale Core

Experiment #	Pyrolysis Conditions				Reactor Pyrolysis			TGA of Spent Shale	
	Temperature (°C)	Pressure (psig)	Sample Mass (g)	Wt loss (%)	Oil Yield (wt%)	Gas Yield (wt%)	Unreacted Organics (wt%)	Coke (wt%)
Expt-1	500	0	47.71	15.30	10.71	4.59	0.08	0.33
Expt-2	400	0	41.39	16.77	12.46	4.31	2.66	5.78
Expt-3	300	0	50.88	1.08	0.64	0.44	15.52	1.16
Expt-4	500	500	38.5	18.70	10.63	8.07	0.16	1.03
Expt-5	400	500	39.09	9.67	2.06	7.61	7.68	0.22
Expt-6	300	500	44.47	2.46	0.67	1.79	15.16	1.44

TABLE 7.8

Results for the Nonisothermal Pyrolysis of the 1 in. Diameter S2 Oil Shale Core at Ambient Pressure

Experiment #	Heating Rate (°C/min)	Sample Mass (g)	Wt. Loss (%)	Oil Yield (wt%)	Oil Yield/ Wt. Loss
Expt-1	1	145.08	15.34	8.76	0.57
Expt-2	5	144.46	13.50	8.20	0.60
Expt-3	10	145.32	10.41	7.66	0.73

Note: Samples were held at final temperature of 500°C for 2 h.

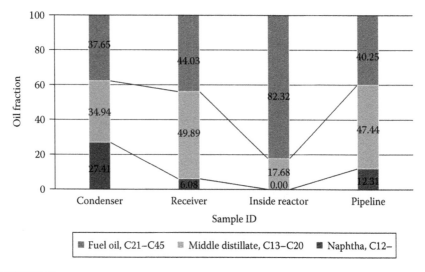

FIGURE 7.11

Grade of oil samples collected during the pyrolysis of S2 1 in. diameter core at a heating rate of 5°C/min to a final temperature of 500°C. The column labels indicate the location in the experimental apparatus where the oil was sampled (see Figure 7.3).

- 350°C and 500°C experiments at ambient pressure—The temperature reading 1 in. from the center (TC-3) is higher than the surface temperature (TC-5); TC-5 is the controlling thermocouple.
- 350°C experiment at ambient pressure—TC-2 has a lower temperature reading than the center of the core.
- 500°C experiment at 500 psig—TC-3 has a higher temperature reading than all of the other thermocouples.

The experiment with an isothermal surface temperature of 350°C and ambient pressure has a steady-state temperature gradient of approximately

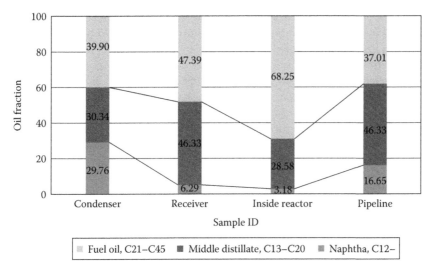

FIGURE 7.12
Grade of oil samples collected during the pyrolysis of S2 1 in. diameter core at a heating rate of 10°C/min to final temperature of 500°C. The column labels indicate the location in the experimental apparatus where the oil was sampled (see Figure 7.3).

80°C/in. This gradient is primarily due to heat losses despite heavy insulation. The lowest temperature (TC-2) is 310°C, which is outside the pyrolysis range of 350°C–400°C. The experiment at 500°C and ambient pressure has a similar gradient, but the lowest temperature (TC-2) is above 400°C, which is in the pyrolysis range.

At 500 psig, there is less temperature gradient in the core. In the experiment, the temperature control was on the surface of the reactor and not on the surface of the core. Under these conditions, higher pressure would lead to better heat transfer through the medium in the annulus and therefore a smaller overall temperature gradient in the system. Heat transport effects are a limiting phenomenon for pyrolyzing kerogen in bulk quantities.

Results from the isothermal experiments at various temperatures ranging from 350°C to 500°C can be seen in Table 7.9. The low weight loss for the experiment at 350°C and ambient pressure can be attributed to the temperature set point itself and to the large temperature gradient in the sample. In contrast, the heat transfer effects are not as prominent in the 500 psig experiments, perhaps due to better heat transfer in the annular space that keeps the surface of the core at a higher temperature. For example, in the experiment at 400°C and 500 psig, each thermocouple measured a temperature above 300°C, the onset of pyrolysis. With higher core temperatures in the pressurized experiments, the weight loss was higher than for the same experiments at ambient pressure. However, the oil yield was lower because more gas was produced at 500 psig than at ambient pressure, as indicated by the oil-yield-to-weight-loss ratio in Table 7.9.

The highest coke formation, 6.06% of the sample's original mass, occurred in the experiment at 500 psig and 500°C. This large amount of coke formation can be attributed to a longer residence time caused by a smaller pressure difference between the core and the reactor's annulus. As the heat moves into the core, the internal core pressure increases due to pyrolysis reactions and thermal expansion. The generated pressure in the pores pushes the products out of the sample. This driving force is reduced if the pressure difference is small, resulting in longer residence times for the pyrolysis products within the core. Longer residence times at high temperatures increase secondary reactions (e.g., cracking and coking).

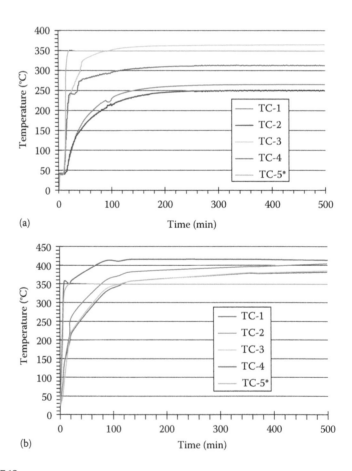

(a)

(b)

FIGURE 7.13

Temperature profiles during pyrolysis of 2.5 in. diameter core sample. (a) 350°C and ambient pressure (b) 350°C and 500 psig (*Control temperature probe.)   (*Continued*)

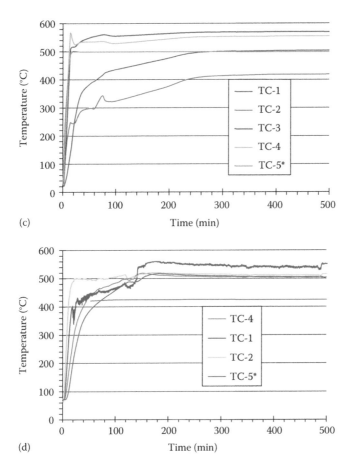

FIGURE 7.13 (*Continued*)
Temperature profiles during pyrolysis of 2.5 in. diameter core sample. (c) 500°C and ambient pressure (d) 500°C and 500 psig (*Control temperature probe.)

7.3.3.5 Pyrolysis of 1 in. Diameter GR Cores

Following the isothermal pyrolysis of the 1 in. diameter GR1, GR2, and GR3 samples, the weight loss and oil yield were measured and the gas produced was calculated by material balance. The results are shown in Figure 7.14. An increase in temperature resulted in an increased weight loss and oil yield. GR1, the organic-rich sample, showed greater weight loss and oil yield than either the GR2 or GR3 samples.

A small amount of spent shale (post-pyrolysis core sample) was further pyrolyzed and then combusted, as described in Section 7.2.2.5, to estimate the unreacted organic remaining in the shale and the amount of coke formed during the pyrolysis, respectively. The results from the TGA analysis of the spent shale are shown in Figure 7.15.

TABLE 7.9

Experimental Results from the Pyrolysis of 2.5 in. Diameter S2 Oil Shale Cores at Different Isothermal Reactor Temperatures and Pressures

Experiment #	Temperature (°C)	Pressure (psig)	Sample Mass (g)	Hold Time (h)	Wt. Loss (%)	Oil Yield (wt%)	Oil Yield/ Wt Loss
Expt-1	350	0	493.58	48	3.67	2.77	0.75
Expt-2	350	500	695.15	48	14.44	8.32	0.57
Expt-3	500	0	961.99	48	21.58	11.71	0.54
Expt-4	500	500	760.00	24	24.52	7.97	0.32

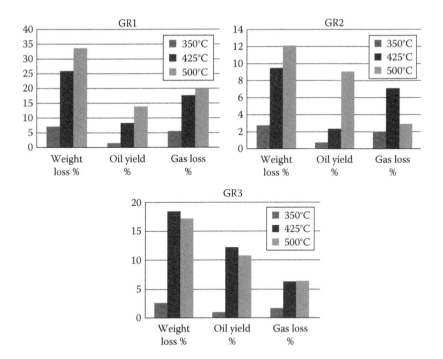

FIGURE 7.14
Weight loss, oil yield, and gas production (e.g., gas loss) during isothermal pyrolysis of GR core sections. The y-axis represents the data in weight percent.

The images of the spent shales after isothermal pyrolysis of the different sections of the three cores are shown in Figure 7.16. During high-temperature (500°C) pyrolysis, the cores with higher organic content (GR1) showed more deformation than at lower-temperature pyrolysis (350°C); they also showed more deformation than organic-lean cores (GR2). Pyrolysis experiments at 350°C resulted in less weight loss and correspondingly low oil yield, an indication that organic decomposition was slow.

The pyrolysis of GR cores showed a trend similar to the powdered TGA and CHNS analyses of the same sections. GR1 samples showed a greater weight loss (33%) at higher temperatures (500°C), but the oil yield (13.7 wt%) did not correspond to the weight loss. One reason for this discrepancy is that during the high-temperature, isothermal pyrolysis (24 h hold time), mineral decomposition may have contributed to the weight loss. The TGA analysis of spent shale showed the presence of significant organic material in pyrolyzed GR1 core samples. This organic matter could either be unreacted organics or heavy oil produced during core pyrolysis. Pyrolysis of GR1 cores also produced more coke relative to organic-lean

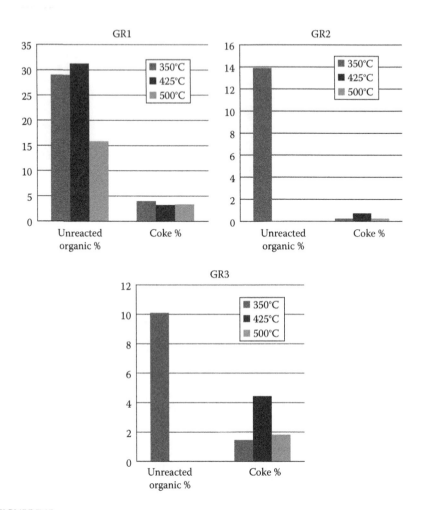

FIGURE 7.15
Unreacted organics and coke in the spent shales from isothermal pyrolysis of GR core sections. The *y*-axis represents the data in weight percent.

samples (GR2 and GR3). The results suggest that it is not only the temperature that influences coke formation but also the organic content of the shale.

Oil and gas samples from the reactor pyrolysis experiment were collected and analyzed using GC. The chromatograms of oils from GR core samples pyrolyzed at 500°C are shown in Figure 7.17. The distribution of hydrocarbons (Figure 7.18) shows that the oils produced from different shales under identical conditions differ in composition. A similar trend is seen in the gaseous products.

FIGURE 7.16
Images of the spent shales from the pyrolysis of GR core samples.

7.3.4 Comparison of Multiscale Core Pyrolysis Results

A comparison of the overall mass balances for the 2.5 and 0.75 in. diameter S2 core experiments at 500 psig is shown in Table 7.10. Because the shale oil produced in the larger core has a longer core residence time, it undergoes secondary reactions of coking and cracking, which decreases oil yield while increasing gas and coke yields. Also in the larger core, the weight loss is higher, which could be beneficial in increasing the flow pathways for the products.

For all the cores, most of the higher temperature experiments yielded higher weight loss (as seen in Tables 7.7 and 7.9 and in Figure 7.14). The grade and the yield of the oil were dependent on sample size, pressure, final temperature, and heating rate. A lower heating rate yielded a higher percentage of oil than higher heating rates. Higher pressure and higher temperature increased the rate of secondary coking and cracking reactions by increasing the residence time of the shale oil in hot zones; the 2.5 in. diameter core at high pressure and temperature yielded the most coke and gas. One exception to this trend was the 0.75 in. diameter core; more coke formed at 400°C than at 500°C.

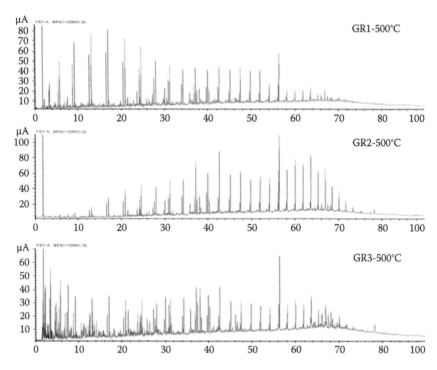

FIGURE 7.17
Gas chromatograms of the oil produced at 500°C from GR1, GR2, and GR3 oil shales.

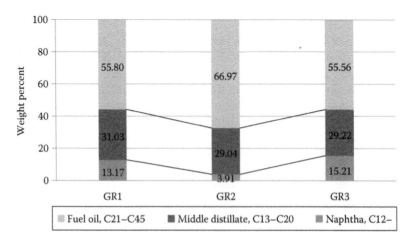

FIGURE 7.18
Oil fractions based on single carbon number distribution of the shale oils produced from pyrolysis of GR core samples at 500°C. Single carbon number distribution is a way of representing the GC spectra with the help of an internal standard.

TABLE 7.10

Comparison of Overall Mass Balance Results for the Isothermal Pyrolysis
of 2.5 in. and 0.75 in. Diameter S2 Cores at 500°C and 500 psig

Material Balance	2.5 in. Diameter Core	0.75 in. Diameter Core (%)
Wt. loss %	24.52	18.69
Oil yield, wt%	7.96	10.63
Coke, wt%	6.06	1.03
Gas, wt%	16.56	8.06
Unreacted organics (wt %)	0.05	0.43

7.4 Summary and Conclusions

The TGA and multiscale pyrolysis experiments were used to obtain intrinsic
reaction rates of Green River oil shale and to study the effects of core size,
heating regime, pressure, and final temperature on product distribution and
oil yield. It was found that larger core sizes produced more coke and gas per
unit mass of core sample, indicating that secondary reactions were occurring
inside the core samples. The oil production in nonisothermal heating regimes
was dampened by heat transport effects in core sizes with diameters larger
than 1 in. It was found that higher pressure lowered the temperature gradient
in larger core samples, producing a higher weight loss but a lower oil yield.

Pyrolysis experiments on larger core samples or pilot-scale in situ experiments are needed to provide data for model validation. Such experiments
will provide researchers with information on how the oil shale rock will crack
and fracture due to thermal expansion, kerogen degradation, oil/gas expansion, and pressure changes, potentially creating macrochannels for product
removal. More research is also required to determine if higher temperatures
and pressures generated from pyrolysis reactions could create enough back-
pressure to move products through the rock to a collecting well.

The fundamental characterization of Green River oil shale described in
Chapter 5 together with models that describe how oil forms during pyrolysis
(Chapter 6 and this chapter) and moves through the rock (Chapters 8 and 9)
are steps toward creating new technologies that make oil shale an economically and environmentally viable source of long-term energy.

References

Franks, A. J. and B. D. Goodier. 1922. Preliminary study of the organic matter of
    Colorado oil shales. *Quarterly of the Colorado School of Mines* 17(4):3–16.
Lee, S. 1990. *Oil Shale Technology*. Boca Raton, FL: CRC Press.

Red Leaf Resources, Inc. 2015. EcoShale® Technology. http://www.redleafinc.com/ecoshaletechnology. Accessed on January 10, 2016.

Ryan, R. C., T. D. Fowler, G. L. Beer, and V. Nair. 2010. Shell's in-situ conversion process—From laboratory to field pilots. In *Oil Shale: A Solution to the Liquid Fuel Dilemma*, eds. O. I. Ogunsola, A. M. Hartstein, and O. Ogunsola, pp. 161–183. Washington, DC: ACS Symposium Series.

Symington, W. A., D. L. Olgaard, G. A. Otten, T. C. Phillips, M. M. Thomas, and J. D. Yeakel. 2006. ExxonMobil's Electrofrac process for in situ oil shale conversion. Paper presented at the *26th Oil Shale Symposium*, Golden, CO. http://www.ceri-mines.org/documents/A05b-BillSymingtonpaper.pdf. Accessed on January 10, 2016.

Tiwari, P. 2012. Oil shale pyrolysis: Benchscale experimental studies and modeling. Ph.D. dissertation, University of Utah, Salt Lake City, UT.

Vyazovkin, S. and C. A. Wight. 2000. Estimating realistic confidence intervals for the activation energy determined from thermoanalytical measurements. Analytical Chemistry, 72(14):3171–3175.

Wellington, S. L., I. E. Berchenko, R. E. Pierre de Rouffignac et al. 2005. In situ thermal processing of an oil shale formation to produce a desired product. U.S. Patent US6880633 B2.

8

Pore-Scale Transport Processes during Oil Shale Pyrolysis

Jan D. Miller and Chen-Luh Lin

CONTENTS

To improve our understanding of transport phenomena during the in situ and ex situ pyrolysis processing of oil shale, pore scale analysis of oil shale samples undergoing pyrolysis at different temperatures is critical. In this study, the pore space of Green River oil shale samples before and after pyrolysis was characterized and digitized using the following multiscale, noninvasive, nondestructive, three-dimensional imaging techniques: x-ray microtomography (XMT), x-ray nanotomography (XNT), and high-resolution x-ray microtomography (HRXMT). Once the digital representation of the pore space was established, the Lattice Boltzmann method was used to estimate flow properties such as absolute permeabilities for the different pore network structures.

This analysis was performed for several oil shale samples over a range of pyrolysis temperatures in order to better understand the nature of the pore networks created during thermal treatment. The sample set included the GR1, GR2, and GR3 Skyline 16 oil shale samples discussed in Chapters 5 through 7.

8.1 Introduction

Unconventional oil resources are defined as extra heavy oils and bitumens associated with oil sand deposits and as kerogen associated with oil shale resources. Most of the world's known oil sand and oil shale deposits are in North America, and the combined liquid fuels potential from these resources far exceeds the world's known conventional oil reserves. The most significant oil shale deposits are in the Green River Formation of Colorado, Utah, and Wyoming with an estimated resource size of 1.5–1.8 trillion barrels. It is expected that oil shale resources will be used primarily for producing transportation fuels. This production will require an understanding of processes that occur over a wide range of length and time scales from the structure of kerogen and how its binds to an inorganic matrix to the fluid flow resulting from in situ processing of an oil shale interval that covers hundreds of acres.

Fundamental issues important to the pyrolysis processing of oil shale in Utah include

1. Kerogen conversion to oil, gas, and coke
2. Nature of the pore space before and after pyrolysis
3. Porous media characteristics after pyrolysis
4. Absolute permeabilities
5. Relative permeabilities

Permeability, or hydraulic conductivity, is the proportionality constant defined from Darcy's law. Absolute permeability refers to the permeability associated with single-phase saturated flow. Relative permeability describes the permeability of two-phase flow as a function of composition.

Item 1, kerogen conversion, was considered in detail in Chapters 6 and 7. This chapter considers the very challenging characterization problems identified in items 2–5. The pore structure and the connectivity of the pore space (e.g., permeability) are important features that determine fluid flow. The pore structure created during oil shale pyrolysis influences the reactions occurring within particles and the flow characteristics of oil and gas products. Hence, study of the nature of the pores and of the corresponding permeability is required.

In previous work, Lee et al. (1994) reported the relation of bulk density of oil shale to the computed tomography (CT) number and to oil yields using x-ray tomography for Australian oil shale. A 2010 study characterized oil shale using x-ray tomography before and after pyrolysis (Mustafaoglu 2010). The application of XMT to the study of thermal cracking in Fushun oil shale under different temperatures was reported by Kang et al. (2011).

8.2 Materials and Methods

8.2.1 Sample Selection

The organic and inorganic compositions of oil shale vary from one geological environment to another. To address this variability in resource composition and its effect on the pyrolysis product distribution, two sets of oil shale drill core samples were used for this study. The first set of samples, dating from 1990, was obtained from the Green River Formation in eastern Utah. These 0.75 in. diameter drill core samples from the Mahogany zone, labeled S2, were used for the preliminary characterization study. These samples are the same 0.75 in. diameter S2 cores described in Chapter 7.

The second set of samples used in this study was obtained from the Skyline 16 core. This set of fresh (drilled in 2012), organic-rich (Mahogany zone) samples of core, described in Chapters 4 and 5, was from the depth range of 461.1–548.9 ft. The samples were identified as GR1 (461.1–462.1 ft), GR2 (485.9–486.9 ft), and GR3 (548.1–549.1 ft). The core samples were 1 in. in diameter.

8.2.2 Pyrolysis of Oil Shale Core Samples

For the first set of cylindrical, S2 oil shale core samples, a 6 in. long, 0.75 in. diameter core was loaded in the pyrolysis reactor described in Section 7.2.2 of Chapter 7. The core was heated from the outside using a band heater. A heating rate of 100°C per min was used to reach the specified reaction temperature (e.g., the experiment was isothermal). The core was then held for 24 h at this temperature. Nitrogen (N_2) gas was flowed at a steady rate of 55 mL/min during the pyrolysis experiment. The condensate was collected in a series of two condensers held at −6°C. The core was cooled to

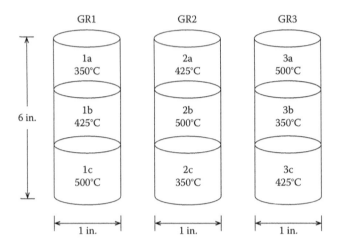

FIGURE 8.1
The GR1, GR2, and GR3 core samples (6 in. long) were sectioned into three portions (each about 2 in. long) and subjected to isothermal pyrolysis at different temperatures.

ambient temperature, removed from the reactor, and subjected to CT analysis as described later. Additional details about the experiment can be found in Section 7.2.2 of Chapter 7.

The second set of 1 in. diameter core samples (GR1, GR2, GR3) was used for a more detailed pyrolysis study. As shown in Figure 8.1, each core (approximately 6 in. long) was divided into three sections to perform the isothermal pyrolysis. The resulting sectioned cores were labeled as GR1a, GR1b, GR1c, GR2a, GR2b, GR2c, GR3a, GR3b, and GR3c. Each section was approximately 2 in. long and weighed 30–60 g. For each experiment, the sample core was packed into a reactor where it could expand both vertically and radially. As with the first set of oil shale samples, a heating rate of 100°C/min was used to achieve the reaction temperature and then the core was held at that temperature for 24 h. These continuous flow, isothermal experiments were conducted at 350°C, 425°C, and 500°C with hot N_2 flow (100 mL/min). The temperature of the reactor surface was used to control the pyrolysis temperature. See Section 7.2.2 for additional details about the pyrolysis experiments.

8.2.3 Multiscale X-Ray Microtomography/Nanotomography

The pore structure and the connectivity of the pore space are important features that determine fluid flow in oil shale during pyrolysis. The XMT/XNT technique is the best noninvasive, nondestructive method available today to characterize complex pore structures. Cone beam XMT systems, introduced commercially a decade ago, are a valuable tool for three-dimensional (3D) visualization, characterization, and analysis of multiphase systems at a voxel (volume element) resolution of 10 μm. This resolution is sufficient to describe mineral structure and composition of multiphase particles having a size on the order of 100 μm (Miller and Lin 2004).

Most of the commercial XMT systems are based on the principle of point projection of an x-ray source through the sample onto a detector. In this design, the achievable resolution is a function of both the x-ray source spot size and the detector resolution. For conventional x-ray systems with large detector pixel sizes, a large geometric magnification combined with a small source spot size is required to achieve high resolution. The highest achievable resolution for these systems is limited by the spot size and the closest distance allowed between the sample and the source. As a consequence, traditional XMT systems use transmission-type sources to achieve a small source spot size combined with a minimal distance between sample and source.

The proximity of the sample to the source imposes a severe restriction on high-resolution tomography. For extended samples, the minimum distance between source and sample can be large, leading to a limited resolution. One solution is the relaxation of the restriction that the sample be close to the source by utilizing an x-ray detector with high resolution. In fact, by using a high-resolution x-ray detector, an imaging resolution better than the x-ray source spot size can be realized for sources with spot sizes larger than the x-ray detector resolution. Such a system is the HRXMT from Xradia (2010), which employs an x-ray detector with submicron resolution combined with a microfocus x-ray source to achieve a voxel resolution of 1 μm. With this resolution, the structure and composition of multiphase particles having a size on the order of 10 μm can be determined. Working distances between source, sample, and detector are typically around 100 mm, so full tomography, even for larger samples, can be achieved.

In addition to the XMT/HRXMT systems, further resolution is possible using the XNT scanner (Xradia, 2010). The XNT provides two key improvements: (1) over one order of magnitude resolution gain to at least 60 nm and (2) a Zernike phase-contrast imaging mode that dramatically enhances the contrast of low-density features.

In this oil shale pyrolysis study, the 3D pore network structures of selected oil shale resources before and after pyrolysis have been characterized noninvasively at varying resolution from tens of microns down to 60 nm using XMT, HRXMT, and XNT imaging techniques.

8.2.4 Lattice Boltzmann Method for Pore Scale Modeling of Single Phase Fluid Flow

Unlike the conventional computational fluid dynamics methods, which involve a "top-down" approach based on discretization of macroscopic continuum equations, the Lattice Boltzmann method (LBM) (Qian et al. 1992,

Chen 1993, Shan and Chen 1993, He and Luo 1997, Stockman 1999, Wolf-Gladrow 2000, Succi 2001, Sukop and Or 2003) is based on a "bottom-up" approach where constructed kinetic models incorporate microscopic model interactions and mesoscopic kinetic equations so that the macroscopic-averaged properties of the flow obey the desired macroscopic equations. Hence, the macroscopic dynamic behavior is the result of the collective behavior of the microscopic particles in the system.

Due to several attractive features, LBM has received increasing attention in the area of fluid flow simulation in porous media (Martys and Chen 1996, Lin and Miller 2004). These features include its ability to incorporate molecular level interactions and a structure that facilitates code parallelization. Over the last decade, LBM has become an emergent mathematical technique capable of handling the complex boundary conditions for flow in porous structures such as oil shale samples in a reasonable span of time. This technique is also popular for its ability to incorporate additional physical complexities such as multicomponent and multiphase flow (Chen and Doolen 1998). Computer simulation can then be used to calculate macroscopic variables of the flow such as absolute and relative permeabilities (Videla et al. 2008).

8.3 Characterization of S2 Oil Shale Core Samples

Based on results from thermogravimetric analysis (TGA), the estimated organic content of the S2 oil shale drill core samples from the Mahogany zone was 17.5 wt%. Further information can be found in Chapter 7 and in Tiwari (2012).

The S2 oil shale core samples were very hard and fine-grained. They showed a laminated structure composed mainly of dolomite, calcite, and quartz in addition to clay minerals in different percentages. The clay minerals were represented by illite and kaolinite. These clay minerals exhibited good crystallinity as indicated from x-ray diffraction analysis. The scanning electron microscope analysis indicated the presence of gypsum and pyrite minerals.

Optical microscopy analysis of thin sections of an S2 oil shale sample, shown in Figure 8.2, confirms the lamellar structure in which different minerals are distributed in very thin and parallel laminae. These laminae include alternating layers of clay minerals and microcrystalline carbonate minerals. The iron oxides and organic matter give color to the banding structure. While both iron oxide and organic matter are mostly associated with the clay mineral layers, they are occasionally associated with the carbonate layers. The clay mineral layers range in thickness

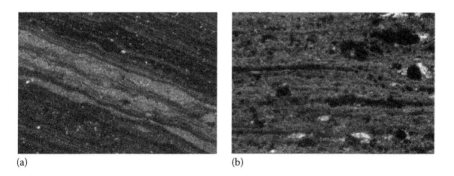

(a)   (b)

FIGURE 8.2
Micrograph of the S2 oil shale drill core samples. (a) Alternating carbonate layers and clay layers with kerogen distribution, especially in the clay layers (cross polarized light). (b) Organic matter is mostly associated with the clay layers (plane polarized light).

from 20–30 µm, while the carbonate mineral layers range in thickness from 10–20 µm. The quartz mineral is of silt size and is found in the carbonate layer as elongated particles parallel to the banding structure of the oil shale.

An S2 core sample was subjected to pyrolysis at 400°C (N_2 flow). Using a combination of XMT/XNT and specialized software, the 3D network of pores, kerogen/mineral phases, crack networks, and flow channels of the oil shale sample were imaged before and after pyrolysis. The porous network structures that evolved during pyrolysis were obtained via image digitalization.

8.3.1 Multiscale X-Ray Microtomography/Nanotomography before Pyrolysis

Three different S2 oil shale core samples were imaged at three resolutions: XMT at 39 µm voxel resolution, HRXMT at 1 µm voxel resolution, and XNT at 60 nm voxel resolution. Figure 8.3 shows the 3D, volume-rendered images from the reconstructed multiscale x-ray CT data of the samples before pyrolysis. On the left, the gray scale level indicates variations in the x-ray attenuation coefficients that depend on the density and atomic number of the material within each voxel. Kerogen-rich (dark gray) and silicates-rich (light gray) lamellar structures are observed. On the right are the distributions of the kerogen phase. As expected, the XNT image reveals smaller inclusions of kerogen with dimensions of less than one µm in size. These results further validate the results obtained from optical microscopy. At a voxel resolution of 60 nm (XNT), individual grains can be identified easily.

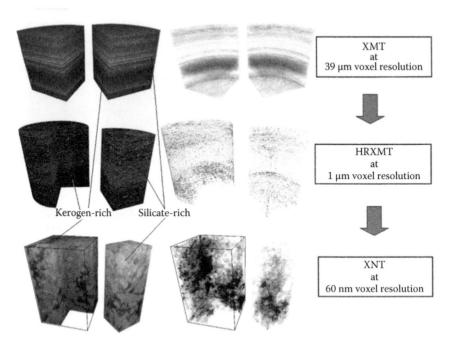

FIGURE 8.3

Volume-rendered images of S2 oil shale drill core from the reconstructions of multiscale x-ray CT data including XMT at 39 μm voxel resolution, HRXMT at 1 μm voxel resolution, and XNT at 60 nm voxel resolution. Left column: The gray-scale level indicates variations in the x-ray attenuation coefficients. Middle column: The gray-scale level shows the distribution of the kerogen phase.

8.3.2 Multiscale X-Ray Microtomography/ Nanotomography after Pyrolysis

Figure 8.4 shows the 3D, volume-rendered images from the reconstructed multiscale x-ray CT data after pyrolysis of the three S2 oil shale core samples shown in Figure 8.3. Crack networks that developed during the pyrolysis process are evident and well-defined. The two distinct regions with different sizes of cracks and voids are identified as region A and region B. Cracks and voids as small as 100 nm (from XNT images) are observed inside region A (silicate-rich, lamellar structure). However, larger, anisotropic cracks and voids developed inside region B (kerogen-rich, lamellar structure from HRXMT images) of Figure 8.4. Figure 8.5 shows the triplanar and volume-rendered images of the residual product after pyrolysis (region A).

8.3.3 Pore Scale Modeling of Post-Pyrolysis S2 Oil Shale Core Sample

As indicated previously, the cracks and voids inside region A (silicate-rich, lamellar structure) of S2 after pyrolysis are small and are created due to

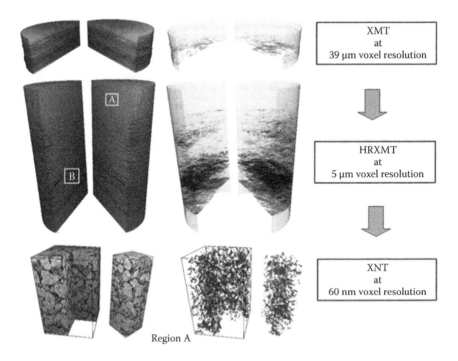

Region A

FIGURE 8.4

Volume-rendered images of S2 oil shale drill core after pyrolysis (400°C, N_2 flow) from the reconstructions of multiscale x-ray CT data including XMT at 39 μm voxel resolution, HRXMT at 5 μm voxel resolution, and XNT at 60 nm voxel resolution. Left column: The gray-scale level indicates variations in the x-ray attenuation coefficients. Middle column: The gray-scale level shows the distribution of the kerogen phase.

the thermal expansion of grain boundaries. An LBM simulation was performed using the post-pyrolysis pore space of the oil shale in region A. Figure 8.6 illustrates the 3D view of the LBM simulation for saturated flow through this pore space. On the right-hand side of Figure 8.6, the solid phase is removed to show the nature of the flow channels. The velocity scale is color-coded as shown by the color bar in Figure 8.6. Solids are white and solution velocity ranges from black for no flow through blue, green, yellow, and, finally, red for the highest flow rate. Based on this simulation, the estimated absolute permeability in region A is 0.00363 μm² or 0.363 millidarcy (mD). It is noted that the absolute permeability in all regions is highly anisotropic.

Figure 8.7 shows the 3D views of LBM-simulated flow along the x-axis (bedding plane) through the reconstructed HRXMT image of region B (kerogen-rich, lamellar structure) of the S2 sample after pyrolysis; stratigraphic up is to the left. The estimated absolute permeability is 3.87×10^{-8} cm² or 3.87 darcy, four orders of magnitude higher than that in region A.

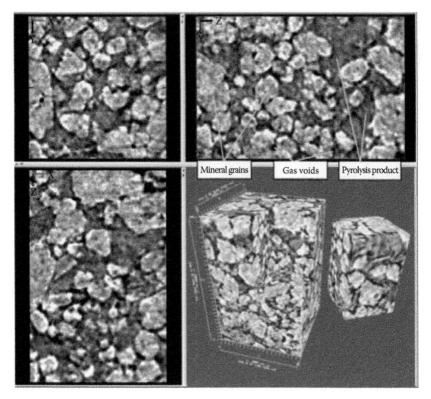

FIGURE 8.5

Triplanar (upper left and right, lower left) and volume-rendered (lower right) images of S2 oil shale sample (20 × 20 × 30 μm³, region A) after pyrolysis at 400°C, obtained from XNT at 60 nm voxel resolution.

8.4 Characterization of Skyline 16 Oil Shale Core Samples

TGA tests were performed on the three GR samples from the Skyline 16 oil shale core (see Table 7.4, Chapter 7). As shown in Table 8.1, the organic carbon contents are 21.13 wt%, 7.20 wt%, and 11.16 wt% for GR1, GR2, and GR3, respectively. These differences clearly indicate the heterogeneity of the GR core samples.

These three samples were each divided into three subsections (see Figure 8.1), and the resulting nine samples were then pyrolyzed at various temperatures as described in Section 8.2.2. The weight loss results after pyrolysis are also shown in Table 8.1. As expected, weight loss increases with temperature, and the cores with the highest organic content show the highest weight loss. In pyrolysis experiments at 425°C and 500°C, weight loss exceeds the weight

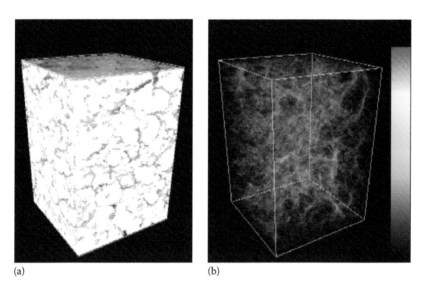

(a) (b)

FIGURE 8.6

3D views of LBM–simulated flow through the reconstructed XNT image (60 nm voxel resolution) of region A of S2 oil shale sample after pyrolysis at 400°C. (a) Solid phase is white. (b) Transparent solid phase to reveal flow channels colored by solution velocity.

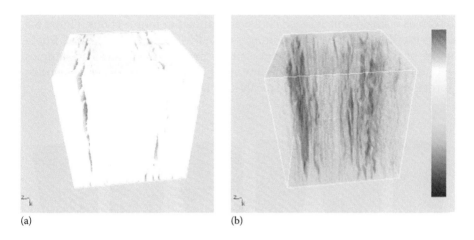

(a) (b)

FIGURE 8.7

3D views of LBM–simulated flow along x-axis (bedding plane) through the reconstructed HRXMT image (1 μm voxel resolution) of region B of S2 oil shale sample after pyrolysis at 400°C. (a) Solid phase is white. (b) Transparent solid phase to reveal flow channels colored by solution velocity.

TABLE 8.1

Weight Loss of GR Core Samples during Pyrolysis

Oil Shale Core Samples (Depth, ft)	ID	Initial Weight (g)	Organic Carbon (Wt%)	Temperature (°C)	Weight Loss (%)
GR1 (461.1–462.1)	1a	40.32		350	7.08
	1b	40.49	21.13	425	25.91
	1c	41.63		500	33.74
GR2 (485.9–486.9)	2c	50.78		350	2.84
	2a	61.82	7.20	425	9.61
	2b	52.92		500	12.26
GR3 (548.1–549.1)	3b	51.75		350	2.54
	3c	47.88	11.16	425	18.40
	3a	51.20		500	17.17

Measurements were normalized based on initial weight.

Source: Tiwari, P., Oil shale pyrolysis: Benchscale experimental studies and modeling, Ph.D. dissertation, University of Utah, Salt Lake City, UT, 2012.

percent organic matter in the sample. Loss of moisture and of bound water in clays and minerals in the shale may have resulted in this phenomenon.

8.4.1 Multiscale X-Ray Microtomography/Nanotomography before and after Pyrolysis

The 6 in. long, 1 in. diameter GR1, GR2, and GR3 core samples were too long to scan as a single unit, so each core was scanned in four sections. Figure 8.8 shows the 3D, volume-rendered images from the reconstructed x-ray CT data of one of the sections of each core sample prior to pyrolysis. The samples were imaged with XMT at 41.67 μm voxel resolution. Lamellar structures (kerogen-rich in dark gray and silicate-rich in light gray) are observed.

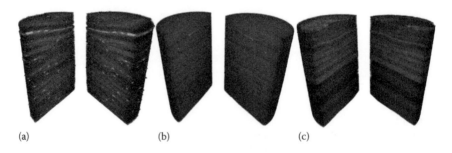

(a)   (b)   (c)

FIGURE 8.8

Volume-rendered images of sections of (a) GR1, (b) GR2, and (c) GR3 core samples before pyrolysis from the reconstructions of XMT data at 41.67 μm voxel resolution.

FIGURE 8.9
Photographs of spent shale samples after pyrolysis of GR1 (1a, 1b, 1c), GR2 (2a, 2b, 2c), and GR3 (3a, 3b, 3c) core samples at different reaction temperatures (325°C, 425°C, and 500°C).

Photos of the subdivided GR1, GR2, and GR3 core samples after isothermal pyrolysis are shown in Figure 8.9. Due to their higher organic content, the GR1 cores exhibited greater deformation during high-temperature (500°C) pyrolysis than either of the leaner cores (GR2 and GR3). The cores with the highest organic content, namely, GR1b and GR1c, essentially crumble due to decomposition of the organic matter. The leanest set of core samples, GR2, remains intact and retains its original appearance. The core with the intermediate organic content, GR3, displays intermediate characteristics with respect to deformation. The overall patterns are consistent with those observed by Duvall et al. (1983). More detailed analysis of the GR pyrolysis experiments is available in Tiwari et al. (2013).

8.4.2 Porosity Changes during Pyrolysis

Selected pieces of the nine post-pyrolysis GR samples were scanned using HRXMT. Characterization of the textural change during pyrolysis was analyzed based on the comparison of the same sections of cores before and after pyrolysis. Figure 8.10 shows the triplanar image for pyrolyzed GR1b

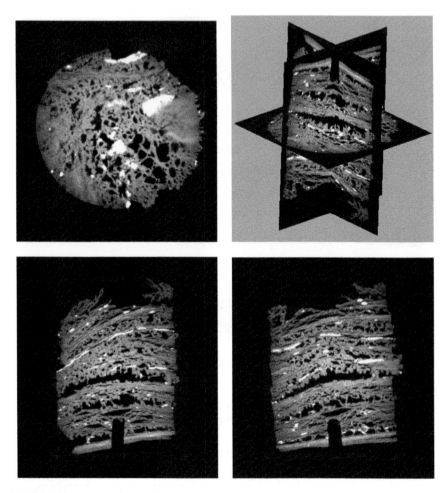

FIGURE 8.10
Triplanar images of GR1b sample (20% organic matter) after pyrolysis at 425°C. Weight loss was 25.91% (see Table 8.1).

(pyrolysis temperature of 425°C). Large amounts of pore space were generated by kerogen conversion to liquids and gases during the pyrolysis process, as evidenced by the dark patches observed in the images. Adjacent gray regions in the pore spaces appear to be reaction products. The distribution and size of the pore space generated is uneven and is consistent with the initial kerogen distribution along the bedding plane. As expected, pores were generated along the kerogen-rich layers.

Based on the relative position of mineral grains inside core GR1b, as shown in Figure 8.11, two section pairs from the 3D images were identified at the same positions along the core before and after pyrolysis at 425°C. Before pyrolysis, the distance between the top (747) and bottom (698) layers was

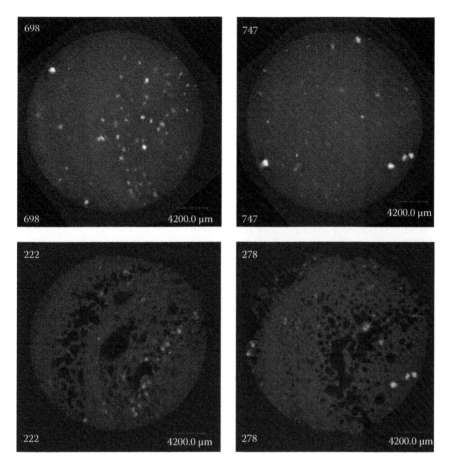

FIGURE 8.11
Effect of pyrolysis at 425°C on GR1b sample. Based on the relative position of mineral grains inside GR1b, two section pairs from 3D CT images were identified at the same positions along the core before and after pyrolysis. The length of the core increased by 291.69 μm or 14.3%.

49 voxels or 2041.83 μm (voxel resolution of 41.67 μm). After pyrolysis, the distance between the top (278) and bottom (222) layers increased to 56 voxels or 2333.52 μm (voxel resolution of 41.67 μm) due to the creation of pore volume during pyrolysis. Overall, pore volume creation increased the length of the core by 291.69 μm, a 14.3% increase. This increase is higher than the increases reported by Duvall et al. (1983), except for their richest sample perpendicular to the bedding plane, which saw a 20% increase in length. Figure 8.11 also shows the heterogeneous distribution of the organic matter in the shale (dark gray color in images) that corresponds to the pore space created in the core after pyrolysis.

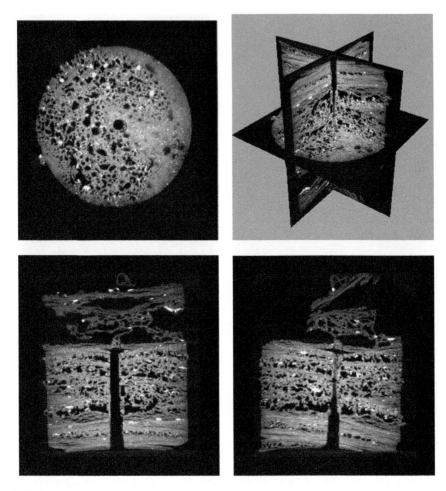

FIGURE 8.12
Triplanar image of GR1c sample (20% organic matter) after pyrolysis at 500°C. The weight loss was 33.74% (see Table 8.1).

Figure 8.12 shows the triplanar image for pyrolyzed core GR1c (pyrolysis temperature of 500°C). The pore space generated is observed to be patchy despite the higher pyrolysis temperature. The two section pairs from the 3D images were identified at the same positions along the core before and after pyrolysis at 500°C as seen in Figure 8.13. Before pyrolysis, the distance between the top (235) and bottom (227) layers was 8 voxels or 333.36 µm (voxel resolution 41.67 µm). After pyrolysis, the distance between the top (143) and bottom (123) layers was 20 voxels or 833.4 µm (voxel resolution 41.67 µm). The length of the core increased by 500.04 µm or 24% during the pyrolysis, which is consistent with the observations of Duvall et al. (1983) for the richest

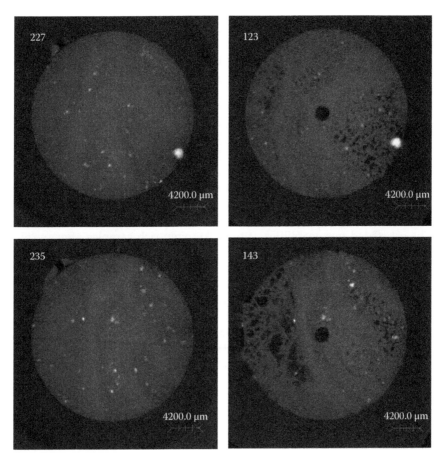

FIGURE 8.13
Effect of pyrolysis at 500°C on GR1c sample. Based on the relative position of mineral grains inside the GR1c, two section pairs from 3D CT images were identified at the same positions along the core before and after pyrolysis. The length of the core increased by 500.04 μm or 24%.

sample perpendicular to the bedding plane. The larger thermal expansion with sample GR1c is consistent with the expectation of higher thermal expansion at higher temperatures in unconfined conditions. Qualitatively, there is no significant difference between the images at 500°C and at 425°C. One would expect to observe a greater amount of pore space at 500°C based on the higher amount of oil generated, but that is not the case. At this temperature, the coke formed might be obscuring the pore space. Additionally, the kerogen content may differ between the GR1b and GR1c samples due to sample heterogeneity.

8.4.3 Pore Scale Modeling of Green River Core Samples

The absolute permeabilities of the GR1b and GR1c cores after pyrolysis were determined based on coupling with LBM simulation using a subset of the HRXMT data in a region that exhibited maximum permeability along the bedding plane (e.g., a region containing more cracks). The pyrolyzed sample was a 40 mm long, 26 mm diameter core, while the computational domains for the LBM simulations were $12.6 \times 12.6 \times 12.6$ mm^3 and $16.8 \times 16.8 \times 16.8$ mm^3 for the GR1b and GR1c samples, respectively. For comparison, Figure 8.14 shows the 3D views of LBM-simulated flow along the x-axis (bedding plane; stratigraphic up is to the left) through the reconstructed HRXMT image of cores GR1b and GR1c samples. The transparent solid phase on the right side of the figure reveals flow channels, e.g., the pore network structure after pyrolysis. The velocity scale is color-coded as shown by the color bar.

The estimated permeabilities of these kerogen-rich samples after pyrolysis are 2.88×10^{-5} cm^2 (2919 darcy) and 1.71×10^{-6} cm^2 (173 darcy) for GR1b (425°C) and GR1c (500°C), respectively. The higher permeability of the GR1b sample, despite being pyrolyzed at a lower temperature, can be attributed to the heterogeneity of the core. These permeabilities are several orders of magnitude higher than the permeability of 3.87 darcy measured for the kerogen-rich region of the S2 core after pyrolysis at 400°C.

Figures 8.15 and 8.16 show the triplanar images for the GR2b and GR3c samples, respectively, after pyrolysis at 500°C. Only a small extent of pore generation (e.g., cracks) is seen in the organic-lean (or mineral-rich) GR2b sample, even though the pyrolysis temperature was high. In the GR3c sample (intermediate organic content), directional cracks were created during pyrolysis that can be seen in Figure 8.16. The nonuniform pore space that was created has a resultant impact on permeability, although post-pyrolysis permeability was not computed for either sample.

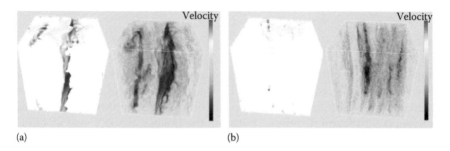

(a)   (b)

FIGURE 8.14

Three-dimensional views of LBM–simulated flow along the x-axis through the reconstructed HRXMT image of core samples GR1b and GR1c, pyrolyzed at temperatures of 425°C and 500°C, respectively. (a) Solid phase is white. (b) Transparent solid phase to reveal flow channels colored by solution velocity.

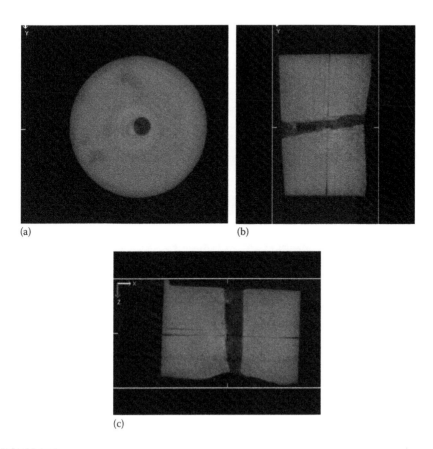

(a)   (b)

(c)

FIGURE 8.15
Triplanar image of GR2b sample after pyrolysis at 500°C. Sections are from (a) the X–Y plane, (b) the Y–Z plane, and (c) the X–Z plane. This sample was the leanest of the GR samples, containing about 10% organic matter. The pore space created is not easily seen at this 42 μm voxel resolution.

8.5 Conclusions

Using multiscale x-ray CT, detailed 3D imaging of oil shale core samples before and after pyrolysis was done to establish the pore structure of the core after reaction. For the S2 core samples, the pore structure of the unreacted material was not well defined at a voxel resolution of 39 μm. After pyrolysis at 400°C, selected images of the S2 core were obtained at voxel resolutions from 39 μm to 60 nm. It is evident that XNT imaging will be required to provide satisfactory pore structure information for the silicate-rich zone. The pore structure deduced from the images was used for LBM simulations to calculate the permeability in the pore space. The permeability of the silicate-rich

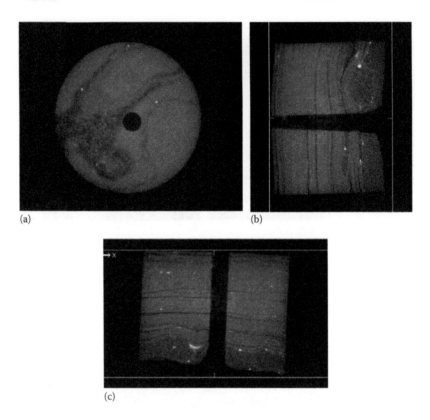

FIGURE 8.16
Triplanar image of GR3a sample after pyrolysis at 500°C. Sections are from (a) the X–Y plane, (b) the Y–Z plane, and (c) the X–Z plane. This sample contained about 15% organic matter. Concentration of the organic matter in certain locations and creation of localized pore space in those locations are observed.

zone of the pyrolyzed sample was on the order of millidarcies, while the permeability of the kerogen-rich zone was very anisotropic and about four orders of magnitude higher.

For the second set of Skyline 16 GR core samples, the results show that the samples within a single 1 ft section of core had significant variations in composition. During pyrolysis, different sections of cores released vary-ing amounts of organic matter, creating different amounts of void space. The channels formed during the pyrolysis are likely to affect the ultimate product formation and distribution. Characterization of the textural change during pyrolysis was accomplished from HRXMT analysis based on com-parison of the same core sections before and after pyrolysis for kerogen-rich cores (GR1b and GR1c) at different pyrolysis temperatures (425°C and 500°C). Porosities of 20%–25% were measured using XMT. Absolute permeabilities of the pyrolyzed GR1b and GR1c (kerogen rich) cores were calculated based on the pore network structure from HRXMT analysis coupled with LBM simulation. Based on visual observation of images in Figures 8.10, 8.12, 8.15, and 8.16, pyrolysis of the organic-rich (GR1b and GR1c) cores at high temperature produced larger pore space during thermal treatment than pyrolysis of the leaner GR2b and GR3c samples. The HRXMT scans of these cores at 42 μm voxel resolution (see Figures 8.10 and 8.12) allowed for pore spaces, reaction products, and source rock to be distinguished and spatially resolved. The porosity created was highly patchy and depended on the distribution of kerogen in the shale, resulting in flow channels that may not have been fully connected.

References

Chen, H. 1993. Discrete boltzmann systems and fluid flows. *Computers in Physics* 7:632–637.

Chen, S. and G. Doolen. 1998. Lattice Boltzmann method for fluid flows. *Annual Review of Fluid Mechanics* 30:329–364.

Duvall, F.E.W., H.Y. Sohn, C.H. Pitt, and M.C. Bronson. 1983. Physical behaviour of oil shale at various temperatures and compressive loads. *Fuel* 62(12):1455–1461.

He, X. and L.-S. Luo. 1997. Theory of the Lattice Boltzmann method: From the Boltzmann equation to the Lattice Boltzmann equation. *Physical Review E* 56:6811–6817.

Kang, Z., D. Yang, Y. Zhao, and Y. Hu. 2011. Thermal cracking and corresponding permeability of Fushun oil shale. *Oil Shale* 28:273–283.

Lee C., R.G. Mclver, and R. Chang. 1994. X-ray computed tomography of Australian oil shales: Non-destructive visualization and density determination. *Fuel* 73:1317–1321.

Lin, C. L. and J. D. Miller. 2004. Pore structure of particle beds for fluid transport simulation during filtration. *International Journal of Mineral Processing* 73:281–294.

Martys, N. and H. Chen. 1996. Simulation of multicomponent fluids in complex three-dimensional geometries by the Lattice Boltzmann model. *Physical Review E* 53:743–750.

Miller, J.D. and C.L. Lin. 2004. Three-dimensional analysis of particulates in mineral processing systems by cone-beam x-ray microtomography. *Minerals and Metallurgical Processing* 21(3):113–124.

Mustafaoglu, O. 2010. *Characterization and Pyrolysis of Oil Shale Samples: An Alternative Energy Option*. Saarbrücken, Germany: LAP Lambert Academic Publishing.

Qian, Y., D. d'Humieres., and P. Lallemand. 1992. Lattice BGK models for Navier-Stokes equation. *Europhysics Letters* 17:479–484.

Shan, X. and H. Chen. 1993. Lattice Boltzmann model for simulating flows with multiple phases and components. *Physical Review E* 47:1815–1819.

Stockman, H. 1999. A 3D Lattice Boltzmann code for modelling flow and multicomponent dispersion, Sandia Report SAND99–0162. Albuquerque, NM and Livermore, CA: Sandia National Laboratories.

Succi, S. 2001. *The Lattice Boltzmann Equation for Fluid Dynamics and Beyond*. New York: Oxford University Press.

Sukop, M.C. and E. Or. 2003. Invasion percolation of single component, multiphase fluids with Lattice Boltzmann models. *Physical B* 338:298–303.

Tiwari, P. 2012. Oil shale pyrolysis: Benchscale experimental studies and modeling. Ph.D. dissertation, University of Utah, Salt Lake City, UT.

Tiwari, P., M. Deo, C.L. Lin, and J.D. Miller. 2013. Characterization of oil shale pore structure before and after pyrolysis by using x-ray micro CT. *Fuel* 107:547–554.

Videla, A. R., C. L. Lin, and J.D. Miller. 2008. Simulation of saturated fluid flow in packed particle beds—the lattice-Boltzmann method for the calculation of permeability from XMT images. *Journal of the Chinese Institute of Chemical Engineers* 9:117–128.

Wolf-Gladrow, D. 2000. *Lattice-Gas Cellular Automata and Lattice Boltzmann Models.* New York: Springer.

Xradia. n.d. ZEISS Xradia 3D x-ray microscopes: Defining XRM. http://www.xradia. com/zeiss-products/. Accessed March 28, 2016.

9

Geomechanical and Fluid Transport Properties

Thang Q. Tran and John D. McLennan

CONTENTS

This chapter describes an experimental program that was developed to demonstrate how the mechanical and transport properties of representative oil shales from the Green River Formation in the Uinta Basin evolve up to, through, and after in situ pyrolysis. The main goal of the program was to provide information related to mechanical properties of oil shale at temperatures and in situ stresses characteristic of in situ extraction. Beyond this, the program sought to advance the understanding of drive mechanisms (compaction and expansion), subsidence/heave, and transport/storage (permeability/porosity) prior to and after pyrolysis.

Oil or gas is moved from a reservoir to a wellbore due to imbalances in pressure or similar mechanisms (drive mechanisms). For example, expansion drive results from exsolution of dissolved gas in conventional oils. Volumetric expansion of the gas "drives" the oil to the wellbore. Alternatively, in uncon-solidated, high-porosity formations, in situ effective stress increases with reduction in reservoir pressure. This tends to "compact" pores and "drive" fluid to the wellbore. In oil shale, expansion drive could result from genera-tion of oil and gas; compaction drive could result from pore creation and the concurrent reduction of material integrity.

With in situ oil shale production, there is a potential for subsidence/heave, which is a reduction/increase in surface elevation due to a compaction/thermal expansion of a formation at depth. Consider that pore volume is created during pyrolysis and is initially filled with liquid and gas that can-not be drained because the permeability of the surrounding medium has not yet increased. Thermal expansion may cause heave of the surface and of formations abutting the reservoir. Eventually, drainage may be possible with development of a "pathway" from a heating well to a production well. The fluid will then move to the production well and the pore pressure will reduce, potentially leading to subsidence.

9.1 Introduction

Oil shale has been described as a complex, indurated, heterogeneous, and laminated sedimentary rock that contains organic matter, mineral matri-ces, and a small amount of bound and/or unbound water. The main organic constituent is kerogen, a heterogeneous mixture of materials including res-ins, cuticles, and fragments of wood (Chong et al. 1976). Kerogen is notably insoluble in ordinary organic solvents such as toluene. When subjected to retorting or thermal treatment, kerogen undergoes chemical decomposition to form a variety of products. If the thermal treatment occurs in the absence of oxygen, the process is called pyrolysis.

Currently, the most common method of hydrocarbon recovery from oil shales is ex situ extraction. This process involves mining the oil shale and treating it at surface processing facilities. Mining, surface retorting, and disposal of the spent shale create a large surface disturbance that must be reclaimed. In addition, the retorting process may release sulfur dioxide, lead, and nitrogen oxides into the atmosphere (Herro 2013). Regardless, with good planning and environmental precautions, commerciality is possible. There are commercial-scale, ex situ operations in Estonia, Brazil, and parts of China as well as pilot-scale operations in the Uinta Basin of eastern Utah.

Conversely, in situ extraction involves pyrolyzing kerogen-rich formations in place—below ground—and using production wells to collect the gaseous and liquid hydrocarbon products (Trimmer and Heard 1980). Due to the lack of oxygen, pyrolysis of the source rock produces different products than combustion. Decomposition of the organic material accelerates with heating, producing materials ranging from organic gases to bitumen to char. The process is an analog to the natural generation of oil and other hydrocarbons from organic material in a source rock, although those processes occur at lower temperatures and over protracted geologic time (Eseme et al. 2007, Herro 2013, Tiwari 2013).

In situ treatment of oil shale generally requires aggressive thermal exposure. Even so, in situ heat transfer likely occurs slowly by conduction (Islam and Skalle 2013). Heat transfer may be supplemented by convection, with the development of expulsion-induced, microcracking networks and the eventual communication of the pyrolysis zone with a production wellbore.

9.2 Geomechanical Issues Related to In Situ Treatment

Mechanical and transport properties of oil shales subjected to high pressure and temperature, particularly temperatures required for pyrolysis (400°C–450°C), are complex and evolve over time as temperature increases and/or as time-dependent creep occurs. With the exception of several key legacy publications and ongoing proprietary measurements (White et al. 2015), the thermomechanical response of representative oil shales is speculative. The public domain literature largely describes testing at temperatures of approximately 200°C–300°C (Chong et al. 1976, 1987, Trimmer and Heard 1980, Eseme et al. 2007, Zhao et al. 2011, Tiwari 2013, Islam and Skalle 2013) with little data available above 400°C (Leavitt et al. 1987, Eseme et al. 2007, Islam and Skalle 2013). At ambient temperatures, prior to pyrolysis, oil shales tend to have a high Young's modulus (the resistance of an elastic material to deformation when an increment of stress is applied), high strength (load bearing capacity), and low permeability (facility for transport of fluids). During heating for pyrolysis, solid kerogen decomposes into bitumen; with further heating, oil and various gas species are produced. This process creates porosity (liquid- or gas-filled voids in a medium), which, if interconnected, causes the oil shale to become more permeable, has low resistance to in situ loading, and deforms in a more ductile fashion after initial yielding occurs (for example, Leavitt et al. 1987).

Because thermal stresses are generated and the integrity and character of the oil shale are impacted during thermal treatment, geomechanical issues that arise include the following:

- Permeability and porosity evolution with changes in the kerogen.
- Expansion in conjunction with the heating schedule with an associated potential for heaving of surrounding and overlying material

and possible ramifications at the surface. With any subsequent drainage, in situ compaction and subsidence at the surface could be anticipated.

- Mechanical deformation of the bulk reservoir associated with changes in the pore-filling material and phase, as kerogen is converted to liquid, gas, and char.

During heating and pyrolysis, permeability may change substantially with time. Initial thermal expansion likely leads to microfracturing and initial permeability increases. Permeability may later be reduced due to additional compressive deformation (some of the original matrix is now porosity) that is experienced. The time for this permeability degradation depends on the compressive stress, the richness of the oil shale, and the heating rate (World Energy Council 2010). However, with increasing temperature and kerogen conversion, pore pressure is anticipated to increase as the generated liquids and gases and the associated pressure may not be able to dissipate, a condition known as undrained behavior (Crawford et al. 2008). If drainage develops, as might occur when a pyrolysis front reaches a production well, undrained behavior may transform to drained behavior. Here, drainage refers to the ability for a created liquid to be discharged away from a particular location with the consequent reduction in generated pore pressure to an equilibrium value. The resistance to deformation may decrease during drainage because liquid-filled porosity decreases (filled only with more compressible gases with a resulting reduction in pressure). Furthermore, with drainage, the effective stress increases and consequently, the potential for shear failure and pore collapse increases.

As indicated, the inherent reactions occurring in oil shale at anticipated in situ treatment temperatures turn kerogen into soluble bitumen, char, hydrocarbon liquids, and gas. All phases, including the more inert mineral matrix, attempt to expand in opposition to the in situ stresses in the reservoir. The gaseous components attempt to expand the most. A net expansion can result, particularly if the created fluids are unable to drain from the shale. This expansion can affect the surrounding area with heaving of overlying material or fracturing of over- and underlying leaner formations. Whether fracturing that occurs within the heated material heals subsequently by creep (time-dependent deformation) is uncertain, particularly recognizing that residual, pyrolyzed laboratory specimens do maintain some structure. After the kerogen is converted and fluids flow out of the shale, the resultant porosity could lead to in situ compaction and correlative surface subsidence. The compaction as well as any heave occurring earlier in the deformation history could jeopardize seal integrity. As this complex temporal behavior suggests, a transient reservoir history is envisioned.

In addition to bulk physical expansion or contraction, increasing temperature impacts the overall strength and resistance to deformation at in situ stress conditions. When the organic matter is heated to more than 450°C, about 90% of the organic volume is no longer solid (Trimmer and Heard 1980). Lean oil shales maintain their structure due to the higher volume of more inert minerals and the development of fewer pores during pyrolysis. In contrast, kerogen-rich oil shales lose much of their strength when the organic matter is decomposed due to a substantial increase in porosity. Strength and resistance to deformation in geologic materials are normally associated with frictional interactions among solids and cementitious material between the solids. Since fluids cannot carry shearing stresses, fluid-filled porosity will not resist shear deformation and shear failure. Subsequently, even before fluid drainage is ultimately enabled, the porous oil shale will be susceptible to "pore collapse." Once drainage has been enabled by a communication with the production well, this pore collapse is further encouraged due to the accompanying effective stress increase.

9.3 Deformation Model for Heated Oil Shale

An experimental program was undertaken to provide mechanical properties data to at least intuitively support or refute the transient model of oil shale response to pyrolysis. This geomechanical transience as a consequence of heating correlates with kerogen decomposition. The result is a multiphase material comprised of minerals, char, kerogen, and pore-filling fluids (Eseme et al. 2007).

These kerogen-rich shales may initially contain natural fractures, including microfractures. In addition, generated pressure may create supplementary fractures and fluid migration may occur. At the very least, microfracturing resulting from pyrolysis of kerogen is anticipated to initially increase permeability and ultimately promote drainage. Porosity is also evolving as the kerogen reacts under thermal treatment. As the temperature increases, volumetric evolution of liquid and gas will attempt to exceed volumetric porosity generation and pressure will gradually increase. When the pressure in the pores becomes greater than the minimum total in situ stress and a thermally reduced tensile strength, further microfracturing will ensue. Exploiting these microfractures, fluid will migrate vertically due to buoyancy or in the direction of the negative pressure gradient if there is some communication to a production well. The pores are now being voided, cohesion associated with solid kerogen has been eliminated, and frictional resistance to deformation may be jeopardized due to the generated porosity.

The consequence of heating, phase conversion, microfracturing, and fluid movement is significant strength degradation with grain rearrangement as well as pore collapse. Wellbore integrity and the integrity of overlying rock acting as a pseudo-seal to out-of-zone gas movement may both be jeopardized as well. In situ compaction in rich layers may become significant enough to manifest finite surface subsidence. Experimental data presented in this chapter will show that strength and resistance to deformation are greatly reduced beginning at temperatures well below those where significant pyrolytic activity is anticipated.

Experimental verification and support from published literature for the hypothesized behavior associated with pyrolysis should include the following:

- Demonstration of nominally elastic behavior at ambient temperature (until near failure) when representative material is subjected to appropriate in situ stress conditions.
- Demonstration of increased permeability (if drainage is allowed), decreased load bearing capacity, and accelerated deformation as temperature and kerogen pyrolysis increase.
- Identification of post-peak deformation—it could be compactive or dilatant to large volumetric strains, depending on drainage conditions.

9.4 Literature Review

There is some literature on laboratory-scale oil shale retorting that outlines various efforts for determining mechanical properties along with some analysis of the materials before and after pyrolysis (Chong et al. 1976, 1987, Trimmer and Heard 1980, Leavitt et al. 1987, Eseme et al. 2007, Zhao et al. 2011). Mechanical properties can refer to directionally dependent values of Young's modulus, Poisson's ratio, shear modulus, uniaxial and triaxial compressive strength, and standard failure loci (envelope) properties, the simplest of which could be cohesion and angle of internal friction. There are also many publications on the effect of elevated temperature and pressure on the composition of the organic material and on the mechanisms of compositional change. There is a smaller body of public-domain research characterizing the mechanical properties of oil shales at in situ pyrolysis temperatures and realistic stress conditions.

9.4.1 Oil Shale Mechanical Properties

Eseme et al. (2007) described how organic material (i.e., kerogen) affects oil shale's mechanical properties. They performed testing that indicates differences in mechanical properties between lean and rich oil shale samples (see Table 9.1). Since desirable hydrocarbon products form when kerogen reaches

TABLE 9.1

Mechanical Properties for Two Grades of Western U.S. Oil Shales at Various Temperatures

Property	Temperatures							
	25°C		150°C		300°C		450°C	
Organic carbon content (%)	20	50	20	50	20	50	20	50
Oil yield (L/t)	63	210	63	210	63	210	63	210
Grain density (g/cm³)	2.4	1.8	—	—	—	—	—	—
Young's modulus, GigaPascals (GPa)	16 ± 2	4.5 ± 0.5	6.5 ± 0.5	5	1	0.5	<1	<1
Porosity (%)	0	0	—	—	—	—	—	—
Unconfined compressive strength (MPa)	125 ± 25	50 ± 30	45 ± 5	35	20	15	18	—
Poisson's ratio	0.2	0.35	—	—	0.3	0.3	—	0.4?
Friction angle (°)	40.5 ± 0.5	20	30	25	30	25	30	10?
Cohesion (MPa)	28 ± 7	28 ± 7	10	7	7	5	5?	—
Tensile strength (MPa)	13 ± 1	9.5 ± 1.5	4	2	2	1	—	—

Source: Modified after Eseme, E., J. L. Urai, B. M. Krooss, and R. Littke. 2007. Review of mechanical properties of oil shales: Implications for exploitation and basin modelling. Oil Shale 24(2):159–174.

temperatures in the 300°C–550°C range, they ran tests on lean and rich oil shales at 150°C and 350°C. Because they did not have access to devices that could reach 450°C, they estimated the properties at that temperature (Table 9.1). Even at room temperature, the differences associated with the grade of the oil shale are noticeable. The amount of organic material directly correlates with Young's modulus, Poisson's ratio, and compressive strength of the samples. Chong et al. (1987) found that the mechanical properties of oil shale under uniaxial compression (a laboratory test where a sample is not subjected to hydrostatic confining pressure) are influenced by the mineralogical variations and distribution, the volumetric organic content, and the depositional setting. Variations in the organic and mineral contents can change mechanical properties by 100% or possibly more. Islam and Skalle (2013) developed a model for empirical estimation of certain mechanical properties of oil shale at pyrolysis temperatures. They performed triaxial compression testing on samples that were oriented parallel and perpendicular to bedding.

Capabilities of testing equipment have limited public-domain empirical research on oil shale properties although some measurements have been made at high temperatures (up to 550°C) and modest pressures (up to

20 MPa). Baharia et al. (2011) are among the few authors who devised an experimental procedure for high-temperature and high-pressure Green River oil shale testing. They fitted each sample into a stainless steel jacket, attached acoustic transducers to each end, and installed this fixture into a pressure vessel. The system was equipped with an electric heater and confining pressure (to simulate in situ stress) was provided by argon gas. Measurements were planned at temperatures between 300°C and 550°C. Due to equipment limitations, only measurements at temperatures up to 250°C were achieved.

Lankford (1976) assessed how the strain experienced during sample loading was influenced by the applied strain rate. In similar research, Chong et al. (1987) assessed the influence of strain rate on Young's modulus and Poisson's ratio. Both Lankford and Chong et al. recorded a substantial increase (on the order of 40%) in the initial Young's modulus in a lean oil shale that was cored perpendicular to bedding (horizontal plug) when the strain rate was increased by five orders of magnitude at ambient temperature.

From archived measurements on oil shale at elevated temperature such as those by Baharia et al. (2011) and other researchers, stress–strain behavior can often be conveniently approximated by well-accepted soil mechanics procedures. Duncan and Chang (1970) developed a model for the calculation of stress–strain characteristics in soil that can be effectively used in representing the stress–strain characteristics of oil shales. In soils, as in heated oil shales, stress–strain behavior depends on a number of factors including density, water/fluid content, structure, drainage conditions, boundary conditions (i.e., uniaxial strain, triaxial compression), duration of the loading, stress history, confining pressure, and shear stress (Duncan and Chang 1970). Accounting for all these factors is difficult when assuming strictly linear elastic behavior. Therefore, a model that adopts a nonlinear stress–strain curve is often implemented using an equation proposed by Kondner and Zelasko (1963):

$$\sigma_1 - \sigma_3 = \frac{\varepsilon}{a + b\varepsilon} \tag{9.1}$$

where
 σ_1 and σ_3 are the major and minor principal stresses, respectively
 ε is the axial strain
 a and b are constants whose values are determined experimentally.

This equation accommodates a hyperbolic stress–strain curve and is useful in representing the nonlinear stress–strain behavior. Data from one measurement are shown in Figure 9.1. The method is particularly useful for linearizing data and allows for effective comparison of multiple experiments where strain hardening is prevalent.

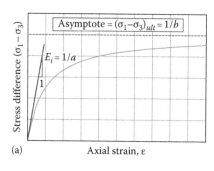
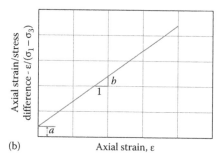

(a)   Axial strain, ε (b)   Axial strain, ε

FIGURE 9.1
Schematic linearization of axial stress difference and axial strain data. (Modified after Duncan, J. M. and Chang, C.-Y., *J. Soil Mech. Found. Div.*, 96(5), 1629, 1970.) The panel (a) shows an idealized set of laboratory data, plotting the axial stress difference versus the axial strain. The axial strain asymptotically approaches an ultimate stress difference. Using Equation 9.1, these data are linearized in the panel (b) by plotting the axial strain divided by the axial stress difference ($\varepsilon/(\sigma_1-\sigma_3)$) on the ordinate.

9.4.2 Permeability and Porosity

The mechanical response of oil shale at high temperature and pressure is intimately coupled with (1) the evolution of storage (porosity) due to kerogen decomposition to form liquids and gases; (2) the development of pressure due to the generation of liquids and gases from higher-density, solid material; and (3) the movement of multiple phases of fluid in response to pressure gradients (permeability).

Much of the available porosity is created by the phase/compositional changes associated with elevated temperatures. As porosity develops, so does pore pressure. This porosity development may initially be unconnected, resulting in undrained behavior where the pore pressure is unable to dissipate. Drained conditions develop when spatial porosity evolution is accompanied by at least a temporary increase in permeability. As fluid drains, effective stress increases and is borne by residual inorganic mineral matter.

Eseme et al. (2007) found that porosity increased with increased temperatures as kerogen steadily decomposed. However, they could not perform measurements at temperatures above 350°C. Thomas (1966) demonstrated an increase in porosity and permeability with hydrocarbon yield when retorting at temperatures up to 538°C at a constant confining pressure of 6.89 MPa. Thomas found a minimum retorting temperature for shale oil production of 371°C for unstressed shale. This temperature systematically decreased to 332°C and 318°C when the applied confining pressures were 13.79 and 17.25 MPa, respectively. Based on a restricted number of tests, this suggests that—at least above confining pressures of 6.89 MPa—the minimum temperature required to create pore structure decreases as the in situ mean stress increases. From an intuitive perspective, this result is consistent with natural methanogenesis that is favored by both temperature and pressure.

Other papers have considered the permeability and porosity of oil shale samples before and after exposure to high temperatures when confining pressure was applied during the testing (confining pressure simulates native, in situ stresses). Few of these evaluations were above 300°C and, in many cases, produced fluids were extracted prior to the measurements. Removal of produced fluid before testing implies that drained conditions could be achieved and likely underestimates the amount of char developed (Glauser 2015).

Tiwari et al. (2013) reported on the effect of temperature on oil shale pyrolysis and the creation of pore volume during thermal treatment. As discussed in Chapter 8, 1 in. diameter plugs from different depths of the Uinta Basin Skyline 16 core were characterized with computed tomography before and after exposure to elevated temperatures to analyze pore space. These tests were performed under drained conditions, where generated pore pressure was allowed to dissipate either by draining, by unconstrained thermal expansion, or by hydraulically induced microfracturing. The samples were not subjected to in situ stresses during pyrolysis, and substantial permeability developed due to unconstrained expansion. When heated to 300°C, weight loss of the samples was minimal. In some experiments at 500°C, weight loss exceeded the weight of organic content measured in the samples. The sample with the highest organic content exhibited the greatest thermal expansion at 500°C. Greater pore space was created during pyrolysis of oil shales with higher organic content. Channels formed during pyrolysis were hypothesized to also affect the final gaseous, liquid, and solid product formation and distribution.

It has long been recognized that the permeability of most oil shale at ambient conditions is extremely small. Consequently, porosity and permeability have sometimes been inferred indirectly using a technique known as the GRI method, after the Gas Research Institute (Core Laboratories 2015). With this protocol, shale is disaggregated to a grain-size distribution that is adequately small to be unaffected by microcracking. Permeability and porosity are determined by monitoring transient gas pressure decay in an isolated vessel at ambient stress conditions. Tinni et al. (2012) conducted experiments on oil shales using the GRI method. While they achieved consistent results, critics questioned whether the measurements were representative of intact samples.

Regardless of whether unsteady pressure decay (GRI method), pressure pulse, or steady-state measurements are carried out, a great deal of variability has been reported in permeability measurements in the literature. Tisot and Sohns (1970) conducted permeability experiments on oil shale samples from the Bureau of Mines experimental mine near Rifle, CO. Each bedding-parallel core plug was heated to temperatures of 385°C or 441°C at 1°C/min. A constant load of 80, 200, or 325 psi was maintained to approximate

TABLE 9.2

Experimental Data from 240 L/t Oil Shale Heated to 385°C

Stress (MPa)	Compression at Zero Permeability (%)	Time Elapsed to Zero Permeability (min)	Time Elapsed to Structural Collapse (min)	Maximum Compression (%)	Organic Matter Loss (wt%)
0.552	37.5	85	170	44.5	34.4
1.38	36.6	55	140	44.1	33.0
2.47	33.8	22	100	41.5	28.9

Source: Modified from Tisot, P.R. and Sohns, H.W., *J. Chem. Eng. Data*, 15(3), 425, 1970.

conditions at 100, 250, and 400 ft depths in an in situ environment where overlying rock had been rubblized mechanically. The induced permeability and porosity were assessed along with determination of the yield temperature, e.g., the temperature at which an in situ structure would collapse, as evidenced by accelerated reduction in load-bearing capacity. After the sample had distinctly failed (substantial reduction in load-bearing capacity), permeability to gas was determined with a 3 psi differential pressure. Experimental data from the literature for 240 L/t oil shale heated to 385°C are shown in Table 9.2. The permeability and porosity increased in proportion to increases in the temperature and pressure.

While these data are useful for understanding the structural response of oil shales that have been heated to modest temperatures, there are fewer results at retorting temperatures. Trimmer and Heard (1980) performed experiments to measure the change in bulk volume and permeability of crushed oil shale aggregate subjected to high temperatures (upwards of 700°C) and uniaxial strain boundary conditions. That testing matrix included samples with axial stresses ranging from 0.69 to 6.2 MPa. Samples were exposed to a constant stress and then heated at 10°C/h up to a maximum temperature of 700°C. Values of volumetric strain, the volumetric deformation of a material normalized by its initial volume, up to 34% were recorded. The permeability was commensurately dependent on how much volumetric deformation occurred. These samples originally had permeabilities characteristic of compacted particulate aggregate (100–200 darcies). After exposure to stress and high temperatures, permeability decreased to levels of 0.3–14 darcies. The largest decrease in both porosity and permeability occurred at temperatures between 300°C and 450°C, characteristic of temperatures required for pyrolysis. In comparison, intact, virgin oil shale will typically have permeabilities in the sub-nanodarcy range. Table 9.3 shows Trimmer and Heard's experimental data for various grades of oil shale at 500°C and higher. While Trimmer and Heard showed that there is no simple relationship between strain and porosity/permeability, their data are of limited value for in situ, nonrubblized recovery because of the aggregated/crushed nature of the samples.

TABLE 9.3

Permeability and Compaction Data for Various Grades of Oil Shale Collected in Trimmer and Heard's Experiments (1980)

Grade (L/t)	Applied Axial Stress (MPa)	Equivalent Particle Diameter (cm)	Heating Rate (°C/h)	Maximum Temperature (°C)	Permeability at 500°C (Darcies)	Volumetric Strain at 500°C	Initial Porosity	Final Porosity (Corrected) at 500°C
101	4.14	—	9.7	500	—	0.331	0.464	0.278
110	4.14	—	10.2	500	—	0.296	0.411	0.265
102	2.07	1.1 (I)	6.7–250	500	—	0.322	0.486	0.314
110	0.69	—	11.0	510	(14.1)	0.242	0.494	0.403
107	1.38	—	13.5	500	(3.1)	0.324	0.489	0.317
110	0.69	—	11.4	500	(2.4)	0.296	0.530	0.386
99	2.07	—	10.0	680	(2.0)	0.256	0.509	0.405
(104)	2.07	—	10.1	690	(7.8)	0.285	0.519	0.386
69	2.07	1.0 (I)	9.3	680	—	0.337	0.505	0.336
102	2.07	—	10.5	530	11.1	0.322	0.498	0.331
98	2.07	—	10.3	530	5.8	0.317	0.452	0.316
95	2.07	—	9.9	520	3.4	0.292	0.542	0.377
100	2.07	0.57 (I)	10.1	500	7.3	0.321	0.486	0.301
100	4.14	0.65 (I)	9.8	500	1.1	0.330	0.486	0.306
100	6.2	—	9.4	500	0.38	0.300	0.430	0.276
100	4.14	1.3 (II)	9.9	520	1.6	0.331	0.440	0.262
109	4.14	0.35 (IV)	25.2	520	0.34	0.315	0.407	0.254
100	4.14	1.3 (II)	9.9	500	0.76	0.337	0.436	0.251
98	4.14	0.60 (III)	9.7	500	0.44	0.333	0.426	0.238
100	4.14	0.55 (III)	9.5	500	0.50	0.345	0.431	0.237

A common problem evident in legacy research is the discrepancies that occur due to loading, unloading, heating, and cooling of the sample after pyrolysis. Zhao et al. (2011) overcame this problem by constructing an apparatus that could provide stress, heat the sample to retorting temperatures, cool the sample, and measure sample permeability before and after pyrolysis without removing the sample from the pressure vessel. They reported on two Chinese oil shales. Below a threshold temperature, permeability initially increased slightly with temperature and then decreased to a low magnitude as pyrolysis and sample drainage occurred. Above a temperature threshold, the permeability increased with temperature. Zhao et al. noted that in situ oil shale technologies should consider permeability variation with temperature so that a relatively high permeability could be attained and maintained.

9.5 New Experimental Measurements

In order to supplement oil shale measurements in the public domain, two experimental programs have been used to characterize mechanical properties and permeability of oil shale:

1. Measurement of oil shale deformation in a high-pressure, high-temperature, triaxial vessel under realistic in situ pressure, temperature, and stress conditions
2. Measurement of steady-state permeability of samples pyrolyzed under realistic in situ conditions.

These experiments were designed to overcome the deficiencies of previous experiments where thermal treatment did not reach extremely high temperatures or where crushed shale rather than solid core samples were used for permeability assessment.

9.5.1 Triaxial Testing of Mechanical Properties

9.5.1.1 Triaxial Testing Apparatus

A high-pressure, high-temperature vessel, shown in Figure 9.2, was fabricated to measure representative oil shale response (mechanical deformation as a function of stress) under realistic in situ pressure and stress conditions when high temperature is applied. Figure 9.3 is a design drawing for the pressure vessel. The internal diameter of the vessel is 0.46 m and the height with the vessel lid attached is roughly 2 m. A suite of measurements on oil shale samples was made in this vessel. These measurements delineate key

FIGURE 9.2
Photograph of the high-pressure, high-temperature vessel. For scale, the chain link fence is 1.8 m high.

mechanical properties and their evolution with time and temperature during in situ thermal processing. All data acquisition and control used LabVIEW™ or Opto 22©.

The apparatus was designed to accommodate samples up to 10.2 cm in diameter and 20.4 cm long. The in situ geologic environment was simulated by applying vertical (e.g., axial) stress with a 181 tonne hydraulic actuator (hydraulic jack) reacting against the upper end cap. This actuator was controlled with a feedback loop where axial deformation was measured with linear variable displacement transducers (LVDTs) and hydraulic pressure in the jack was manipulated to maintain a prescribed stress or strain rate using a Teledyne ISCO pump. The LVDTs measured linear displacement by converting a position from a mechanical reference into a proportional electrical signal that contained amplitude for distance and phase for direction. This allowed for precise readings of sample deformation under the high temperatures and confining pressures that were imposed. Force, stress, and deformation were recorded.

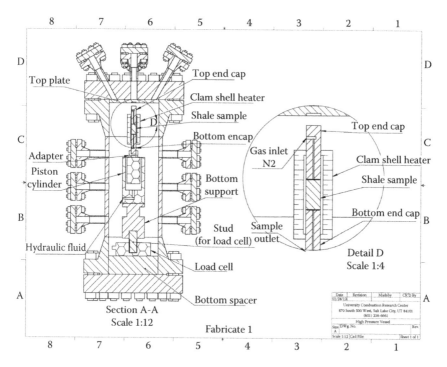

FIGURE 9.3

Schematic diagram of the pressure vessel. The circled area is enlarged at the right and a schematic of the sample (up to 10 cm diameter) and its end caps is shown.

The original hydraulic jack, shown in Figure 9.4, was sized for tests on 10.2 cm diameter samples. Since the samples tested were only 3.81 cm in diameter, a smaller jack was substituted. The load cell, visible in Figure 9.4 at the top of the hydraulic jack, measured force. This load cell, along with pressure transducers for the pump that drives the jack, was used to determine the axial stress.

Radial confining pressure (pneumatic pressure to 10.34 MPa) was provided with nitrogen (N_2) since typical liquid confining fluids would break down at the temperatures being applied. Heating to 538°C was provided by an electrical clamshell heater positioned around the sample. As indicated, axial and radial strains were measured with LVDTs.

Figure 9.5 shows a sample enshrouded with the clamshell heater. In this annotated photograph, "1" denotes the upper end cap, which rested on top of the sample. Confining pressure was applied using N_2 from a sixteen-cylinder package. The sample was jacketed in 0.0343 mm thick copper foil, which was slid over the sample and clamped to the end caps to prevent the N_2 confining fluid from penetrating into the sample; see Figure 9.6. While shrink fit Teflon

FIGURE 9.4
Hydraulic jack with load cell on top. Neither the jack nor the load cell is exposed to elevated temperatures.

or polyurethane is commonly used for this purpose, copper was chosen here because of the elevated temperature. The sample (3.81 cm diameter by 7.6 cm long) is not visible in Figure 9.5. It is hidden by the electrical clamshell heater ("2") that was used to heat the sample to a controlled temperature up to 538°C. A fixture ("3") was clamped to the upper ("1") and lower ("4") end caps. This fixture enabled the mounting of three axial LVDTs at 120° spacing; two arms of this three-armed mount, each containing a hole to fit an LVDT, are visible near the bottom in Figure 9.6. Three LVDTs were used so that an average measurement could be made to account for nonuniform axial deformation. Finally, an adaptor ("5") connected this LVDT-mounting fixture to the axial load cell immediately below it (not shown in this photograph but seen in Figure 9.6).

The clamshell heater had an outside diameter of 15 cm and a height of 23 cm. It was custom-designed to allow for the measurement of the sample's radial strain. Four horizontally oriented LVDTs with stainless steel extension arms, oriented at 90° to each other, extended through the holes in the heater and mutually contacted the sample and the LVDTs. The four-armed holder

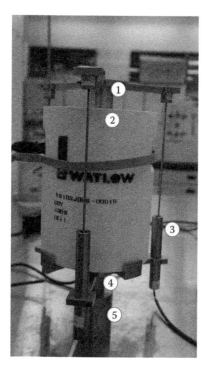

FIGURE 9.5
Heater and axial strain measuring devices. The labels are described in the text.

for the heater is visible in Figure 9.6. With this arrangement, samples could be tested at pyrolysis temperatures and radial deformation could be measured in two orthogonal directions.

9.5.1.2 Triaxial Testing Procedure

A typical procedure for running a triaxial test inside the vessel shown in Figure 9.2 is as follows:

- Fit the sample with end caps and then wrap the sample with a copper jacket.
- Place the sample in the electric clamshell heater and then install the strain-measuring devices.
- Place the entire fixture inside the triaxial vessel on top of the hydraulic jack with a load cell between the jack and the sample fixture.
- Install the upper closure for the vessel. Tighten the bolts on the top closure to ensure a hydraulic seal.
- Heat the sample to the desired temperature at a rate of 10°C/min.

FIGURE 9.6
Photograph of the sample inside a copper jacket with two end caps. A mounting bracket for the heater is visible. The entire fixture is installed in the pressure vessel.

- While sample is heating, apply confining pressure using N_2 gas.
- Increase the axial stress, σ_1, using a syringe pump to drive the hydraulic ram. The confining pressure, σ_3, is maintained constant. The axial stress, σ_1, and the axial deviatoric stress, $\sigma_1-\sigma_3$, are increased in a prescribed and controlled fashion.
- Record axial and radial deformation using three axial LVDTs and four radial LVDTs.

The triaxial testing apparatus was used to duplicate in situ conditions when pyrolysis occurs. The applied stresses likely prevented or at least inhibited the "swelling" and fracturing that are seen in tests run without confinement. Without confining and axial stresses, laboratory simulations of thermal treatment likely result in significant sample expansion and thermally induced fracturing.

The acquired data were used to analyze and model response to high temperatures, in situ stresses, and internally generated pressures resulting from phase conversion in an initially undrained environment. Standard relationships were used to calculate the mechanical properties for each sample, including Young's modulus, Poisson's ratio, bulk modulus, and shear modulus. Young's modulus (E) is a measure of the stiffness of an elastic material. It is defined as the ratio of the stress (force per unit area) along an axis, σ_{zz}, (in this case axial stress) to the strain (ratio of axial deformation to initial length) along that same axis, ε_{zz}, in the range of stress for which Hooke's law holds (elasticity, fully recoverable deformation on unloading) as seen in Equation 9.2:

$$E = \frac{\sigma_{zz}}{\varepsilon_{zz}} \tag{9.2}$$

Poisson's ratio, v, is the negative of the ratio of the radial strain ε_{yy} or ε_{xx} to the axial strain, ε_{zz}; see Equation 9.3.

$$v = -\frac{\varepsilon_{yy}}{\varepsilon_{zz}} \tag{9.3}$$

Since measurements of strain were made in two orthogonal radial directions, two values of Poisson's ratio could be determined. In an isotropic material, these two values would be identical.

Bulk modulus, K, is a measure of the resistance to uniform, hydrostatic compression and is obtained from Young's modulus and Poisson's ratio, presuming elasticity and isotropy:

$$K = -\frac{E}{3(1-2v)} \tag{9.4}$$

Shear modulus, G, characterizes the material's response to shear stress and is also computed from Young's modulus and Poisson's ratio, presuming elasticity and isotropy:

$$G = -\frac{E}{2(1+v)} \tag{9.5}$$

9.5.2 Permeability and Porosity Testing

The original plan to continuously monitor the evolution in permeability did not turn out to be practical. Consequently, sample permeability was measured before and after pyrolysis. A high-pressure, high-temperature, core-flooding system, capable of measuring absolute and relative permeabilities of low-permeability formations, was used. Relative permeability is the ratio of the effective permeability of one fluid when two phases are present in the sample to the absolute permeability when the flowing fluid permeates a sample saturated with the same fluid. The system measures relative permeabilities by flowing two immiscible fluids concurrently in metered proportions. During a relative permeability measurement, flow of each fluid phase inhibits flow of the other phase (and consequently reduces permeability).

The apparatus determines absolute permeability using Darcy's law:

$$q = -\frac{k}{\mu}\frac{\Delta P}{L}A \tag{9.6}$$

Darcy's law relates the flow rate, q, to the permeability, k, the cross-sectional area, A, the dynamic viscosity, μ, the pressure drop, ΔP, and the length, L, of the sample. The measurement allows for calculation of permeability, k, using a liquid with known viscosity (usually water) flowing at a specified rate through a sample with known dimensions. The system measures the pressure differential over the length of the sample and when the pressure and/or rate reach steady state, permeability can be calculated.

Porosity was measured using an Ultra-Pore 3000 porosimeter, manufactured by Core Laboratories. This is an effective porosity, ϕ:

$$\phi = \frac{v_{voids}}{v_{total}} \equiv \frac{\text{Acessible pore volume}}{\text{Bulk (total) volume}} \tag{9.7}$$

9.6 Results

9.6.1 Sample Testing

Green River Formation plug samples were subcored from the Skyline 16 core, as described in previous chapters, and from the White River Mine waste heap (samples provided by the Utah Geological Survey). Both locations are in Uintah County, approximately 97 km southeast of Vernal, UT. For the Skyline 16 core, 305 m of continuous 10 cm diameter core was recovered, slabbed, plugged, archived, and allocated for various tests as described in Chapter 4.

Table 9.4 is a summary of the samples that were tested. The GR samples came from the Skyline 16 core: GR1 (462–463 ft), GR2 (486–487 ft), and GR3 (548–549 ft). An additional sample, GR4, was obtained from a depth of 467 ft. A further description of the GR1, GR2, and GR3 samples can be found in Chapters 4 and 5. The WR designation indicates a White River Mine sample. A designation of V or H in the sample identifier indicates that the plug was taken vertically and nominally perpendicular to bedding or horizontally and

TABLE 9.4

Matrix of Measurements Performed

Sample ID	Depth (m)	Orientation	Testing Method	Confining Pressure (MPa)	Temperature (°C)
GR1-V	140.8	Vertical	Triaxial	6.89	400
GR2-V	148.1	Vertical	Triaxial	6.89	400
GR2-H	148.1	Horizontal	Triaxial	6.89	400
GR3-V	167.3	Vertical	Creep	5.52	121
GR3-H	167.3	Horizontal	Creep	6.55	400
GR4-V-3	142.3	Vertical	Triaxial	3.45	400
GR4-V-1	142.3	Vertical	Porosity/ permeability	20.7	30
GR4-V-2	142.3	Vertical	Porosity/ permeability	20.7	400
WR-13-V-1	—	Vertical	Porosity/ permeability	20.7	30
WR-13-V-2	—	Vertical	Porosity/ permeability	20.7	400

parallel to bedding, respectively. The three generic testing protocols adopted in the testing program were (1) triaxial testing at temperature, including testing at temperatures where pyrolysis occurred, (2) measurement of time-dependent deformation during creep testing with temperature applied at constant stress conditions, representative of virgin in situ conditions, and (3) pre- and post-pyrolysis porosity and permeability measurements. The permeability measurements were designed to assess how pyrolysis under stress affects pore evolution and integrity and to infer if there was a correlation between depth (stresses increase with depth) and the amount of organic material lost due to pyrolysis. There were other measurements carried out during the development and refinement of the equipment that are not reported here, including measurements to assess thermal conductivity and the distribution of applied temperature throughout a sample.

9.6.2 Mechanical Properties

9.6.2.1 Axial Deviatoric Stress versus Axial Strain

Figure 9.7 is an example of stress–strain–time data obtained from the testing performed on a vertical (perpendicular to bedding) sample from the moderately kerogen-rich GR2 zone. This figure demonstrates typical testing chronology and sample behavior, as is summarized in the following steps:

- The temperature was increased to approximately 400°C, reaching this value after about 50 min. Significant pyrolysis was initiated approximately 20 min after the heater temperature reached 400°C.

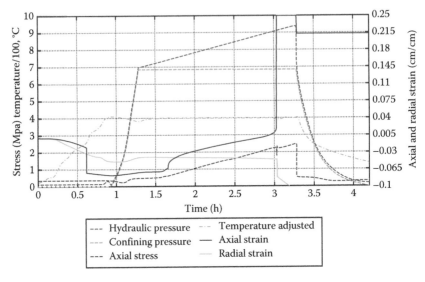

FIGURE 9.7

Chronologic variation of temperature, confining pressure, and axial deviatoric stress as well as axial and radial strains on sample GR2-V. This was a vertically oriented core plug with a grade of 117 L/t. The stabilized testing temperature was 400°C and the confining pressure was 6.9 MPa.

While the temperature was increasing, the axial and radial strains were decreasing, consistent with sample volumetric expansion in both the axial and radial directions (compressive strain is considered to be positive and unrestrained thermal expansion will incrementally cause a reduction in strain). The rapid drop in axial strain at approximately 0.6 h suggests adjustment of LVDT positions; subsequent measurements appear to be unaffected by this realignment.

- After equilibration of the temperature and the strains, hydrostatic confining pressure (equal in all directions, axial and radial) was applied to the sample using N_2 as the pressurizing medium. Controlled application of the confining pressure through a solenoid valve started at approximately 0.9 h. After approximately 1.3 h, the confining pressure reached a target value of 6.9 MPa and was maintained constant for the remainder of the test. In other tests, the confining pressure was applied before the temperature was increased from ambient levels.

- Next, the hydraulic jack applied supplementary axial differential stress so that the overall axial stress linearly increased (compression is positive) until catastrophic failure (catastrophic loss of load-bearing capacity) occurred. The load cell sensed initial sample deviatoric loading (difference between the axial stress and confining

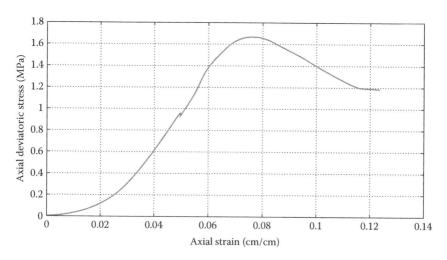

FIGURE 9.8

Axial stress versus axial strain in triaxial testing on sample GR2-V. This was a vertically oriented core plug with a grade of 117 L/t. It is the same sample that is depicted in Figure 9.7. The stabilized testing temperature was 400°C and the confining pressure was 6.9 MPa.

pressure) at approximately 1.6 h, at which point the initial axial deviatoric stress and strain were zeroed (Figure 9.8). Catastrophic failure was evidenced by accelerated strains at approximately 3 h. Testing was terminated at approximately 3.25 h.

Figure 9.8 is a more conventional view of triaxial testing data using the sample shown in Figure 9.7. Figure 9.8 shows axial deviatoric stress versus axial strain. Radial strain can be similarly visualized. Nonzero axial deviatoric stress is incurred after the confining pressure has been applied and maintained constant; axial strain increases after approximately 1.6 h in Figure 9.7, indicating that the sample is bearing a nonhydrostatic axial load. The stresses and strains in Figure 9.8 have been zeroed to reflect deviatoric conditions only.

Mechanical behavior for a laminated material such as oil shale strongly depends on the orientation of the sample (parallel or perpendicular to bedding) and consequently the directionality of the applied axial stress difference. To determine the difference in strength between bedding-perpendicular and bedding-parallel samples, consider Figure 9.9. Figure 9.9 is a comparison plot for "twin" samples from the moderately organic-rich GR2 section of the Skyline 16 core. One sample was subcored in a vertical direction perpendicular to bedding and the other was plugged in a horizontal orientation. The sample orientation correlates with differences in the stress and strain behavior shown in Figure 9.9. The horizontal sample required a higher stress to deform an equivalent amount to strains incurred by the vertical sample; the horizontal sample also failed more catastrophically as evidenced

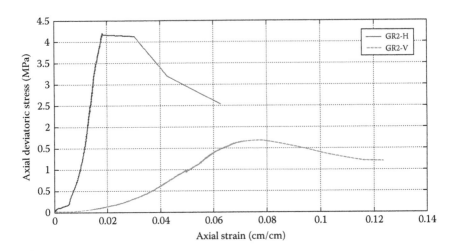

FIGURE 9.9
Comparison of the axial strain versus axial deviatoric stress in triaxial testing for vertical and horizontal GR2 samples from the Skyline 16 well. The samples were nominally twins with the same moderately organic-rich grade. The confining pressure was 6.89 MPa and the temperature for the testing was 400°C.

by the rapid loss of load-bearing capacity after a brief period of near perfectly plastic deformation (nonrecoverable deformation with constant axial stress). The vertical sample experienced ultimate failure in a more ductile manner. The role of laminations (lean and rich layers) can also be seen in the differences in peak stresses between the two samples with the vertical sample much weaker than the horizontal sample. These behavioral differences are consistent with behavior of laminated material at lower temperatures; see Vutukuri et al. (1974).

As shown, the orientation of the applied stress is important and has significant relevance in modeling where transverse isotropy (different mechanical properties normal to bedding from those in the plane of the bedding) can be readily incorporated. Similarly, mechanical performance is a function of the oil shale grade. Figure 9.10 compares triaxial stress–strain behavior for comparable samples from a premium grade (250 L/t) zone, GR1, and a moderately organic-rich (117 L/t) zone, GR2. With the organic material (kerogen) being modified, displaced, and converted during heating, the higher-grade sample accommodated less stress (e.g., failed more easily) than the lower-grade material due to generation of pores and drainage as well as viscoelastic changes in the kerogen itself.

9.6.2.2 Radial and Axial Strains

Up to this point, figures have highlighted axial deviatoric stress versus axial strain. Figure 9.11 shows both average radial and axial strain versus axial

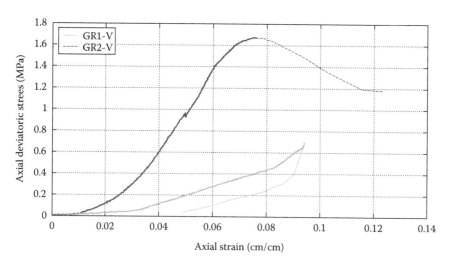

FIGURE 9.10
Comparison of the axial strain versus axial deviatoric stress in triaxial testing of two vertical samples with different grades: GR1-V with a grade of 250 L/t and GR2-V with a grade of 117 L/t. In each case, the confining pressure was 6.9 MPa and the stabilized temperature was 400°C.

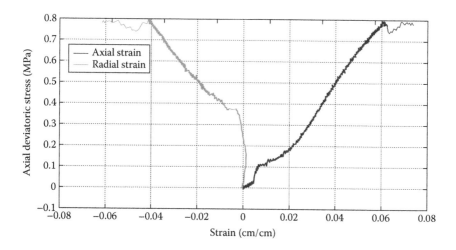

FIGURE 9.11
Axial and radial strain versus axial deviatoric stress during triaxial testing of a vertical sample (GR4-V-3) with a grade of 104 L/t. Notice the initial resistance to radial deformation followed by deformation with a Poisson's ratio of about 0.5 and ultimately by plastic behavior.

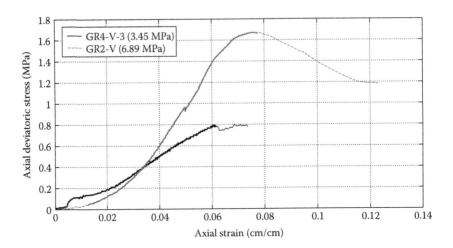

FIGURE 9.12
Comparison of the axial deviatoric stress versus axial strain during triaxial testing of two vertical samples (GR4-V and GR2-V) with similar grades but two different confining pressures. The temperature was 400°C.

deviatoric stress for the GR4 sample. The sample was plugged vertically, normal to bedding. Initially, there was relatively little radial strain, with the axial deformation being accommodated by the softer, organic-rich zones and the leaner zones inhibiting overall radial expansion. However, with only a small axial deviatoric stress applied, radial deformation accelerated dramatically. At an axial deviatoric stress of only ~0.8 MPa, relatively plastic deformation is evident.

The stress–strain behaviors of the GR4 sample (grade of 104 L/t) shown in Figure 9.11 and of a GR2 sample (grade of 117 L/t) are compared in Figure 9.12. Both samples were tested at 400°C but at different confining pressures: 6.9 MPa for the GR2 sample and 3.45 MPa for the GR4 sample. Ignoring the early flat parts of the curves, Young's modulus increases significantly with confining pressure as expected; see Vutukuri et al. (1974) or Jaeger and Cook (1976). For these two samples, doubling the confining pressure doubles Young's modulus at the same temperature and nearly the same grade. Additionally, and consistent with soil and rock mechanics experience, strength (peak axial stress) depends on the effective confining stress.

Table 9.5 summarizes the mechanical properties of four samples with different orientations and/or grade. The directionality of the axial deviatoric loading impacted the strength and modulus of each sample; at the same confining pressure and temperature, the horizontal sample required a higher stress to fail. Grade was also a first-order controlling factor. The richer shale sample (from zone GR1) had a lower Young's modulus and strength than the leaner shale (from zone GR2).

TABLE 9.5

Comparison of Mechanical Properties for Samples with Different Grades and Orientations at a Testing Temperature of 400°C

Property	GR2-H	GR2-V	GR1-V	GR4-V
Grade (L/t)	117	117	250	104
Sample orientation	Horizontal	Vertical	Vertical	Vertical
Testing temperature (°C)	400	400	400	400
Confining pressure (MPa)	6.89	6.89	6.55	3.45
Young's modulus (MPa)	45.1	36.0	8.03	15.5
Poisson's ratio	—	0.05	0.07	—
Bulk modulus (MPa)	—	13.2	3.10	—
Shear modulus (MPa)	15.1	17.2	3.76	—
Peak stress (MPa)	3.48	1.93	0.90	1.0

9.6.3 Permeability and Porosity

Permeability and porosity were measured for four samples before and after pyrolysis. Two lean samples from the White River Mine had a grade of 21 L/t and the more kerogen-rich GR4 samples had a grade of 104 L/t. Before pyrolysis, steady-state, absolute permeability to water was determined at 21 MPa confining pressure and ambient temperature (25°C). After pyrolysis, the steady-state, absolute permeability to water was measured again under the same temperature and confining pressure conditions.

Figure 9.13 synthesizes data from the permeability testing of a low-kerogen-content sample (WR-13-V) after pyrolysis. A water flow rate of 0.05 mL/min resulted in a pressure difference of ~12 MPa at 21 MPa confining pressure and 25°C. This pressure difference was used to calculate an absolute permeability of the sample of approximately 1.2 microdarcies. Prior to pyrolysis, the permeability of the sample was less than 2 nanodarcies.

Pre- and post-pyrolysis porosity and permeability measurements for the four samples that were evaluated are shown in Table 9.6. Note the remarkable change in both porosity and permeability with pyrolysis. These results would be anticipated to vary if the stress conditions at pyrolysis were different.

With the organic-rich samples, there were complications in running the porosity and permeability tests due to swelling of the plug during pyrolysis, which prevented the sample from fitting into the testing vessel without surface grinding. Based on the swelling that occurred in these confined samples during pyrolysis, it is clear that without the establishment and maintenance of drainage pathways (interconnected pores, hydraulic fractures, etc.), gas generation and porosity evolution will encourage swelling in situ and possible surface heave. While swelling will be inhibited by in situ stress, any hydraulically induced channels can close due to post-pyrolysis drained deformation, which is counterproductive to moving liquid and gas to a production well.

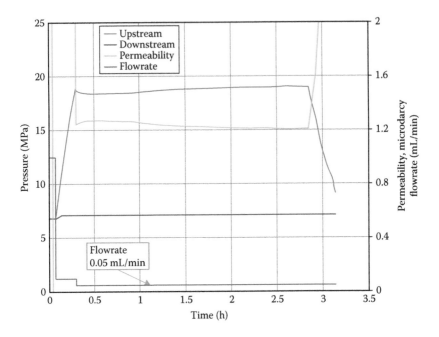

FIGURE 9.13

Post-pyrolysis absolute permeability measurements on a lean (21 L/t) oil shale sample, WR-13-V-1. The differential pressure of approximately 11.72 MPa along the length of the sample was substantial, even at a low flow rate. The measurement was carried out at ambient laboratory temperatures and the sample was subjected to a total confining pressure of 21 MPa.

TABLE 9.6

Pre- and Post-Pyrolysis Permeability and Porosity Measurements for Lean and Rich Oil Shale Samples

Sample ID	Grade (L/t)	Pyrolysis	Porosity (%)	Absolute Permeability (nanodarcy)
WR-13-V-1	21	Before	2.1	<2
WR-13-V-2	21	After	22.18	1.21×10^3
G4-V-1	104	Before	2.238	<2
GR4-V-2	104	After	34	1.04×10^6

Permeability depends on the orientation of the sample (vertical or horizontal). As is well known for layered media, intact vertical plugs are typically less permeable than horizontal plugs due to laminations. Following pyrolysis of oil shale, the liquid and gaseous products preferentially follow the laminae (which usually are nominally horizontal in the settings evaluated). Without well-developed, laterally extensive laminae, removal of produced liquids and gases will be very difficult unless an extensive set of steeply dipping fractures develops or is present.

The relationship between permeability and the richness of the sample is uncertain. During pyrolysis, a sample likely undergoes swelling as the solid-phase kerogen is decomposed into liquids and gases. While the sample being heated should be relatively stable due to constraint from in situ stresses, this expansion may or may not create cracks. Although thermal cracking is anticipated, it may be preceded by a stage of high ductility. It can be inferred that ductility is associated with partially converted, semisolid organic material with or without liquids and gases present. If a reasonable volume of relatively inert mineral matter is present and the sample is undrained, internal pore pressure could cause localized hydraulic fracturing (the real life analog is expulsion of petroleum from geologic source rock). As the liquid flows out of the sample, pathways develop that regulate the degree of porosity and permeability after pyrolysis. Longer-term integrity of these pathways is uncertain; they may heal because of deformation as the pore fluid ultimately drains from the sample. With drainage, effective stress will increase and the integrity of the residual framework may be jeopardized.

9.7 Conclusions

In situ extraction relies on high temperatures to decompose the native kerogen into a bitumen intermediate and then into liquids, gases, and some pore-filling char. This complex behavior is coupled with the generation of microfractures inside the oil shale where the gas may migrate. The expulsion of gas into these microfractures is expected to increase permeability, promoting drainage. Generated liquids and gases are anticipated to affect the porosity of the pyrolyzing oil shales, since the evolution of these fluids will exceed the volumetric porosity generation and increases in pressure will result. When the pressure in the pores exceeds the minimum total in situ stress and a rather small tensile strength (typically tensile strength will be on the order of 10% of the unconfined compressive strength), further microfracturing will occur and the liquid/gas will migrate, usually vertically or in the direction of the negative pressure gradient toward a production well. However, porosity creation resulting from kerogen decomposition dramatically increases the risk of compressive deformation associated with pore collapse, particularly when drainage leads to reduced pore pressure. This deformation may jeopardize the integrity of the wellbore and overlying rock and also reduce permeability.

Based on the novel experiments described in this chapter, the evolution of permeability and porosity under in situ confinement at elevated temperatures has been measured. Multiple drive mechanisms are operational for in situ oil shale extraction. Initially, an expansion drive operates as liquids and gases are generated. Even if conductive pathways are not initially available to a production wellbore, creation of hydraulic microfractures can result from phase changes during pyrolysis. Evidence for this conclusion comes from the laboratory triaxial testing described, which indicates that pyrolysis can result in substantial axial deformation (initially expansion and subsequently compression). On the presumption that this expansion and compression correlate with significant porosity generation and collapse (Table 9.6), these pores are prone to significant compaction once a continuous connection evolves to a production wellbore, and liquids and gases can be moved by this compaction drive. Both expansion and compaction drives occur once drainage is enabled. However, if compaction continues, interconnected porosity and permeability can degrade with time and increasing effective stress. Concurrently, on a broader scale, depending on the domain that is heated and its depth, heave may initially result and subsidence may follow. In this work, mechanical properties and their evolution with temperature have been developed in order to populate conventional geomechanical simulators to assess such possibilities.

References

Baharia, M., M. Batzle, and G. Radziszewski. 2011. Temperature and pressure dependence on mechanical and elastic properties of oil shale as revealed by ultrasonic and micro x-ray CT. Paper presented at *45th U.S. Rock Mechanics/Geomechanics Symposium*, San Francisco, CA.

Chong, K. P., J. S. Harkins, M. D. Kuruppu, and A. I. Leskinen. 1987. Strain rate dependent mechanical properties of western oil shale. Paper presented at *28th US Symposium on Rock Mechanics*, Tucson, AZ.

Chong, K. P., J. W. Smith, B. Chang, P. M. Hoyt, and H. C. Carpenter. 1976. Characterizations of oil shale under uniaxial compression. Paper presented at *17th U.S. Symposium on Rock Mechanics*, Snowbird, UT.

Core Laboratories. 2015. Unconventional reservoir services. http://www.corelab.com/ps/gri-analysis. Accessed on June 20, 2015.

Crawford, P., K. Biglarbigi, A. Dammer, and E. Knaus. 2008. Advances in world oil shale production technologies. Paper presented at *SPE Annual Technical Conference and Exhibition*, Denver, CO.

Duncan, J. M. and C.-Y. Chang. 1970. Nonlinear analysis of stress and strain in soils. *Journal of the Soil Mechanics and Foundations Division* 96(5):1629–1653.

Eseme, E., J. L. Urai, B. M. Krooss, and R. Littke. 2007. Review of mechanical properties of oil shales: Implications for exploitation and basin modelling. *Oil Shale* 24(2):159–174.

Glauser, W. J. 2015. Simulating evolution of poroelasticity and subsidence of Green River oil shale under in-situ pyrolysis. MS thesis, University of Utah, Salt Lake City, UT.

Herro, A. 2013. Plenty of shale, plenty of problems. WorldWatch Institute. http://www.worldwatch.org/node/5167. Accessed on June 25, 2015.

Islam, M. A. and P. Skalle. 2013. An experimental investigation of shale mechanical properties through drained and undrained test mechanisms. *Rock Mechanics and Rock Engineering* 46(6):1391–1413.

Jaeger, J. C. and N. G. W. Cook. 1976. *Fundamentals of Rock Mechanics*, 2nd edn. London, U.K.: Chapman and Hall.

Kang, Z., D. Yang, Y. Zhao, and Y. Hu. 2011. Thermal cracking and corresponding permeability of Fushun oil shale. *Oil Shale* 28(2):273–283.

Kondner, R. and J. Zelasko. 1963. Hyperbolic stress-strain response: Cohesive soils. *Journal of the Soil Mechanics and Foundations Division* 89(1):115–143.

Lankford, J. 1976. Dynamic strength of oil shale. *Society of Petroleum Engineers Journal* 16(1):17–22.

Leavitt, D. R., A. L. Tyler, and A. S. Kafesjian. 1987. Kerogen decomposition kinetics of selected Green River and eastern U.S. oil shales from thermal solution experiments. *Energy & Fuels* 1(6):520–525.

Terzaghi, K., R. B. Peck, and G. Mesri. 1996. *Soil Mechanics in Engineering Practice*, 3rd edn. New York: John Wiley & Sons.

Thomas, G. W. 1966. Some affects of overburden pressure on oil shale during underground retorting. *Society of Petroleum Engineers Journal* 6(1):1–8.

Tinni, A., E. Fathi, R. Agarwal, C. Sondergeld, I. Y. Akkutlu, and C. Rai. 2012. Shale permeability measurements on plugs and crushed sample. Paper presented at *SPE Canadian Unconventional Resources Conference*, Calgary, Alberta, Canada.

Tisot, P. R. and H. W. Sohns. 1970. Structural response of rich Green River oil shales to heat and stress and its relationship to induced permeability. *Journal of Chemical Engineering Data* 15(3):425–434.

Tiwari, P. 2013. Oil shale pyrolysis: Bench-scale experimental studies and modeling. Ph.D. dissertation, University of Utah, Salt Lake City, UT.

Tiwari, P., M. Deo, C. L. Lin, and J. D. Miller. 2013. Characterization of oil shale pore structure before and after pyrolysis by using x-ray micro-CT. *Fuel* 107:547–554.

Trimmer, D. and H. C. Heard. 1980. Compaction and permeability of oil-shale aggregates at high-temperatures. *Society of Petroleum Engineers Journal* 20(2):95–104.

Vanden Berg, M. 2015 (May). Core center news: Skyline 16 Green River Formation core—World class lacustrine teaching tool. Utah Geological Survey, Salt Lake City, UT.

Vutukuri, V. S., R. D. Lama, and S. S. Saluja. 1974. *Handbook on Mechanical Properties of Rock: Testing Techniques and Results, Vol. I*, Aedermannsdorf, Switzerland: Trans Tech Publications.

White, J. A., A. Burnham, A., and D. Camp. 2015. A thermoplasticity model for oil shale. Paper presented at *ISRM GeoProc 2015*, Salt Lake City, UT, unpublished.

World Energy Council. 2010. Oil shale. In *2010 Survey of Energy Resources*, pp. 93–122. London, U.K.: World Energy Council.

10

Modeling of Well Arrangement and Its Effect on Energy Ratio for In Situ Thermal Treatment of Oil Shale in the Uinta Basin

Michal Hradisky and Philip J. Smith

CONTENTS

While commercial development of any in situ oil shale production process is multifaceted, the overall success always depends on the underlying technology and its in situ implementation. This chapter focuses on the fundamental heat transfer problem that is common to all in situ technologies. A suite of 242 simulations is performed to study the effect of well arrangement on energy ratio for an in situ process scenario employing horizontal wells located in the Uinta Basin. Simulations incorporate implementation of a detailed stratigraphic delineation and associated physical properties and capture a wide range of well configurations. As such, simulation results can be used to determine the most sensitive design parameters that affect the energy ratio and thus the viability of well arrangement designs.

10.1 Introduction

As of 2015, there was not a single commercial in situ oil shale production facility in the world. While multiple patents and technologies have been proposed and filed for in situ thermal retorting (BLM 2006, 2011, Burnham et al. 2008, Symington et al. 2006, Wellington et al. 2005), none have yet been proven to be commercially viable. In contrast, ex situ thermal processing of oil shale has occurred at a commercial scale for more than a century and continues today. In countries around the world such as Estonia, Brazil, and China, oil shale is used for electricity generation and liquid fuel production. Nevertheless, in situ oil shale retorting offers an appealing alternative to ex situ oil shale processing as it has the potential to reduce the adverse environmental aspects often associated with ex situ technologies such as surface disturbance, carbon dioxide (CO_2) emissions during retorting, and disposal of spent shale. For an in situ process, the resource stays in place and a series of horizontal or vertical wells is drilled that is then used to supply heat to the retorting region and to extract the products. Heat may be supplied by electrical heaters; fuel cells; internal combustion; circulating, externally generated hot gases; or microwaves (BLM 2006, 2011, Burnham et al. 2008, Fox 1983, Symington et al. 2006, Wellington et al. 2005). Another benefit of in situ technology is its potential to target shale zones that would be difficult to mine or zones with leaner resources that may not be economical to mine.

While in situ technology offers possible benefits over ex situ processes, the main difference in commercial viability between these two processes is the maturity of technology used to convert solid kerogen into products and to extract products from the source rock. In situ oil shale retorting technology is still considered experimental, as it is appreciably more difficult to study than ex situ processes.

Oil shale retorting, whether ex situ or in situ, is a very simple concept: it is an application of heat to the oil shale rock over a period of time, which in turn breaks down the kerogen and converts it into gaseous and/or liquid products that are then extracted. For an ex situ process, the heat transfer occurs within a matter of minutes or hours, since the size of the crushed oil shale pieces that are processed is relatively small (on the order of centimeters). The predominant mode of heat transfer for an ex situ process in horizontal or vertical retorts is convection, which allows for a relatively fast and even heat distribution.

For an in situ process, the heat needs to be applied significantly longer (on the order of months or years) and a large portion of the heat can be wasted to the surrounding rock by conduction. Prior to heating, fractures may be introduced in the retorting region through the use of explosives (Fox 1983) or fracking (Symington et al. 2006, Wellington et al. 2005). The fractures increase source rock permeability, promoting faster and more even heat distribution and enhancing product recovery. Other processes rely on thermomechanical fracturing (Burnham et al. 2008), which may occur during heating due to the geomechanical properties of oil shale, as it tends to expand and become brittle once the kerogen locked within the source rock is produced and extracted, as detailed in Chapter 8. Both convective and conductive modes of heat transfer can be present in the fractured (or rubblized) portions of the retorting region near wellbores. However, farther away from these regions of increased permeability, the conductive mode of heat transfer dominates simply because of the natural decrease in the occurrence and size of fractures. Therefore, for both ex situ and in situ processes, it is the heat transfer from the heating medium to the source rock that determines product quantity, quality, and production rate.

While kinetic mechanisms describe the thermal decomposition of kerogen into products, these mechanisms rely on an accurate time–temperature history of the oil shale rock. If the heat transfer is not accurately captured, any subsequent estimates of production rates of oil and gas and of their quality will also be incorrect. While the various companies focused on in situ development (BLM 2006, 2011, Burnham et al. 2008, Symington et al. 2006, Wellington et al. 2005) rely on their specific patented technology to transfer heat from the heating medium to the oil shale formation, it is ultimately the rate of heat transfer throughout the formation that determines the overall efficiency and economic viability of these technologies.

10.2 Simulation Approach

There are a few simulation packages that are commonly used to study in situ heating of oil shale, including STARS, a thermal and advanced process reservoir simulator developed by the Computer Modelling Group Ltd. (Arbabi et al. 2010, Bauman et al. 2009, Hazra 2014); the Finite-Element-Heat-Mass (FEHM) program developed by Los Alamos National Laboratory (Kelkar et al. 2011); the COMSOL multiphysics modeling software (Wallman and Burnham 2008); the STAR-CCM+ comprehensive engineering package developed by CD-adapco (Hradisky et al. 2014); and other research programs (Brandt 2008, Fan et al. 2010). The prediction of correct temperature profiles, pyrolysis kinetics, and product recovery in the formation being heated is difficult with any simulation tool. Model validation, in which simulation results are compared with experimental data, is always desirable prior to applying the simulation tool in a predictive mode. Otherwise, the uncertainties associated with the model outputs are unknown. However, the experimental data for any in situ process are extremely scarce and often proprietary and confidential.

The work described in this chapter uses STAR-CCM+ (CD-adapco 2015). To study the underlying heat transfer problem because of its high-performance parallel computing capability and because detailed oil shale stratigraphy and user-programmed subroutines can be implemented. STAR-CCM+ was used to accurately match the spatially dependent, time-temperature history of an industrial in situ heater test over a multimonth time period (Hradisky et al. 2014). Therefore, this work builds on a proven (i.e., validated) methodology and uses information provided by other authors in this book to construct a high-performance computing model of in situ thermal retorting in a Uinta Basin location.

10.2.1 Numerical Setup

In the validation study of long-term, in situ heating (Hradisky et al. 2014), only the conductive component of heat transfer was needed for the model to match the experimental data. Therefore, convective heat transfer, if indeed present, was not significant in comparison to the predominant conductive heat transfer mode. Consequently, only the conductive component of heat transfer is modeled in the current study.

STAR-CCM+ solves the following energy equation for solid heat transfer:

$$\frac{\partial}{\partial t}\int_V \rho c_p T dV + \oint_A \rho c_p T v_s \cdot da = -\oint_A q'' \cdot da + \int_V s dV \qquad (10.1)$$

where
  ρ represents the solid density
  c_p is the specific heat
  T is the temperature
  v_s is the solid convective velocity
  q'' is the solid heat flux vector
  s is the source term (heat source or heat sink)
  a represents the normal area vector
  A represents the area
  V represents the volume.

For the purposes of this simulation, this equation reduces to

$$\frac{\partial}{\partial t}\int_V \rho c_p T dV = -\oint_A q'' \cdot da \qquad (10.2)$$

since both the solid convective velocity and the source/sink terms are zero. The solid heat flux, q'', is computed using Fourier's law as

$$q'' = -\lambda \cdot \nabla T \qquad (10.3)$$

where
  λ represents the thermal conductivity tensor
  ∇T represents the temperature gradient.

Equation 10.2 is solved using an implicit iterative approach with second-order discretizations for both the temporal and spatial terms. STAR-CCM+ v10.02.010 was used for the simulations, and all simulations were run on the University of Utah Ash cluster (CHPC 2015). Each simulation was run using 360 computing cores and, based on the number of computational cells, as well as the heterogeneous nature of Ash compute nodes, took anywhere from 4 to 24 h to complete.

10.2.2 Uinta Basin Study Area

The study region is based in eastern Utah in the Uinta Basin of the Green River Formation. As noted in previous chapters, the Green River Formation spans Utah, Colorado, and Wyoming and holds an estimated 4.3 trillion of barrels of oil in oil shale resources. However, only a portion of the actual estimated resource is potentially viable for commercial development (Birdwell et al. 2013, Vanden Berg 2008). As discussed in Chapter 4, the majority of the Uinta Basin is comprised of relatively lean oil shale zones interspersed with thin, rich zones. While it may be economical to process the richest zones using either ex situ or in situ technologies, only in situ technologies provide a potential solution for extracting resources from the relatively lean oil shale zones.

In the Uinta Basin, the thickest and richest oil shale deposits are located in the northeast zone. In order to characterize this zone, the U.S. Geological Survey (USGS) drilled multiple oil shale cores, including the Coyote Wash 1 well, one of the deepest oil shale cores in the Uinta Basin (Vanden Berg et al. 2013). It extends down to a depth of 3460 ft. The detailed stratigraphic delineation of the oil shale zones for the Coyote Wash 1 well, obtained using Fischer assay in 1 ft increments, is shown in Figure 10.1. Only the interval from 1742 to 2562 ft (531 to 781 m) in depth is shown because it is the interval used in the in situ retorting simulations. The average richness over this interval is 13.3 gallons per short ton (GPT). The maximum richness of 76.8 GPT occurs at 2236 ft (681 m) below the surface in the Mahogany zone. Below 2562 ft (781 m), the average richness decreases to less than 10 GPT. Further details regarding the geologic model of the Green River oil shale in the Uinta Basin are provided in Chapter 4.

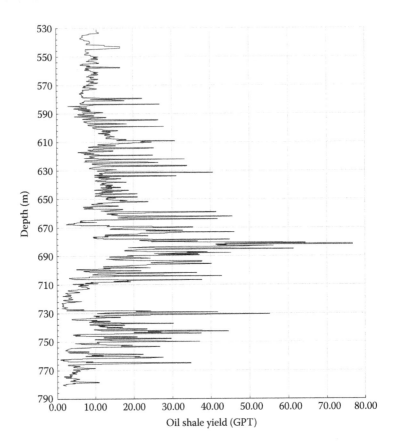

FIGURE 10.1
Oil shale yield for the Coyote Wash 1 core over the studied depth interval. Data adapted from USGS log database for well U044. (From Vanden Berg, M.D. et al., Utah oil shale database: UGS Open File Report 469, CD-ROM, 2006.)

10.2.3 Design of Experiments

To examine the effect of well arrangement on the energy ratio for an in situ heating process, a design of experiments was set up using six design parameters: lateral well spacing (h), vertical spacing between rows of wells (v), vertical well angle offset (θ), vertical location (f_r) of the wells within the 250 m oil shale interval captured in the simulation, well radius (r), and number of well rows. These parameters are shown in Figure 10.2. Table 10.1 lists these design parameters and their ranges used in the simulations. The first four design parameters are continuous and could take any value in the specified range, including the minimum and maximum values. The last two design parameters, well radius and number of well rows, are discrete parameters and can only take on the values specified in the table.

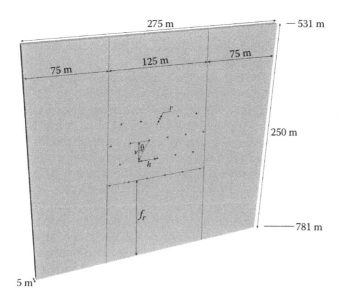

FIGURE 10.2
Simulation geometry showing a representative well distribution with simulation domain dimensions and design parameters.

TABLE 10.1

Geometric Design Parameters and Their Ranges

Design Parameter	Range	
Horizontal well spacing (h)	Min: 5.09 m	Max: 24.96 m
Vertical well row spacing (v)	Min: 2.55 m	Max: 24.61 m
Vertical well offset (θ)	Min: 0.11°	Max: 63.39°
Vertical location (f_r)	Min: 50 m	Max: 150 m
Well radius (r)	0.1016 m, 0.1143 m, 0.127 m, 0.1397 m, 0.1524 m	
Number of well rows	1, 2, 3, 4, 5, 6, 7, 8, 9, 10	

For the design of experiments, a latin hypercube (LH) statistical sampling technique was used to span the parameter space of all six variables simultaneously. This sampling technique is often used to sample multidimensional parameter space and provides a more accurate estimate of means, variances, and distribution functions of an output compared to random sampling. Furthermore, LH sampling assures that there is no preexisting correlation between the input design parameters (Wang 2003). Using LH sampling, 242 designs were constructed using the six design parameters. The values of the six parameters for each of the 242 designs can be found in Table 10A.1 (see Appendix 10A). A simulation was performed for each design.

10.2.4 Simulation Domain and Boundary Conditions

The geometric domain of the in situ retort for all simulations is 5 m in the longitudinal (horizontal) direction, 275 m in the transverse (lateral) direction, and 250 m in the vertical direction, as shown in Figure 10.2. The domain size in the vertical direction corresponds to the depth interval from 1742 to 2562 ft (531–781 m). The simulation domain uses periodic boundary conditions in the longitudinal direction, making this geometry, as well as the results, scalable in the longitudinal direction for well lengths on the order of hundreds or thousands of meters. However, by using the periodic boundary condition, it is assumed that the oil shale stratigraphy does not vary as a function of well length. In the transverse direction, only the middle 125 m of the domain contains wells. The remaining 150, 75 m on each side of the retorting section, is used as a buffer region to ensure that the heat transfer occurring in the center portion is not influenced by the side boundaries. All vertical and transverse domain boundary conditions are set to be adiabatic, i.e., no heat can enter or leave the domain through these boundaries. This boundary condition represents an optimal heating scenario where no heat is lost to the surrounding formation beyond the simulation domain.

While still subject to simplifications, the simulation domain shown in Figure 10.2 was chosen so that well arrangement designs more closely representing real in situ conditions (including a buffer region) could be incorporated within the domain. Modeling only a limited subset of the full well arrangement, without including the surrounding buffer region, idealizes the model scenario and ignores the fact that for any commercial-scale, in situ production process, there are always heat losses to the surrounding formation. This type of simulation leads to favorable temperature profiles and therefore overemphasizes product generation for a given energy input.

The well arrangement scenarios that were simulated in this study are listed in Table 10A.1. These scenarios capture the well arrangement periodicity that may occur for an industrial in situ retorting process without ignoring the inevitable heat losses that occur at the edge of the retorting region, where the heat flows from the retorting region into the surrounding buffer region. Figure 10.3 shows two representative designs (58 and 226) out of the 242 completed for this study. The design parameters for these simulations are shown in detail in Table 10A.1.

Since the geometry captures only the horizontal portion of the wells, all portions of the wells from the surface to the horizontal section are assumed to be unheated. It is also assumed that any or all of the horizontal wells can serve as producer wells at any point between the start and end of heating. Therefore, no specific producer wells are included in the simulation.

The heater wells are specified with a prescribed temperature boundary of 675 K. This heater temperature is within the range required to convert the kerogen into products (Biglarbigi et al. 2008). This temperature is applied

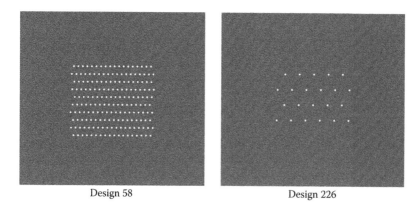

Design 58   Design 226

FIGURE 10.3
Illustrative simulation domains for two specific design scenarios (58 and 226) out of the 242 total designs. The well radius has been increased by 1500% for illustrative purposes.

instantaneously at the start of the simulation and held constant throughout the 7 years of simulated heating.

10.2.5 Simulation Considerations

Not all assumptions mentioned in the previous section would necessarily be applicable to an actual industrial-scale in situ process. For example, the heated section would not be limited to a width of 125 m and a height of 250 m. Rather, it could be continuously extended by drilling new wells and introducing new heaters.

Furthermore, the heat supplied by heaters would not be instantaneous. Instead, heater temperatures would gradually increase over the course of days or weeks, reducing the large energy demand seen in the simulations at the onset of pyrolysis. Also, while well heating in the simulations begins after all the wells are drilled, well heating in an actual process could begin as soon as a single well or group of wells is drilled and the heater(s) is placed in the well(s). Possibilities for staging (i.e., optimizing) the heat supply are almost endless and subject to change for each specific in situ technology. While optimized heat delivery could possibly increase the efficiency of any given process, it plays only a secondary role to heat efficiencies established by well arrangements. If a particular well arrangement is subject to large energy penalties, optimizing the heat supply will not dramatically alter the feasibility of the design.

Lastly, the adiabatic boundary conditions prescribed on the vertical and transverse boundaries create an idealized situation, which is non-existent in real-life applications. Since no heat can leave the idealized simulation domain, all heat applied through heater wells heats the source rock, pyrolyzing the oil shale if the temperature becomes high enough for a sufficient length of time. While changing any of these factors could increase or decrease the overall energy ratio, these factors are considered to play a less significant role than the heat transfer generated by the well arrangement and distribution throughout the formation.

10.2.6 Oil Shale Properties

Oil shale is, by nature, a heterogeneous material, and therefore its properties vary based on the orientation of the bedding planes, which are formed by layers of deposits. Oil shale properties such as density, specific heat, and thermal conductivity play a crucial role in heat transfer throughout the formation and therefore have a direct effect on the resulting temperature field and product yields.

10.2.6.1 Density

The stratigraphic oil shale grade variation for the Coyote Wash 1 core is used to estimate the total organic content (TOC) and kerogen in place for each distinct oil shale layer. Equation 10.4 is used to estimate the TOC based on oil shale grade (Cook 1974):

$$TOC = \frac{(G+0.7714)}{2.216} \tag{10.4}$$

where
  TOC is the initial total organic carbon in wt%
  G is the oil shale grade in GPT

The composition of a typical kerogen is 80 wt% carbon, 10 wt% hydrogen, and 10 wt% nitrogen, sulfur, and oxygen (Atwood 2006). Therefore, the following equation is used to obtain the total weight fraction of kerogen available in each specific oil shale layer:

$$K_i = \frac{TOC}{0.8} \tag{10.5}$$

where K_i is the initial kerogen (wt%). The initial density of oil shale is then determined using following relationship:

$$\frac{1}{\rho_{T,i}} = \frac{(K_i/100)}{\rho_K} + \frac{(1-K_i/100)}{\rho_M} \tag{10.6}$$

where
  $\rho_{T,i}$ is the initial density of the source rock
  ρ_K is the average density of kerogen
  ρ_M is the average density of the remaining mineral matrix.

Kerogen density can vary between 1.0 and 1.3 g/cm³ (Eseme et al. 2007); in this work, it is assumed to be 1.06 g/cm³ (Smith 1969). The density of the mineral matrix is assumed to be 2.72 g/cm³ (Smith 1969). This density implementation does not account for the shale water content as the source rock contains, on average, less than 1.3 wt% water over the study interval (Vanden Berg et al. 2006).

During the simulations, the density of the source rock changes as kerogen is converted into products and extracted. Because STAR-CCM+ is based on a finite volume approach, density is adjusted according to the following relationship:

$$\rho_{eff} = \left(\frac{f_k\, m_{k,i} + m_m}{m_{k,i} + m_m} \right) \rho_{T,i} \tag{10.7}$$

where

ρ_{eff} is the effective oil shale density

f_k represents the fraction of kerogen remaining in the oil shale

$m_{k,i}$ represents the initial mass of kerogen and is directly related to K_i

m_m represents the mass of the mineral matrix, which remains constant throughout the simulation and is directly related to the quantity $1-(K_i/100)$.

The initial value of the kerogen fraction, f_k, is one. As kerogen is converted into products, this fraction decreases until it reaches a value of zero. Therefore, f_k is computed according to the following relationship:

$$f_k = \frac{m_k}{m_{k,i}} \tag{10.8}$$

where m_k is the remaining, or available, mass of kerogen.

10.2.6.2 Specific Heat

The specific heat implementation uses empirical relationships developed for Green River oil shales (Shaw 1947) and accounts for property changes that occur as the kerogen is converted during pyrolysis. For raw shales, the following equation is used:

$$c_{p,r} = 0.172 + (0.067 + 0.00162\, G)10^{-3}\, T \tag{10.9}$$

where

$c_{p,r}$ is the specific heat of the original, raw source rock in Btu/lb-°F above 25°C

G is the oil shale grade in GPT

T is the temperature in °R.

For spent shale, the specific heat is computed by

$$c_{p,s} = 0.174 + 0.051 \cdot 10^{-3} T \tag{10.10}$$

where
   $c_{p,s}$ is the specific heat of spent shale in Btu/lb-°F above 25°C
   T is temperature in °R.

To account for property changes as the kerogen is converted into products, the specific heat properties of the raw and spent shale are combined according to the following relationship:

$$c_{p,\text{eff}} = f_k\, c_{p,r} + (1 - f_k)\, c_{p,s} \tag{10.11}$$

where $c_{p,\text{eff}}$ is the effective specific heat of the oil shale as it is retorted.

10.2.6.3 Thermal Conductivity

Thermal conductivity in oil shale is reported in two predominant directions: parallel and perpendicular. The parallel thermal conductivity is measured in the direction of the shale bedding planes, while the perpendicular thermal conductivity is measured in the direction perpendicular to the shale bedding planes. The simulations reported in this chapter use anisotropic thermal conductivity (parallel as well as perpendicular) to capture the heterogeneity of oil shale. As with density and specific heat, the thermal conductivity property changes as the kerogen is converted from its original solid state to products. The general form of the equation is as follows (Tihen et al. 1967):

$$\lambda = c_1 + c_2\,G + c_3\,T + c_4\,G^2 + c_5\,T^2 + c_6\,GT \tag{10.12}$$

where
   λ represents the thermal conductivity in W/m-°C
   c_1 through c_6 are empirical constants
   G is the Fischer assay of shale in GPT
   T is the temperature in °C.

Coefficients obtained from experiments conducted on Green River oil shale (raw and spent shales) for both parallel and perpendicular directions are listed in Table 10.2.

TABLE 10.2

Coefficients for Parallel and Perpendicular Thermal Conductivity of Green River Oil Shale

Coefficients	Raw Shale		Spent Shale	
	Parallel	Perpendicular	Parallel	Perpendicular
c_1	2.0670	1.8081	1.5355	1.8246
c_2	-5.7790×10^{-2}	-3.698×10^{-2}	-5.923×10^{-2}	-4.4844×10^{-2}
c_3	1.572×10^{-3}	1.980×10^{-3}	3.03×10^{-5}	-2.309×10^{-4}
c_4	5.686×10^{-4}	3.056×10^{-4}	6.25×10^{-4}	3.652×10^{-4}
c_5	-4.585×10^{-6}	-5.184×10^{-6}	-2.935×10^{-9}	0
c_6	-1.470×10^{-5}	-1.872×10^{-5}	2.698×10^{-6}	1.067×10^{-5}

Source: Tihen, S.S. et al., Thermal conductivity and thermal diffusivity of Green River oil shale, in *Thermal Conductivity: Proceedings of the Seventh Conference*, Gaithersburg, MD, 302, pp. 529–535, 1967.

Equation 10.13 is used to determine the effective thermal conductivity (parallel or perpendicular) as a function of kerogen conversion:

$$\lambda_{\text{eff}} = f_k \lambda_r + \left(1 - f_k\right) \lambda_s \tag{10.13}$$

where

λ_{eff} is the effective thermal conductivity
λ_r represents the thermal conductivity for raw oil shale
λ_s represents the spent shale thermal conductivity.

10.2.7 Oil Shale Pyrolysis Kinetic Model and Product Distribution

Kerogen in the Green River oil shale is a heterogeneous mixture of organic material derived from algae, woody fragments, resins, and spore exines (Hubbard et al. 1950). Pyrolysis of this kerogen is a very complex process as discussed in Chapters 6 and 7. There are numerous kinetic formulations of kerogen decomposition, such as first-order reactions (Shih and Sohn 1978), consecutive (Hubbard et al. 1950) or parallel first-order reactions (Leavitt et al. 1987), as well as variable reaction order models (Al-Ayed et al. 2010). Kinetic formulations for oil shale pyrolysis are discussed in more detail in Chapters 6 and 7.

For this study, a first-order kinetic model is used to model oil shale pyrolysis:

$$\frac{dm}{dt} = -km \tag{10.14}$$

where
  m is the remaining mass of kerogen
  t is time
  k is the chemical reaction rate constant, which is defined as

$$k = -Ae^{-E/RT} \tag{10.15}$$

where
  A is the preexponential factor
  E is the activation energy
  R is the universal gas constant
  T represents the temperature.

For this pyrolysis model, the kerogen is directly converted into oil, gas, or coke (a carbonaceous residue) based on the kinetic rate defined by Equation 10.15. Values for the preexponential factor and the activation energy were obtained from experiments on the GR3 oil shale sample as detailed in Chapter 6. From Table 6.1, the preexponential factor is 9.5×10^{13} 1/s and the activation energy is 221 kJ/mol.

Because the TGA data were collected at experimental conditions that differ greatly from in situ pyrolysis conditions, the actual product yield as well as product distribution under in situ conditions may differ significantly from the TGA data. Furthermore, even under *ideal* in situ retorting conditions where all kerogen is converted into products, not all of those products can be recovered as liquid fuel and/or gas. Additionally, there may be disparities between product yield based on oil shale grade, which is often based on Fischer assay analysis, and actual product yield, which may be less than 70% of Fischer assay (Le Doan 2013). To account for the possible differences between experimental tests and true in situ conditions, the recovered and unrecovered fractions of the overall converted kerogen vary according to the following equation:

$$c_r + c_{unr} = 1 \tag{10.16}$$

where
  c_r represents the fraction of recoverable products (oil and/or gas)
  c_{unr} represents the unrecoverable fraction (i.e., kerogen that is converted to coke, product yields that are less than Fischer assay, and products that are not extractable due to process limitations).

Given that oil and gas are the only two recoverable products accounted for in this analysis, Equation 10.16 can be rewritten as

$$c_r\, m_{f,o} + c_r\, m_{f,g} + c_{unr} = 1 \qquad (10.17)$$

where
$m_{f,o}$ is the mass fraction of oil
$m_{f,g}$ is the mass fraction of gas in the recovered products.

Intrinsically, the summation of $m_{f,o}$ and $m_{f,g}$ is always unity. The product distribution defined by Equation 10.17 is possible because of the single-step, first-order kinetic model implemented in the simulations. If a more complex kinetic model were used, such differentiation may not be possible. Using this relationship, no constraints are applied on the recoverable fraction of the converted kerogen nor on the product distribution for any of the 242 designs.

10.2.8 Energy Metrics

To evaluate the performance of each of the 242 designs, the primary metric is the external energy ratio (EER). In its conceptual form, EER is defined as

$$\mathrm{EER} = \frac{E_{out}}{E_{in}} \qquad (10.18)$$

where
E_{out} represents the energy contained in the oil and gas products
E_{in} represents the energy supplied to the formation.

However, this definition of EER requires the product distribution and the associated energy content to be known.

In this study, the variability of product distribution and recovery associated with Equation 10.17 produces a range of EER values, even for a single design. To incorporate this wide range of possible energy contents across the studied designs, the EER definition is modified as

$$\mathrm{EER} = \mathrm{EER}_{n,y}\, p_f \qquad (10.19)$$

where
$\mathrm{EER}_{n,y}$ is the normalized EER
y represents the year at which the normalized EER is computed
p_f represents the process factor.

The normalized EER is calculated according to the following relation:

$$\text{EER}_{n,y} = \left(\frac{m_{p,y}}{E_{h,y}}\right) \bigg/ \left(\frac{m_{p,7}}{E_{h,7}}\right)_{max} \tag{10.20}$$

where

$m_{p,y}$ represents the mass of kerogen converted into products during a specified time interval denoted by y

$E_{h,y}$ represents the amount of energy supplied by all heaters to the formation during the same time period y

$m_{p,7}$ represents the amount of kerogen converted during 7 years of heating

$E_{h,7}$ represents the energy supplied by all heaters to the formation during the 7 years of heating.

The mass of converted kerogen is calculated using Equation 10.14, while the energy input from heaters during heating is calculated using Fourier's law, as described by Equation 10.3.

The denominator of Equation 10.20 represents the maximum kerogen-to-energy ratio from the 242 designs after 7 years of heating and is the common scaling factor for all normalized energy ratios, no matter the year. The EER continuously changes throughout the 7 years of heating since it captures both the cumulative energy input and the cumulative mass of kerogen converted into products. At the end of the 7-year simulation heating interval, $\text{EER}_{n,7}$ varies between zero and one for all designs, with one being the design for which the amount of kerogen converted per unit of energy input is the largest. All other normalized energy ratios from years 0 through 6 can reach values larger than one but can never be lower than zero.

The process factor, p_f, is calculated according to the following equation:

$$p_f = \left(c_r m_{f,o} \text{HV}_o\right)\left(\frac{m_{p,7}}{E_{h,7}}\right)_{max} + \left(c_r m_{f,g} \text{HV}_g\right)\left(\frac{m_{p,7}}{E_{h,7}}\right)_{max} \tag{10.21}$$

where

HV_o represents the heating value of produced oil

HV_g represents the heating value of produced gas.

By allowing the oil and gas heating values to vary, this definition of the process factor further widens the potential EER range.

The EER can be considered as the energy ratio at the wellhead. It pays no regard to how the heat is actually delivered to the formation nor does it account for all energy inputs needed in order to generate and process the products, such as energy required to drill the wells, energy required to extract products from the ground, or energy for transporting products to processing facilities. The net energy ratio (NER) is a better metric of the overall energy output to input and is highly dependent on the total number of wells that need to be drilled, the location of the process with respect to the existing infrastructure, and the actual amount and type of products, among other considerations. NER is explored in more detail in Chapter 12.

The overall energy input, the way heat is delivered to the formation, and the energy penalties are specific to each respective in situ technology. However, the underlying heat transfer within the formation and the energy needed to heat the formation to generate products are comparable across all technologies. The EER provides a metric to comparatively evaluate performance of in situ designs subject to predominantly conductive heating by calculating heat requirements needed to generate products and their resulting energy content, no matter the underlying differences in the proposed in situ technology.

10.2.9 Computational Mesh and Time Settings

Following the implementation of the detailed oil shale stratigraphy and oil shale properties into the simulations, a computational-mesh and time-refinement study was conducted in order to ensure that simulation results were not mesh- and time-sensitive. For a representative simulation scenario shown in Figure 10.4, the computational time step was varied between

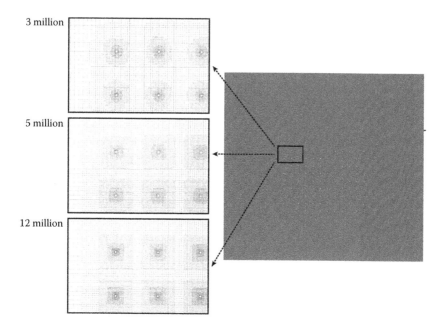

FIGURE 10.4
A representative simulation domain with three different mesh resolutions used for the time and grid refinement study. The well radius has been increased by 1000% for illustrative purposes.

TABLE 10.3

Errors in Mass of Converted Kerogen and Cumulative Energy Percent for Mesh and Time Refinement Study

Mesh Resolution (Million of Cells)	Time Step (s)	Mass of Converted Kerogen (% Error)	Cumulative Energy (% Error)
12	900	0.00	0.00
12	1800	0.11	1.47
12	3600	0.26	3.52
5	7200	0.03	2.20
3	1800	0.08	3.17
3	3600	0.07	2.83
3	7200	0.11	2.07

900 and 7200 s and three different mesh resolutions were used: 3, 5, and 12 million computational cells. Adaptive mesh refinement was used to increase mesh resolution in the vicinity of the heater wells while continuously increasing the cell size with increasing distance from the heater wells to a maximum cell size of $1 \times 1 \times 1$ m^3. The minimum cell volume in the vicinity of the heater was $2.5 \times 2.5 \times 2.5$ cm^3 for the 12 million-cell mesh and $5 \times 5 \times 5$ cm^3 for the 3- and 5-million-cell meshes. The only distinction between the 3- and 5-million-cell meshes was the thickness of the region in which the fine-to-coarse mesh transition occurred (narrower for the 3-million-cell mesh, larger and more gradual for the 5-million-cell mesh). The differences among the three mesh resolutions can be seen in Figure 10.4. Results from the mesh- and time-refinement study are summarized in Table 10.3. Results from the simulation with the finest mesh (12 million cells) at the smallest time step (900 s) were taken as the benchmark for comparison to other simulations with different time and mesh refinements.

As can be seen from Table 10.3, percent errors in mass of converted kerogen and cumulative energy required to heat the formation for 7 years are minimal across the range of mesh refinements and time steps considered. Based on these results, the increased computational cost of simulations with larger meshes and smaller time steps overwhelms the marginal benefit of these simulations. Hence, all simulations in the design of experiments were run on a 3-million-cell mesh with a time step of 3600 s. The exact cell count varies slightly among the 242 designs due to the number of wells in the actual design. The higher the well count, the larger the mesh size to accommodate the fine surface mesh requirements.

This mesh resolution is more than adequate to resolve the detailed oil shale stratigraphy provided in the USGS logs, as discussed in Section 10.2.2. Figure 10.5 shows the implementation of the detailed oil shale stratigraphy for a representative simulation domain, design 58. The oil shale grade shown in Figure 10.5 corresponds to the oil shale grade of the Coyote Wash 1 core shown in Figure 10.1.

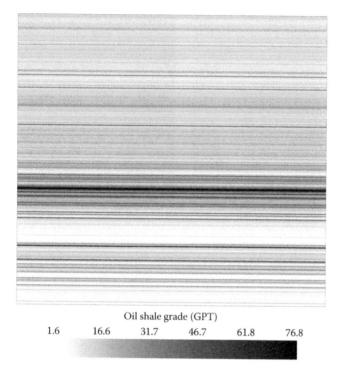

Oil shale grade (GPT)

| 1.6 | 16.6 | 31.7 | 46.7 | 61.8 | 76.8 |

FIGURE 10.5

Detailed oil shale grade implementation inside a simulation for a representative design scenario (58). Data adapted from USGS log database for well U044 (Vanden Berg et al. 2006) as shown in Figure 10.1.

10.3 Simulation Results

Since the computational domain employs periodic boundary conditions in the longitudinal direction, both the energy input and the amount of kerogen converted into products are on a per-meter basis and can be linearly scaled to extend to designs with an arbitrary length of horizontally heated wells. The temperature distribution and kerogen fraction for a representative design, design 58, are shown in Figure 10.6. The left side of the figure shows the temperature distribution and the right side shows the kerogen fraction still remaining (as defined by Equation 10.8) after 7 years of heating. The temperature distribution field throughout the retorting region is fairly uniform. The small temperature variations between rows are due to the differences in thermal conductivities for rich and lean oil shale zones. Intrinsically, lean oil shale zones have higher thermal conductivities and therefore are better at conducting the heat than the rich oil shale zones,

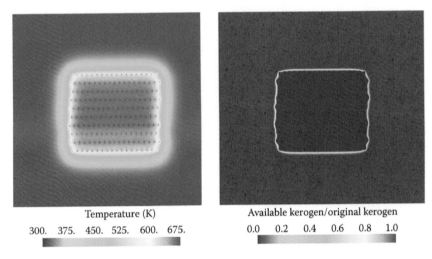

FIGURE 10.6
Temperature and kerogen fraction distributions for design scenario 58 after 7 years of heating.

which tend to have lower thermal conductivities. For this design, all kerogen in the vicinity of heater wells has been converted into products after 7 years of heating.

10.3.1 Normalized Energy Ratio

The energy input and the amount of kerogen converted into products varies significantly across the 242 simulation designs. Figure 10.7 shows the amount of kerogen converted into products after 7 years of heating with respect to the total cumulative heat applied during the same time for all designs. Figure 10.7 further shows the normalized energy ratio after 7 years of heating, $EER_{n,7}$, with both the marker size and the shading representing the magnitude of $EER_{n,7}$ from zero to one. While there is a trend of increasing kerogen conversion with increasing heat input, a relatively large number of well configuration designs have significant heat input but very little production. Therefore, the amount of heat input does not guarantee increased kerogen conversion within the 7 years of heating captured in the simulations. Furthermore, designs with the highest $EER_{n,7}$ values do not necessarily result in the largest kerogen conversion.

The normalized energy ratio also varies significantly for the 242 well configuration designs. Figure 10.8 shows the distribution of the normalized energy ratio after 7 years of heating ($EER_{n,7}$) with respect to the design number. As can be seen, there is a wide distribution of $EER_{n,7}$ values for the different well arrangement designs, with design 211 providing the highest $EER_{n,7}$ value of one. However, $EER_{n,7}$ is not the only metric by which the

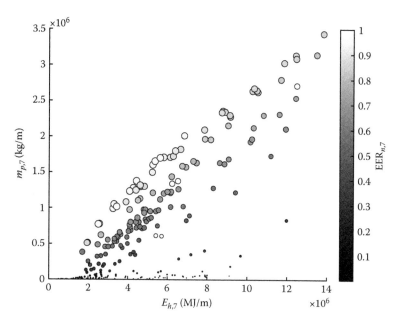

FIGURE 10.7
Amount of kerogen converted into products ($m_{p,7}$) with respect to both the heat applied by heaters during 7 years of heating ($E_{h,7}$) and the normalized energy ratio after 7 years of heating ($EER_{n,7}$) for all 242 designs. Because of the simulation domain periodicity in the longitudinal direction, the results are presented on a per-meter basis.

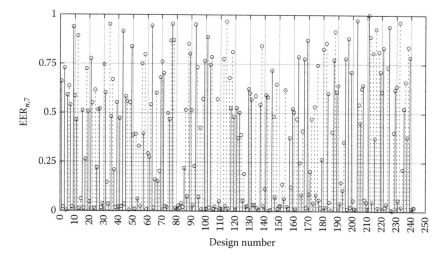

FIGURE 10.8
Normalized external energy ratio ($EER_{n,7}$) after 7 years of heating for all 242 designs.

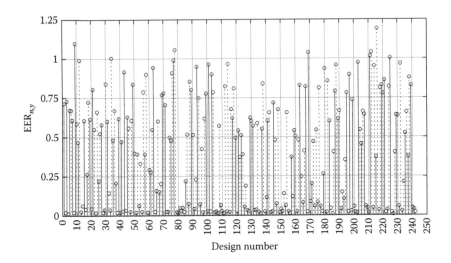

FIGURE 10.9

The maximum normalized external energy ratio (EER$_{n,y}$) for all 242 designs between onset of heating and the end of the 7-year heating interval.

feasibility of a design can be determined. Other metrics include product quality and distribution, as well as costs associated with each design and its geographic location relative to available infrastructure.

Figure 10.9 shows the maximum normalized energy ratio achieved for each design anytime between the start of heating and the end of the simulation (EER$_{n,0}$ to EER$_{n,7}$). As can be seen in the figure, there are eight designs for which the normalized energy ratio is greater than one, including design 211, with design 216 achieving the highest normalized energy ratio sometime during the 7-year simulated heating period. This figure illustrates the fact that the energy ratio continuously evolves, which makes it difficult to compare any two individual designs just on the basis of the energy ratio. While one design may have a higher energy ratio at the onset of heating, another design may need more time to reach its maximum. In fact, it is not guaranteed that the 7-year-long heating interval captures the maximum possible energy ratio for every design; for some designs, a longer heating interval may be required.

This evolution of the normalized energy ratio is shown in Figure 10.10, which depicts the distribution of designs with respect to the normalized energy ratio after 1, 3, 5, and 7 years as defined by Equation 10.20. At the onset of the heating process, most of the heat is used to bring the formation temperature closest to the heaters up to pyrolysis temperature. Therefore, EER$_{n,1}$ is very low, with almost 100% of the designs in the lowest range. As the formation continues to be heated, the normalized energy

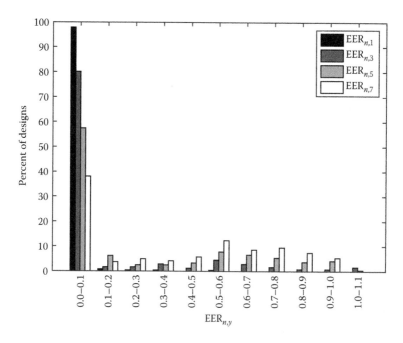

FIGURE 10.10

Distribution of designs based on the normalized energy ratio after one ($EER_{n,1}$), three ($EER_{n,3}$), five ($EER_{n,5}$), and 7 years ($EER_{n,7}$) of heating.

ratio improves. After 3 years of heating, more kerogen is converted into products, leading to a decrease in the number of designs in the lowest category, 0.0–0.1, and an increase in the number of designs in higher categories increases. This redistribution of designs toward increasing ranges of $EER_{n,y}$ continues after 5 and 7 years of heating, signifying that the longer the formation is heated, the better the chance that a particular design achieves a higher normalized energy ratio. However, longer heating does not guarantee the highest normalized ratio as was shown in Figure 10.9 and is evident in Figure 10.10, where the highest normalized energy ratios, 1.0–1.1, occurr in years 3 and 5 rather than year 7.

10.3.2 Process Factor

The magnitude of the process factor (p_f), defined in Equation 10.21, provides the scaling factor for $EER_{n,y}$ necessary to determine the actual energy ratio for any particular design. Its magnitude depends on the fraction of products recovered and on the distribution of products between oil and gas and their respective heating values, as shown in Figure 10.11. In this figure, the heating values range from 20–50 MJ/kg to capture heating values of both the

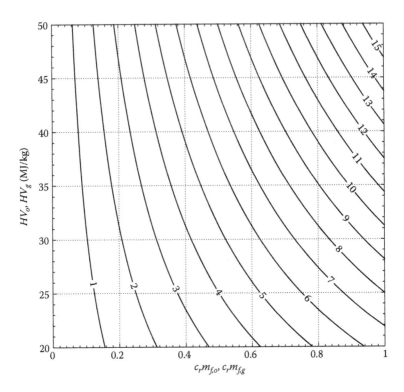

FIGURE 10.11

Magnitude of the process factor for the recoverable fractions of both the oil $(c_r m_{f,o})$ and gas $(c_r m_{f,g})$ with respect to their heating value (HV_o, HV_g).

oil and the gas. The contour lines represent the magnitude of the process factor, which vares based on the product recovery fraction and the product heating value. The magnitude of the process factor increases with increasing recovery ratio $(c_r m_{f,o}, c_r m_{f,g})$ and increasing quality of products, i.e., increasing heating values of oil or gas. A particular scenario with high product recovery and low product quality can have the same process factor as a scenario with low recovery but high product quality.

The maximum process factor of about 15 is achieved for 100% product recovery with all product recovered as a very high-energy oil or gas. While this maximum value is almost unattainable for any industrial in situ process, the process factor can still be relatively high for moderate recovery rates. For instance, for 70% recovery $(c_r = 0.7)$, 70% oil fraction $(m_{f,o} = 0.7)$ with a heating value of 43 MJ/kg, and 30% gas $(m_{f,g} = 0.3)$ with a heating value of 35 MJ/kg, the magnitude of the combined process factor is slightly greater than eight. With this value of process factor, designs with an $EER_{n,y}$ of 0.5 yield values of four for the actual EER, as defined by relationship displayed in Equation 10.19. Therefore, to evaluate the feasibility of any particular design, it is important to consider both the normalized EER and the magnitude of the process factor.

10.3.3 Effect of Design Parameters

Distributions of the normalized energy ratios after 7 years of heating for the 242 well configuration designs with respect to the six input parameters are shown in Figure 10.12. There are four parameters for which there is no distinct correlation between the parameter and the resulting $EER_{n,7}$: horizontal

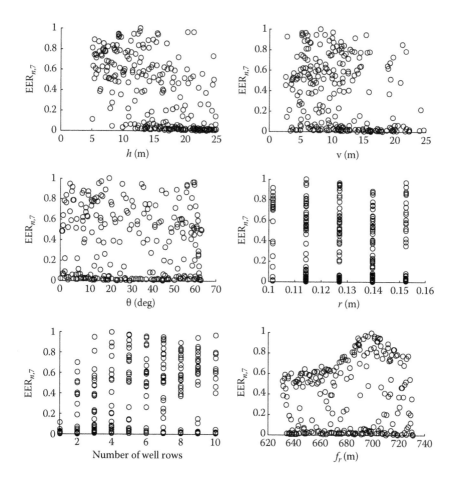

FIGURE 10.12
The effect of six design parameters on the normalized external energy ratio ($EER_{n,7}$) after 7 years of simulated heating. The design parameters are horizontal well spacing (h), vertical well spacing (v), vertical well offset angle (θ), well radius (r), number of well rows, and vertical location of wells within the formation (f_r).

well spacing, vertical well spacing, vertical well offset angle, and well radius. On the other hand, there is a clear correlation for the number of well rows and the vertical location of wells within the formation on the value of $EER_{n,7}$. With only one well row, the maximum attainable $EER_{n,7}$ is less than 0.2. This value increases to 0.75 for two well rows and can be very close to one for three well rows or more. This correlation illustrates the necessity of having at least two rows of heater wells to achieve higher kerogen conversion per unit of energy input, which results in higher values of $EER_{n,7}$. Furthermore, while locating wells near the richest zone of the formation yields the greatest values of $EER_{n,7}$, the effect of vertical location on the normalized energy ratio illustrates that it is possible to obtain favorable energy ratios even when the wells are located in the leaner portions of the stratigraphy.

For any given design, it is possible to compute the minimum distance between any two closest wells using the horizontal well spacing, vertical well spacing, and vertical angle offset. The normalized energy ratio with respect to the minimum well spacing for each of the 242 designs is shown in Figure 10.13. If the minimum well spacing is greater than about 15 m, the normalized energy ratio is close to zero. If the minimum well

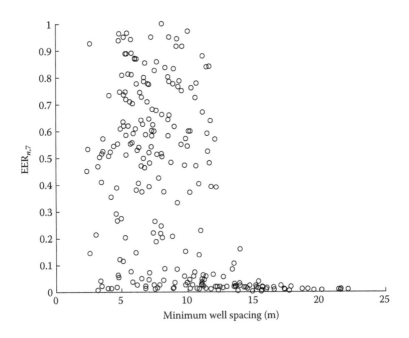

FIGURE 10.13
Normalized external energy ratio ($EER_{n,7}$) as a function of minimum spacing between any two closest wells for all 242 designs.

spacing is decreased to 12.5 m, the maximum normalized energy ratio attainable is less than 0.2. However, if the minimum well spacing is less than 12.5 m, it is possible, though not guaranteed, to achieve high values of the normalized energy ratio, which, given an appropriate magnitude of the process factor, can translate into favorable overall energy ratios. Nevertheless, out of the 242 designs, there are a large number of designs for which $EER_{n,7}$ is relatively low, even with very close well spacing.

10.4 Conclusions

This chapter presents results for simulations of 242 different well arrangement designs for a potential in situ thermal retorting operation located in the Uinta Basin. These simulations build on a proven methodology used previously to accurately model the time- and spatial-temperature history of an in situ test program since product quantity, quality, and distribution ultimately depend on accurately predicting the rate of heat transfer throughout the formation. Rather than focusing on a specific heat delivery system, these simulations capture the time- and spatial-temperature history of heat distribution throughout the study formation by including full stratigraphic information from coring logs to construct a model representative of the study area, including comprehensive implementation of thermal properties. These simulations are able to predict the detailed kerogen conversion at any location in the formation at any given time. High-performance computing was essential in performing this parametric study, given the six design parameters and the detailed properties implementation.

The overall energy ratio (EER) for each design is determined by computing the quantity, distribution, and quality of the recovered products with respect to the amount of heat supplied to the formation. It is shown that the well spacing and placement, the number of heating well rows, and the product recovery, distribution, and quality all play important roles in achieving a favorable energy ratio. As the product recovery and product quality increase, so does the energy ratio. The heating time also plays a crucial role, although increasing the heating time does not guarantee an increased energy ratio. At a certain point, as all kerogen in the vicinity of the wells is converted, the production decreases and additional heat input may not be warranted. Ultimately, these simulations show that given the appropriate well arrangement and product recovery, distribution, and quality, it is possible to achieve favorable energy returns for an in situ oil shale production process in the Uinta Basin.

10A Appendix: Parameter Values for Each of the 242 Simulation Designs

Design Number	$E_{h,7} * 10^6$ (MJ/m)	$m_{p,7} * 10^6$ (kg/m)	h (m)	v (m)	Θ (deg)	f_r (m)	r (m)	Number of Well Rows	Total Number of Wells	$EER_{n,7}$
1	10.8	2.3	8.1	11.6	30.0	109.0	0.1524	9	137	0.66
2	7.4	0.1	20.1	14.1	53.9	126.4	0.1270	10	54	0.02
3	13.5	3.1	6.6	19.7	14.0	67.5	0.1016	8	150	0.73
4	0.8	0.0	22.3	16.1	46.1	70.8	0.1143	1	5	0.00
5	8.6	1.6	5.9	12.5	18.1	129.9	0.1270	6	126	0.59
6	4.8	1.0	5.2	14.3	11.9	110.9	0.1143	3	72	0.64
7	4.9	0.8	8.7	6.4	1.3	138.3	0.1143	5	75	0.54
8	1.9	0.0	17.3	12.4	9.2	114.3	0.1143	2	14	0.02
9	4.3	1.3	9.2	5.1	60.1	88.3	0.1270	6	84	0.94
10	3.0	0.6	13.3	5.6	45.2	109.4	0.1143	4	37	0.59
11	4.0	0.6	18.8	6.9	35.5	58.6	0.1143	6	39	0.46
12	5.2	1.5	14.8	6.0	52.4	69.6	0.1270	8	67	0.89
13	5.9	0.0	23.9	18.7	7.5	99.5	0.1016	10	46	0.01
14	1.8	0.0	24.9	4.9	46.5	98.0	0.1397	4	20	0.06
15	4.9	0.8	5.4	5.4	10.3	141.3	0.1270	5	115	0.51
16	2.5	0.0	17.7	11.1	47.8	96.5	0.1016	4	28	0.01
17	2.5	0.2	22.4	4.7	20.0	115.3	0.1270	5	26	0.27
18	6.2	1.4	10.6	14.2	32.6	68.3	0.1016	5	60	0.72
19	4.4	0.7	7.6	4.3	60.3	137.2	0.1270	6	96	0.51
20	7.8	0.1	21.4	11.7	53.3	66.9	0.1397	10	58	0.05
21	9.1	2.3	8.9	11.4	14.2	54.0	0.1016	8	112	0.78
22	5.4	1.0	16.3	10.0	52.5	101.2	0.1143	7	52	0.55
23	0.9	0.0	21.9	4.3	36.8	82.0	0.1524	1	6	0.01
24	6.1	1.2	9.4	9.1	18.1	125.9	0.1270	6	78	0.62
25	3.6	0.3	10.8	7.4	7.7	53.5	0.1397	3	34	0.22
26	3.4	0.6	20.8	3.7	49.9	130.6	0.1397	7	41	0.52
27	4.2	0.7	7.1	10.0	3.5	139.3	0.1524	3	54	0.52
28	2.1	0.0	24.2	12.8	32.9	131.8	0.1397	3	15	0.02
29	6.0	0.1	18.4	11.6	55.9	82.2	0.1143	8	52	0.04
30	4.1	1.0	5.6	5.0	38.3	73.9	0.1016	5	111	0.75

(Continued)

Design Number	$E_{h,7} * 10^6$ (MJ/m)	$m_{p,7} * 10^6$ (kg/m)	h (m)	v (m)	Θ (deg)	f_r (m)	r (m)	Number of Well Rows	Total Number of Wells	$EER_{n,7}$
31	6.5	1.3	20.2	7.7	54.6	134.9	0.1397	9	56	0.60
32	1.9	0.1	19.5	2.8	62.5	127.4	0.1016	3	19	0.15
33	2.4	0.0	19.3	9.0	9.5	51.0	0.1143	3	18	0.04
34	5.3	1.6	12.4	7.4	36.4	79.8	0.1143	7	70	0.95
35	8.3	1.3	14.3	11.6	0.8	84.5	0.1270	9	81	0.48
36	6.2	1.3	12.6	11.2	11.5	120.5	0.1270	6	59	0.67
37	2.7	0.2	19.6	9.4	60.5	72.2	0.1397	4	25	0.21
38	2.4	0.0	15.2	19.7	23.0	144.4	0.1397	2	16	0.02
39	11.9	2.1	5.7	13.5	20.0	148.4	0.1143	8	174	0.55
40	2.8	0.0	12.9	24.6	6.1	64.6	0.1397	2	19	0.02
41	4.7	0.7	17.8	10.3	58.8	74.9	0.1397	6	42	0.47
42	1.5	0.0	18.2	7.0	44.1	112.3	0.1143	2	14	0.02
43	5.8	1.7	9.6	13.6	5.4	78.1	0.1016	5	65	0.92
44	3.9	0.0	9.9	21.8	3.7	128.0	0.1270	3	36	0.04
45	5.1	1.0	10.1	8.7	34.5	136.2	0.1524	5	61	0.58
46	4.1	0.7	22.7	6.1	55.1	124.4	0.1143	7	38	0.56
47	1.6	0.0	21.7	18.0	31.1	50.0	0.1270	2	11	0.01
48	4.6	0.8	13.1	4.8	34.7	145.6	0.1143	7	66	0.55
49	8.8	2.4	12.4	11.4	2.6	65.2	0.1016	9	90	0.84
50	2.6	0.3	6.3	4.8	48.8	149.2	0.1270	2	39	0.39
51	3.4	0.0	23.2	17.5	24.9	89.7	0.1016	5	27	0.01
52	2.0	0.3	13.4	7.3	43.3	113.3	0.1016	2	18	0.39
53	9.2	0.2	11.1	20.0	8.1	87.7	0.1270	7	78	0.06
54	5.7	0.6	19.1	9.6	47.4	116.4	0.1397	7	45	0.33
55	7.6	0.1	18.6	14.2	34.0	63.6	0.1270	9	59	0.02
56	8.2	2.0	12.0	9.8	40.0	55.5	0.1397	8	83	0.75
57	7.1	0.9	16.5	11.8	17.2	90.6	0.1397	8	60	0.39
58	11.8	3.0	6.7	11.0	15.3	71.5	0.1016	10	184	0.80
59	5.0	0.0	23.7	20.2	19.4	118.7	0.1143	8	37	0.01
60	2.1	0.2	11.8	4.9	62.1	147.0	0.1270	2	21	0.29
61	4.3	0.4	8.2	5.1	41.0	60.3	0.1143	4	60	0.27
62	3.6	0.6	13.7	7.7	59.2	135.1	0.1397	4	37	0.54
63	4.1	1.2	9.2	15.4	4.4	75.7	0.1270	3	40	0.94
64	1.2	0.0	14.6	8.1	57.5	146.5	0.1524	1	8	0.02
65	7.9	0.4	13.9	14.4	22.4	105.7	0.1397	7	63	0.16
66	4.8	0.9	14.1	10.2	24.3	132.1	0.1397	5	44	0.60

(Continued)

Design Number	$E_{h,7} * 10^6$ (MJ/m)	$m_{p,7} * 10^6$ (kg/m)	h (m)	v (m)	Θ (deg)	f_r (m)	r (m)	Number of Well Rows	Total Number of Wells	$EER_{n,7}$
67	7.2	0.3	20.9	10.4	59.4	143.1	0.1397	9	53	0.15
68	3.3	0.2	16.9	8.4	46.8	52.3	0.1397	4	28	0.20
69	7.5	1.6	7.4	8.0	29.7	57.0	0.1524	7	117	0.68
70	4.6	1.1	10.3	12.1	2.5	104.3	0.1143	4	48	0.76
71	1.7	0.4	10.8	6.0	39.4	91.3	0.1016	2	23	0.70
72	7.9	0.1	17.0	17.3	10.3	95.8	0.1524	8	57	0.02
73	6.8	0.0	21.2	13.7	28.9	76.8	0.1270	9	54	0.02
74	4.6	0.7	15.7	7.1	63.4	56.8	0.1397	6	48	0.50
75	4.8	0.7	20.4	3.4	62.8	59.0	0.1143	10	61	0.47
76	3.9	1.1	15.3	6.5	56.7	93.8	0.1524	6	48	0.87
77	3.2	1.0	15.5	5.2	17.3	77.3	0.1270	6	48	0.95
78	4.6	1.3	6.1	6.4	12.4	94.9	0.1397	5	103	0.87
79	7.9	0.0	17.5	15.8	44.6	107.3	0.1143	9	64	0.02
80	4.3	0.0	23.3	21.6	15.3	142.7	0.1270	9	28	0.01
81	2.3	0.0	21.5	11.1	11.4	86.1	0.1270	4	24	0.03
82	0.6	0.0	24.6	9.2	42.5	106.4	0.1397	1	5	0.02
83	3.0	0.0	15.8	15.5	21.8	84.0	0.1524	3	24	0.04
84	4.7	0.0	23.5	12.7	58.4	119.2	0.1270	7	36	0.02
85	12.0	0.8	8.0	24.2	2.2	61.8	0.1143	7	108	0.22
86	6.2	1.0	19.9	8.5	61.3	111.2	0.1270	9	56	0.52
87	2.4	0.1	23.0	6.0	62.5	123.8	0.1143	4	20	0.08
88	6.7	1.8	6.8	21.1	9.9	80.2	0.1397	4	73	0.85
89	10.3	2.6	8.5	12.8	29.3	85.6	0.1143	9	129	0.80
90	3.4	0.6	7.7	19.3	6.9	122.7	0.1270	2	33	0.51
91	2.1	0.0	16.1	11.1	15.9	98.5	0.1143	3	23	0.03
92	2.8	0.2	18.5	11.0	16.5	103.5	0.1397	3	21	0.23
93	2.5	0.8	11.7	8.7	27.8	92.6	0.1270	3	32	0.95
94	2.6	0.6	24.5	4.4	60.7	100.1	0.1524	6	30	0.73
95	2.1	0.0	14.9	7.2	26.9	133.3	0.1143	2	17	0.07
96	5.5	0.7	22.1	8.2	55.5	117.3	0.1524	8	44	0.43
97	7.0	0.0	24.4	21.6	14.4	62.6	0.1524	10	45	0.01
98	4.4	0.8	11.5	3.8	56.3	121.2	0.1397	7	77	0.57
99	4.5	1.1	11.2	9.4	22.7	102.8	0.1143	5	55	0.77
100	1.2	0.0	16.7	11.3	51.3	140.4	0.1524	1	8	0.01
101	4.4	1.3	14.0	5.5	45.6	92.2	0.1397	7	62	0.89
102	4.7	0.0	21.7	14.3	11.8	129.9	0.1397	6	34	0.02

(Continued)

Design Number	$E_{h,7} * 10^6$ (MJ/m)	$m_{p,7} * 10^6$ (kg/m)	h (m)	v (m)	Θ (deg)	f_r (m)	r (m)	Number of Well Rows	Total Number of Wells	$EER_{n,7}$
103	11.7	2.8	5.3	9.6	9.2	68.2	0.1397	10	232	0.75
104	12.4	3.1	6.7	21.7	5.9	66.8	0.1143	7	129	0.79
105	5.1	0.0	21.0	17.8	42.0	116.7	0.1270	6	36	0.01
106	5.6	0.0	15.1	19.9	11.0	106.5	0.1524	5	40	0.03
107	2.0	0.0	19.9	15.7	20.6	65.5	0.1143	2	13	0.01
108	8.9	1.6	12.6	12.5	44.5	145.9	0.1524	8	75	0.57
109	1.6	0.0	12.4	6.7	47.2	87.0	0.1397	2	20	0.05
110	1.2	0.0	13.4	10.6	48.9	145.3	0.1397	1	9	0.01
111	5.1	0.0	24.8	16.2	29.5	142.3	0.1143	7	35	0.01
112	10.5	2.6	7.0	16.0	1.7	63.2	0.1397	7	122	0.78
113	3.0	0.0	21.3	10.2	48.7	134.1	0.1397	4	22	0.03
114	5.4	1.7	21.3	5.7	59.6	79.9	0.1143	10	60	0.97
115	1.3	0.0	23.1	9.4	60.5	137.1	0.1270	2	10	0.01
116	12.5	2.7	7.6	19.2	4.4	85.9	0.1143	8	130	0.68
117	5.1	0.9	8.5	3.8	52.4	149.9	0.1270	8	116	0.53
118	1.9	0.5	14.1	6.1	59.2	95.2	0.1143	3	26	0.81
119	3.4	0.5	9.2	7.4	50.0	149.4	0.1270	3	42	0.48
120	2.0	0.0	19.3	17.1	46.3	55.2	0.1397	2	12	0.01
121	3.5	0.6	12.2	6.4	42.5	128.8	0.1143	4	42	0.53
122	3.7	0.4	23.2	6.7	37.2	57.8	0.1143	6	32	0.37
123	3.3	0.5	24.5	3.6	63.4	127.1	0.1524	7	35	0.50
124	4.0	0.5	12.1	15.8	23.8	112.5	0.1524	3	31	0.39
125	1.7	0.0	15.4	5.0	55.9	108.2	0.1143	2	16	0.06
126	2.1	0.1	14.6	7.9	49.5	144.5	0.1397	2	17	0.19
127	3.9	0.0	23.8	11.4	36.0	78.0	0.1397	6	32	0.02
128	2.9	0.0	22.0	10.9	55.3	71.8	0.1397	4	23	0.03
129	11.6	2.3	7.4	18.9	10.1	118.4	0.1397	10	116	0.62
130	10.2	2.0	10.2	13.9	7.4	143.2	0.1143	10	98	0.60
131	3.6	0.7	9.9	14.1	24.6	97.7	0.1270	3	38	0.57
132	6.3	0.1	13.9	19.3	12.4	123.0	0.1524	5	45	0.03
133	6.4	0.1	13.5	17.2	2.9	83.1	0.1270	6	57	0.03
134	4.1	0.8	18.3	7.3	61.8	117.0	0.1016	6	40	0.59
135	2.0	0.0	18.8	16.0	13.3	133.1	0.1397	2	14	0.02
136	5.3	0.0	19.7	19.7	21.6	144.4	0.1016	6	37	0.01
137	3.5	0.6	10.0	10.5	42.7	67.3	0.1143	3	36	0.54
138	4.8	1.3	13.1	11.8	28.4	82.7	0.1270	5	47	0.84

(Continued)

Design Number	$E_{h,7} * 10^6$ (MJ/m)	$m_{p,7} * 10^6$ (kg/m)	h (m)	v (m)	Θ (deg)	f_r (m)	r (m)	Number of Well Rows	Total Number of Wells	$EER_{n,7}$
139	2.6	0.0	20.8	8.2	53.7	56.9	0.1143	3	18	0.03
140	2.6	0.1	5.1	9.4	8.1	65.6	0.1143	1	25	0.11
141	10.2	1.9	11.5	11.7	4.7	148.1	0.1270	9	97	0.59
142	9.1	1.7	7.3	9.5	10.2	129.4	0.1270	8	135	0.58
143	2.0	0.0	17.4	14.0	31.2	50.0	0.1143	2	15	0.01
144	0.7	0.0	22.2	12.4	25.8	139.5	0.1397	1	5	0.01
145	5.3	1.2	18.8	5.7	58.2	105.3	0.1270	9	59	0.72
146	11.2	1.7	8.8	20.8	0.9	110.4	0.1270	9	98	0.48
147	2.7	0.1	14.3	10.5	1.3	74.8	0.1397	3	27	0.07
148	5.8	1.2	14.7	7.5	61.5	116.4	0.1397	8	64	0.65
149	2.2	0.0	20.1	11.7	3.6	50.7	0.1016	3	18	0.01
150	1.7	0.0	22.8	7.8	54.2	103.4	0.1397	3	16	0.03
151	1.4	0.0	12.3	6.6	61.7	112.6	0.1524	1	10	0.03
152	6.6	0.3	10.8	19.7	28.6	76.4	0.1397	5	55	0.14
153	5.7	0.1	11.9	20.4	9.1	54.9	0.1524	4	42	0.07
154	6.2	1.2	14.5	5.3	40.2	123.7	0.1397	10	85	0.62
155	8.0	0.1	18.5	14.8	29.1	131.3	0.1270	9	55	0.03
156	5.6	0.0	22.0	14.9	41.1	120.7	0.1143	7	40	0.01
157	4.6	0.5	8.2	19.0	10.7	60.9	0.1270	3	46	0.37
158	3.1	0.1	20.5	5.1	52.7	51.9	0.1143	4	26	0.12
159	4.0	0.7	18.1	4.4	59.8	140.2	0.1143	7	48	0.52
160	2.8	0.5	8.1	13.2	31.1	123.4	0.1143	2	30	0.51
161	5.8	0.0	19.2	19.4	35.2	73.2	0.1270	7	45	0.01
162	3.0	0.5	10.8	6.5	14.1	134.6	0.1397	3	35	0.47
163	10.5	2.6	9.4	10.5	5.7	95.5	0.1143	10	132	0.79
164	4.4	0.3	19.9	8.5	62.2	101.2	0.1016	7	43	0.25
165	4.3	0.1	13.3	15.5	40.5	53.7	0.1524	4	36	0.08
166	2.5	0.3	18.1	3.7	59.0	72.0	0.1397	4	26	0.41
167	13.8	3.4	6.2	15.6	16.6	78.0	0.1524	9	182	0.78
168	7.3	0.0	19.7	16.9	38.7	109.1	0.1524	8	50	0.02
169	6.1	1.7	6.0	10.2	5.2	75.2	0.1016	5	104	0.87
170	2.4	0.1	16.9	8.3	1.9	137.8	0.1143	3	22	0.09
171	2.3	0.1	24.2	5.5	46.6	70.3	0.1270	4	20	0.20
172	6.8	1.0	16.1	11.1	55.0	139.8	0.1270	8	56	0.47
173	1.4	0.0	8.2	19.2	19.8	103.0	0.1397	1	16	0.04
174	3.3	0.6	21.2	2.6	61.3	120.1	0.1397	8	46	0.53

(Continued)

Design Number	$E_{h,7} * 10^6$ (MJ/m)	$m_{p,7} * 10^6$ (kg/m)	h (m)	v (m)	Θ (deg)	f_r (m)	r (m)	Number of Well Rows	Total Number of Wells	$EER_{n,7}$
175	2.3	0.1	16.5	9.4	41.6	80.1	0.1397	3	23	0.08
176	9.0	2.2	6.4	11.8	23.1	104.2	0.1143	7	134	0.75
177	4.5	0.1	10.4	21.3	5.4	138.3	0.1270	3	36	0.06
178	7.9	0.0	14.4	24.0	6.9	82.4	0.1397	7	59	0.02
179	4.1	0.3	23.6	8.0	62.2	136.8	0.1397	6	32	0.26
180	8.9	2.3	7.5	10.0	18.9	72.5	0.1270	8	132	0.83
181	9.1	0.1	13.6	22.3	19.3	93.9	0.1143	8	72	0.02
182	6.3	1.7	17.0	8.0	47.6	69.3	0.1524	9	64	0.86
183	4.5	0.9	17.1	7.8	53.6	136.0	0.1143	6	43	0.60
184	2.7	0.0	19.2	10.7	60.7	83.8	0.1016	4	24	0.05
185	1.7	0.0	22.8	3.4	63.3	58.8	0.1016	2	11	0.01
186	3.7	0.5	6.3	12.6	10.9	67.7	0.1270	2	40	0.41
187	12.5	3.1	7.1	14.5	14.9	58.2	0.1016	9	157	0.77
188	6.8	2.0	9.2	9.9	37.3	77.4	0.1270	7	95	0.92
189	4.1	0.8	17.6	6.4	54.5	119.6	0.1524	6	42	0.61
190	6.0	1.2	15.4	6.5	1.3	114.9	0.1270	9	72	0.64
191	1.5	0.0	17.9	3.6	52.2	117.9	0.1397	2	14	0.04
192	2.6	0.1	24.9	6.1	15.3	135.9	0.1143	5	25	0.15
193	9.7	0.3	14.7	13.5	18.0	113.0	0.1143	9	76	0.11
194	5.4	0.6	9.6	4.5	60.6	51.3	0.1524	7	91	0.35
195	7.9	2.0	10.7	13.5	32.2	87.7	0.1270	7	81	0.78
196	1.6	0.0	21.7	7.8	45.9	69.8	0.1397	2	11	0.01
197	6.4	1.8	11.1	11.3	10.3	91.5	0.1524	6	67	0.88
198	5.4	0.0	16.7	15.9	32.0	74.1	0.1270	6	45	0.02
199	6.9	1.6	12.9	7.3	51.6	106.8	0.1016	9	88	0.71
200	5.5	0.1	17.2	12.7	7.1	133.5	0.1524	6	42	0.06
201	0.7	0.0	24.1	4.2	63.1	104.8	0.1270	1	5	0.01
202	4.5	0.0	22.4	22.3	21.3	138.9	0.1524	8	28	0.01
203	4.4	1.4	12.6	10.1	18.4	86.9	0.1524	5	49	0.97
204	5.5	1.0	15.9	4.5	31.8	52.4	0.1397	10	78	0.55
205	3.2	0.5	24.3	2.5	60.2	147.6	0.1270	8	41	0.45
206	4.9	1.0	6.7	8.9	12.1	113.9	0.1143	4	75	0.63
207	10.3	2.1	8.6	18.8	23.4	132.9	0.1270	7	98	0.64
208	3.4	0.1	18.6	9.7	1.6	90.2	0.1270	5	35	0.06
209	3.0	0.0	24.1	7.5	58.6	142.5	0.1270	5	25	0.05
210	1.7	0.0	17.5	11.4	15.0	132.0	0.1016	2	14	0.01

(Continued)

Design Number	$E_{h,7} * 10^6$ (MJ/m)	$m_{p,7} * 10^6$ (kg/m)	h (m)	v (m)	Θ (deg)	f_r (m)	r (m)	Number of Well Rows	Total Number of Wells	$EER_{n,7}$
211	3.3	1.1	12.8	8.1	21.9	82.0	0.1143	4	39	1.00
212	4.6	1.3	5.4	9.5	20.2	92.8	0.1270	4	92	0.89
213	3.4	0.1	13.7	11.1	21.8	76.7	0.1143	4	36	0.05
214	10.3	2.7	5.1	11.4	16.1	78.6	0.1143	8	194	0.81
215	7.0	0.8	16.3	10.2	19.5	146.8	0.1016	8	61	0.37
216	3.4	1.0	9.3	2.8	56.9	91.8	0.1524	7	92	0.93
217	0.6	0.0	25.0	16.2	48.3	88.5	0.1524	1	5	0.02
218	2.0	0.5	13.2	5.8	53.4	97.0	0.1016	3	28	0.81
219	7.3	1.7	5.7	12.5	11.6	59.9	0.1397	5	109	0.71
220	2.5	0.5	5.6	5.0	26.9	105.7	0.1270	2	44	0.61
221	7.8	2.1	9.0	11.4	12.4	64.9	0.1143	7	97	0.83
222	6.8	0.0	19.0	17.8	0.1	141.2	0.1270	9	49	0.02
223	1.6	0.0	16.0	6.2	62.8	122.1	0.1270	2	15	0.04
224	6.7	1.6	5.7	5.4	42.9	102.0	0.1143	9	196	0.74
225	5.6	1.7	15.6	6.0	57.0	75.7	0.1270	9	72	0.94
226	2.7	0.0	21.4	22.2	27.2	94.8	0.1016	4	22	0.01
227	1.1	0.0	23.3	15.0	25.4	61.9	0.1397	1	6	0.00
228	9.4	1.2	16.4	11.1	39.1	121.1	0.1397	10	76	0.40
229	5.6	1.1	16.6	5.8	51.1	125.5	0.1143	9	67	0.62
230	12.4	2.5	11.7	14.5	16.0	115.5	0.1397	10	106	0.64
231	2.4	0.0	13.9	11.4	13.6	86.2	0.1397	3	27	0.06
232	2.5	0.8	20.3	5.2	60.1	85.1	0.1270	5	30	0.96
233	1.3	0.0	12.0	3.7	60.4	110.8	0.1270	1	11	0.02
234	1.9	0.1	18.0	3.2	55.6	112.0	0.1270	3	20	0.21
235	5.2	0.9	14.9	11.6	18.3	55.7	0.1524	5	42	0.52
236	6.5	1.4	17.4	8.8	37.7	98.8	0.1143	9	64	0.66
237	2.9	0.4	13.0	6.2	26.7	68.8	0.1397	3	28	0.38
238	8.7	2.3	8.3	10.7	18.3	91.0	0.1143	8	119	0.84
239	9.1	2.3	8.6	13.1	14.0	62.4	0.1143	7	101	0.79
240	0.5	0.0	23.0	7.0	43.5	94.3	0.1270	1	5	0.01
241	0.9	0.0	15.8	4.9	62.4	88.2	0.1397	1	8	0.02
242	2.0	0.0	22.3	12.2	16.8	89.2	0.1270	3	16	0.02

References

Al-Ayed, O. S., M. Matouq, Z. Anbar, A. M. Khaleela, and E. Abu-Nameh. 2010. Oil shale pyrolysis kinetics and variable activation energy principle. *Appl. Energy* 87:1269–1272.

Arbabi, S., W. Deeg, and M. Lin. 2010. Simulation model for ground freezing process: Application to Shell's freeze wall containment system. Paper presented at the *30th Oil Shale Symposium*, Golden, CO.

Atwood, M. T. 2006. Integration of large scale retorting operations with laboratory testing and analyses. Paper presented at the *26th Oil Shale Symposium*, Golden, CO.

Bauman, J. H., C. K. Huang, M. R. Gani, and M. D. Deo. 2009. Modeling of the in-situ production of oil from oil shale. In *Oil Shale: A Solution to the Liquid Fuel Dilemma*, eds. O. I. Ogunsola, A. M. Hartstein, and O. Ogunsola, pp. 135–146. Washington, DC: ACS Symposium Series.

Biglarbigi, K., H. Mohan, P. Crawford, and H. Mohan. 2008. Economics, barriers, and risks of oil shale development in the United States. Paper presented at the *28th USAEE/IAEE North American Conference*, New Orleans, LA.

Birdwell, J. E., T. J. Mercier, R. C. Johnson, and M. E. Brownfield. 2013. *In-Place Oil Shale Resources Examined by Grade in the Major Basins of the Green River Formation, Colorado, Utah, and Wyoming.* Reston, VA U.S. Geological Survey: Fact Sheet 2012-3145.

Brandt, A. R. 2008. Converting oil shale to liquid fuels: energy inputs and greenhouse gas emissions of the Shell in situ conversion process. *Environ. Sci. Technol.* 42(19):7489–7495.

Burnham, A. K., R. L. Day, and P. H. Wallman. 2008. Overview of American Shale Oil LLC progress and plans. Paper presented at the *28th Oil Shale Symposium*, Golden, CO.

Center for High Performance Computing, University of Utah. 2015. Ash cluster. https://www.chpc.utah.edu/resources/HPC_Clusters.php (accessed October 27, 2015).

CD-adapco. 2015. cd-adapco.com (accessed April 4, 2016).

Cook, E. W. 1974. Green River shale-oil yields: Correlation with elemental analysis. *Fuel* 53:16–20.

Eseme, E., J. L. Urai, B. M. Krooss, and R. Littke. 2007. Review of mechanical properties of oil shales: Implications for exploitation and basin modelling [sic]. *Oil Shale* 24:159–174.

Fan, Y., L. J. Durfolsky, and H. A. Tchelepi. 2010. Numerical simulation of the in-situ upgrading of oil shale. *SPE J.* 15:368–381.

Fox, J. P. 1983. Leaching of oil shale solid wastes: A critical review. Report No. LBL-16496. Berkeley, CA: Lawrence Berkeley Laboratory, University of California.

Hazra, K. G. 2014. Comparison of heating methods for in-situ oil shale extraction. MS thesis, College Station, TX: Texas A&M University.

Hradisky, M., P. J. Smith and J. P. Spinti. 2014. Evaluation of well spacing and arrangement for in-situ thermal treatment of oil shale using HPC simulation tools. Paper presented at the *34th Oil Shale Symposium*, Golden, CO. http://content.lib.utah.edu/cdm/ref/collection/ir-eua/id/3612 (accessed April 4, 2016).

Hubbard, A. B. and W. E. Robinson. 1950. A Thermal Decomposition Study of Colorado Oil Shale. Report of Investigations: 4744. Washington, DC: United States Department of the Interior, Bureau of Mines.

Kelkar, S., R. Pawar, and N. Hoda. 2011. Numerical simulation of coupled thermal-hydrological-mechanical-chemical processes during in situ conversion and production of oil shale. Paper presented at the *31st Oil Shale Symposium*, Golden, CO.

Leavitt, D. R., A. L. Tyler, and A. S. Kafesjian. 1987. Kerogen decomposition kinetics of selected Green River and eastern U.S. oil shales from thermal solution experiments. *Energy Fuels* 1:520–525.

Le Doan, T. V., N. W. Bostrom, A. K. Burnham, R. L. Kleinberg, A. E. Pomerantz, and P. Allix. 2013. Green River oil shale pyrolysis: Semi-open conditions. *Energy Fuels* 27:6447–6459.

Shaw, R. J. 1947. Specific Heat of Colorado Oil Shales. Report of Investigations: 4151. Washington, DC: United States Department of the Interior, Bureau of Mines.

Shih, S. M. and H. Y. Sohn. 1978. A mathematical model for the retorting of a large block of oil shale: effect of the internal temperature gradient. *Fuel* 57:622–630.

Smith, J. W. 1969. *Theoretical Relationship between Density and Oil Yield for Oil Shales.* Report of Investigations: 7248. Washington, DC: United States Department of the Interior, Bureau of Mines.

Symington, W. A., D. L. Olgaard, G. A. Otten, T. C. Phillips, M. Thomas, and J. D. Yeakel. 2006. ExxonMobil's Electrofrac process for in situ oil shale conversion. Paper presented at the *26th Oil Shale Symposium*, Golden, CO.

Tihen, S. S., H. C. Carpenter, and H. W. Sohns. 1967. Thermal conductivity and thermal diffusivity of Green River oil shale. In *Thermal Conductivity, Proceedings of the Seventh Conference*, Gaithersburg, MD, 302, pp. 529–535.

U.S. Bureau of Land Management (BLM). 2006. Oil shale research, development and demonstration project—Plan of operations. San Ramon, CA: Chevron U.S.A., Inc. http://www.blm.gov/style/medialib/blm/co/field_offices/white_river_field/oil_shale.Par.37256.File.dat/OILSHALEPLANOFOPERATIONS.pdf (accessed October 27, 2015).

U.S. Bureau of Land Management (BLM). 2011. Natural Soda Holdings, Inc. Plan of operation. http://www.blm.gov/style/medialib/blm/co/information/nepa/white_river_field/completed_2012_documents.Par.11600.File.dat/NSHI%20BLM%202012%20POO%2004-24.pdf (accessed October 27, 2015).

Vanden Berg, M. D. 2008. *Basin-Wide Evaluation of the Uppermost Green River Formation's Oil-Shale Resource, Uinta Basin, Utah and Colorado.* Special Study 128. Salt Lake City, UT: Utah Geological Survey.

Vanden Berg, M. D., J. R. Dyni, and D. E. Tabet. 2006. Utah oil shale database: UGS Open File Report 469, CD-ROM.

Vanden Berg, M. D., D. R. Lehle, S. M. Carney, and C. D. Morgan. 2013. *Geological Characterization Of The Birds Nest Aquifer, Uinta Basin, Utah.* Special Study 128. Salt Lake City, UT: Utah Geological Survey.

Wallman, P. H. and A. K. Burnham. 2008. Transport model of the AMSO (formerly EGL) oil shale process. Paper presented at the *28th Oil Shale Symposium*, Golden, CO.

Wang, G. G. W. 2003. Adaptive response surface method using inherited Latin hypercube design points. *J. Mech. Des.* 125:210–220.

Wellington, S. L., I. E. Bercenko, E. P. de Rouffignac et al. 2005. In situ thermal processing of an oil shale formation to produce a desired product. U.S. Patent US6880633 B2.

11

Economic Analysis of In Situ Oil Shale Development in the Uinta Basin

Jonathan E. Wilkey, Jennifer P. Spinti, and Terry A. Ring

CONTENTS

This chapter attempts to answer the "how much will it cost" question for every in situ heating scenario discussed in Chapter 10 by performing

1. A discounted cash flow analysis to calculate the supply price (i.e., product price at which sales will generate a specified profit margin) for the produced oil.

2. A design of experiments (DOE) analysis to determine both the range of supply prices that can result from changes in the model input parameters (number of wells drilled, capital and operating cost ranges for the heating system, etc.) and the relative importance of each input parameter to the overall supply price.

All costs in this chapter are in 2014 U.S. dollars (USD).

11.1 Introduction

The production of oil from oil shale in the western United States was first commercially attempted in the mid-1910s (EPA Oil Shale Work Group 1979). In the roughly 100 years since, oil shale has never made the jump from being a "potential" to "proven" source of oil, primarily because no one has demonstrated the economic viability of oil shale relative to other production methods for conventional oil. Consequently, one of the key questions to consider in assessing any oil shale production technology or process is "how much will it cost." Unfortunately, it is difficult to answer this question with any certainty precisely because oil shale is unproven, and as a result, the values of important input parameters and costs are unknown. The reports and studies (Bartis et al. 2005, Bezdek et al. 2006, ICSE 2013, INTEK Inc. 2009, STRAAM Engineers 1979) that have analyzed the costs of developing oil shale do not fully consider the impact of varying their input parameters or costing assumptions. However, as will be discussed in this chapter, much more can be learned about the potential cost range and optimal design of oil shale technologies by rigorously exploring this parameter space.

11.2 Process Description

The in situ production process analyzed here is assumed to be located near Coyote Wash and to have a heating period of 7 years. Coyote Wash is the location of one of the oil shale cores described in Chapter 4 and the location used for the in situ retort simulations discussed in Chapter 10. The major processing steps considered in this cost analysis are shown in Figure 11.1 and discussed in this section. Some of the numbers and values given in the process descriptions are varied as part of the DOE analysis; see Section 11.3.2. These steps only cover the extraction of oil from in situ oil shale. Since oils produced

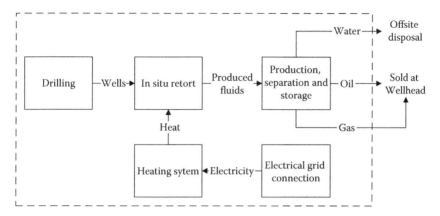

FIGURE 11.1
Process flow diagram for in situ oil shale production. Blocks represent major process steps and pieces of equipment; arrows represent the flow of process inputs/outputs between them. The dotted line indicates the system boundaries. Note that oil and gas are sold as-is at the wellhead and do not include any upgrading or transportation costs.

via in situ retorting will most likely require upgrading and will definitely require transportation to market (neither of which is included in this cost analysis), any oil sold by the process shown in Figure 11.1 would sell at a discount compared to benchmark crudes such as West Texas Intermediate.

11.2.1 Drilling

The first step of the process is drilling the wells for heating and producing the oil shale deposit. Two types of wells are drilled: heating wells and producer wells. Heating wells contain the electrical resistance heaters used to heat the formation (see Section 11.2.2). Producer wells collect retorted oil and gas from the formation. The ratio of heating to production wells is assumed to be 12:1 (Wellington et al. 2003). Both well types require drilling, defined as drilling the actual wellbore and placing and cementing the drill casing. Producer wells are assumed to also require completion, defined as perforating and hydraulically fracturing. As described in Chapter 10, all wells are drilled horizontally.

In order to estimate well parameters (well lengths, costs, drilling times, etc.), the Utah Division of Oil, Gas, and Mining's (DOGM) online database was searched for horizontal wells drilled in the state of Utah in the last 5 years. A total of 132 horizontal wells was identified; these wells form the well sample dataset used in this study (Utah DOGM 2015). Key details about well geometry, drilling and completion time, and capital costs are discussed next.

11.2.1.1 Well Geometry

Based on the geology of the oil shale deposit at Coyote Wash, the middle of the oil shale deposit is assumed to be located at a vertical depth of 2500 ft. All wells, regardless of type, are assumed to have the same geometry: a vertical section that runs from the surface to a depth of 1912 ft, a 90° turn segment that is 924 ft long (made by deflecting each individual 30 ft pipe segment in the turn by 3° from the previous pipe segment), and a lateral segment for either heating or production (depending on well type) that has a variable length. The total length of each well (the sum of the vertical, turn, and lateral segments) is estimated from a log-normal distribution fitted to the total lengths of all wells in the well sample dataset; see Figure 11.2. The length of the lateral (e.g., heating) segment is found by subtracting the lengths of the vertical and turn segments from the total well length. All other aspects of well geometry, such as the number of wells drilled, well spacing, and

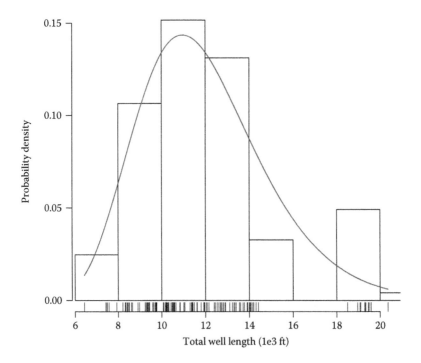

FIGURE 11.2
Log-normal distribution fit of total well lengths from well sample dataset. Histogram and tick marks on *x*-axis show the original data points, while the red line indicates the best log-normal probability distribution function (PDF) fit. Fit parameters: log–mean=9.363, log–standard deviation=0.246. (From DOGM, Data Research Center, http://oilgas.ogm.utah.gov/Data_Center/DataCenter.cfm, 2015.)

vertical offset, are defined by the in situ retorting simulations described in Chapter 10.

11.2.1.2 Well Drilling and Completion Time

All of the wells required for each in situ retorting scenario must be drilled prior to heating. The two factors that determine how long this takes are (1) how many drilling rigs are available (i.e., how many wells can be drilled simultaneously) and (2) how long it takes to drill each well. Multiple drilling rigs are required to drill the wells in a timely fashion. The number of rigs available is assumed to be 14, which is half of Utah's annual average rig count between 2008 and 2013 (Baker Hughes 2015). The length of time it takes to drill each well is estimated based on the elapsed time between the spud dry date (the start date of a well's drilling activity) and the date that the target depth was reached (when drilling activity ceases) as reported in the Utah DOGM database for all wells in the sample dataset. Producers and heaters are assumed to take the same amount of time to drill. While this assumption ignores the additional time that must be spent on completion for producer wells, the DOGM dataset includes time delays during drilling activity such as waiting for a drilling rig to arrive on site that would not likely occur with an in situ oil shale drilling project. Based on the available data, the drilling time is best fit using a log-normal distribution (see Figure 11.3), from which drilling times can be varied as part of the DOE analysis.

11.2.1.3 Well Drilling and Completion Capital Costs

Drilling costs typically increase exponentially with total well length and completion costs with the length of the lateral segment. Unfortunately, cost data were too sparsely reported in the well sample dataset (15 wells had drilling costs, 39 had completion costs) to find a good fit for well costs versus length using regression techniques. As a result, the drilling costs were fitted using a normal distribution (Figure 11.4) and completion costs were fitted using a log-normal distribution (Figure 11.5). The costs for each well were then drawn from these distributions in the DOE analysis.

11.2.2 Heating System

Numerous heating systems have been proposed for in situ oil shale production, ranging from downhole-fired heaters to fuel cells to microwave systems (INTEK Inc. 2011). However, the only in situ heating systems that are commercially available at present are electrical resistance heating systems. Therefore, electrical resistance heaters are used as the heat source in this analysis. The heating system consists of a heat tracing line that converts

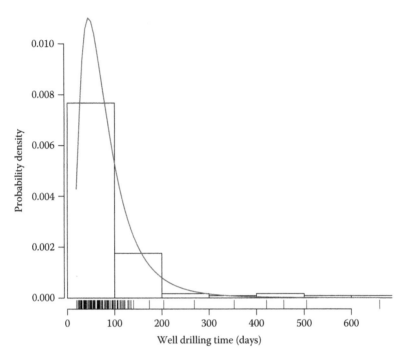

FIGURE 11.3

Log-normal distribution fit of well drilling time from well sample dataset. Fit parameters: log–mean=4.238, log–standard deviation=0.641. (From Utah DOGM), Data Research Center, http://oilgas.ogm.utah.gov/Data_Center/DataCenter.cfm, 2015.)

electricity into heat in the lateral segment of each heating well. The oil shale deposit is then heated through the wall of the drill casing as described in Chapter 10. The capital cost of the heat tracing system (~$50/ft) is based on a case study using mineral-insulated, electric heat tracing lines in California heavy oil wells (McQueen et al. 2009).

11.2.3 Electrical Grid

The electrical power demand for each scenario as a function of time is obtained from the output of the in situ retorting simulations. Assuming a lateral well length for all scenarios of 8500 ft, which is the median lateral well length in the well sample dataset, these power demand curves are plotted in Figure 11.6. Electrical power is purchased from Utah's electrical grid, requiring power lines and a substation. The electrical line is assumed to cost $959,700 USD per mile, equivalent to a 230 kV, 400 MW, single-circuit transmission line, and the substation to cost 10% of the total cost of the electrical line (Black & Veatch 2014). Since the exact length of the shortest route to a suitable grid connection point is unknown, the length of the electrical line is

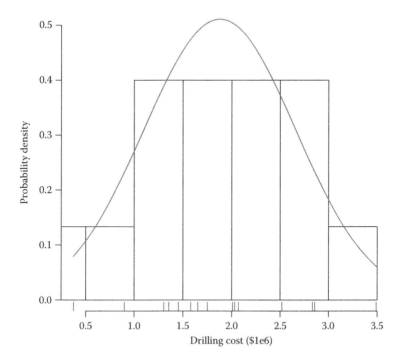

FIGURE 11.4

Normal distribution fit of drilling costs from well sample dataset. Fit parameters: mean = 1.88e6, standard deviation = 0.78e6. (From Utah DOGM), Data Research Center. http://oilgas.ogm. utah.gov/Data_Center/DataCenter.cfm, 2015.)

assumed to be 50 miles, which is roughly 50% longer than the straight-line path from Coyote Wash to Vernal, Utah, the largest city in the Uinta Basin. Based on these assumptions, the capital cost for the electrical grid connection is $53 million USD.

Alternative designs involving the construction of an electrical plant onsite were considered, but given the electrical heating demands of the in situ retorting scenarios, there was no economic justification for building onsite generators. For example, consider the median energy demand curve in Figure 11.6, which has an initial demand of 6.8 GWh/day (283 MW). Demand drops dramatically to approximately 2.8 GWh/day (117 MW) after the first month and then gradually tapers off to less than 1.0 GWh/day (42 MW) after about 1216 days. Given the median curve's cumulative energy requirements (2933 GWh), the total electricity cost is $178 million. In comparison, if a 100 MW, natural-gas, combined-cycle power plant were built onsite, the capital cost would be $196 million, assuming Williams six-tenth's scaling (Williams 1947) and Energy Information Administration (EIA) power plant cost data (U.S. EIA 2013). The electricity purchase option is the economic winner, even before considering other factors such as (1) the time value of

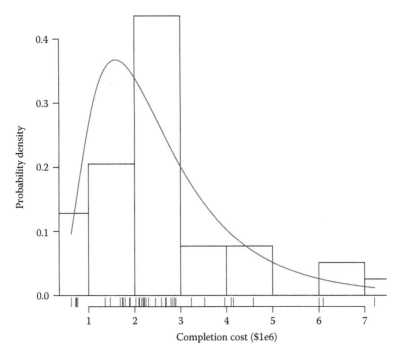

FIGURE 11.5

Log-normal distribution fit of completion costs from well sample dataset. Fit param-eters: log–mean = 14.614, log–standard deviation = 0.577. (From Utah Division of Oil, Gas, and Mining (DOGM), Data research center, http://oilgas.ogm.utah.gov/Data_Center/DataCenter.cfm, 2015.)

money (upfront capital cost versus ongoing future expense spread out over the 7 years of heating/production), (2) the cost of building a pipeline and purchasing fuel to supply the plant before sufficient produced gas is avail-able, and (3) the gap between initial demand and the 283 MW capacity that must still be purchased for the first 3 years. In reality, the more pressing issue is likely whether or not the grid would have the capacity to supply the energy demand (especially the initial demand). As a point of comparison, Utah's annual average power output is 4.9 GW (U.S. EIA 2015d). Therefore, the median initial demand (at median lateral length) is 5% of the state's power output.

11.2.4 In Situ Retort

The simulated in situ retort was described in detail in Chapter 10. It is in the retort that the oil shale kerogen is pyrolyzed as prescribed by the local temperature field and the kerogen kinetics. In all scenarios, heating occurs for 7 years. The results from each in situ retorting simulation specify, as a

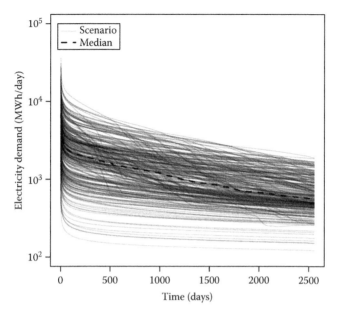

FIGURE 11.6
Daily electrical energy demand curves for all 242 in situ retort scenarios. The y-axis is on a log-scale. The median of all curves is shown as a dotted line.

function of time, the power requirements for heating and the total mass of products from the kerogen pyrolysis. The maximum possible oil production would occur if 100% of the kerogen were converted to oil and were immediately recovered from the deposit. In reality, kerogen decomposes into oil, gas, and coke, and any produced oil would have to travel through the deposit, delaying production and leaving some oil behind.

These issues are addressed by specifying two mass fractions as part of the DOE analysis: x_r, which represents the fraction of converted hydrocarbons (i.e., pyrolysis products) that is recovered from the formation, and x_g, which represents the fraction of the converted hydrocarbons that is gas (the balance is assumed to be oil). The mass fraction of converted hydrocarbons recovered from the formation, x_r, was modeled as a normal distribution based on the reported product recovery values in Chapter 6 and in Ryan et al. (2010). The mass fraction of oil converted into gas, x_g, was also modeled as a normal distribution based on values from the same sources.

By specifying these two parameters, the volume of produced oil (v_o) and gas (v_g) can be calculated using Equations 11.1 and 11.2:

$$v_o = \frac{m_k}{\rho_o} \cdot x_r \left(1 - x_g\right) \tag{11.1}$$

$$v_g = \frac{m_k}{\rho_g} \cdot x_r \cdot x_g \qquad (11.2)$$

where
   m_k is the mass of converted kerogen (as calculated by the in situ retort simulation)
   ρ_o and ρ_g are the densities of the produced oil and gas, respectively.

The density of the oil is assumed to be 843 kg/m³, equivalent to an API gravity of 36°. This density is based on the reported properties of shale oil produced via Shell's in situ conversion process (Beer et al. 2008). The oil and gas heating values, or energy densities, respectively, are taken from property tables for oil and gas produced from 26.7 GPT oil shale (U.S. DOE 1979). The shale oil heating value is 42.55 MJ/kg, the gas density is 1.24 kg/m³, and the gas heating value is 22.78 MJ/kg.

11.2.5 Production, Separation, and Storage System

Any fluids generated from the in situ retort must be produced and then separated and stored at the surface. The design of the production, separation, and storage (PSS) system is based on EIA oil and gas lease equipment and operating costs for primary oil production in the Rocky Mountain region (EIA 2010). EIA's PSS system includes equipment for production by artificial lift with electric motors through a wellhead separator and into a tank battery. Both capital and annual operating costs are presented as a function of well depth. Capital costs are given for a well producing up to 20 barrels of oil per day (BPD) with a water volume fraction of up to 10%. Annual operating costs, on the other hand, assume a production rate of 10 BPD per well. For an 11,000-ft-long well (the median total well length in the well sample dataset) producing at these rates (sizing for 20 BPD for capital costs and charging annual operating costs at 10 BPD), the PSS capital and operating costs are approximately $322,000 and $57,000 per year, respectively, as calculated by linear interpolation using the available well length data points. Since the production rate varies in each scenario, these costs are scaled to meet the maximum production rate of each in situ retort simulation; see Section 11.3.1.1 for a discussion of the cost scaling methods.

   After separation, produced oil and gas are sold at the wellhead as is, while water is sent offsite for disposal. The "break-even" price for produced oil at the wellhead is one of the model outputs. Note that the produced oil would sell at a discount relative to market prices for other crudes because of upgrading and transportation costs required for bringing the oil to market (e.g., a refinery). Produced gas prices are modeled as a normal distribution based on actual natural gas wellhead prices in the Uinta Basin over the last 5 years

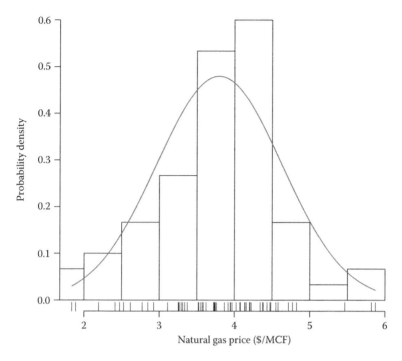

FIGURE 11.7

Normal distribution fit of monthly natural gas wellhead prices in dollars per thousand standard cubic feet ($/MCF) over the January 2010 to December 2014 time period. Fit parameters: mean = 3.80, standard deviation = 0.83. (From U.S. Energy Information Administration (EIA). 2015c. U.S. natural gas prices. http://www.eia.gov/dnav/ng/ng_pri_sum_dcu_nus_m.htm. Accessed on October, 2015.)

(U.S. EIA 2015c); see Figure 11.7. They are varied as part of the DOE analysis. Any water that is produced is gravity-separated in the tank battery system and sent offsite for disposal, a cost that is included in the EIA PSS operating costs model.

11.3 Assessment Methodology

This assessment combines discounted cash flow and DOE analyses to determine the economic viability of each scenario. The first component, a discounted cash flow analysis, is used to determine the oil supply price for each scenario. The second component, a DOE analysis, is used to find the ranges of input parameter values for which a future project is economically viable.

The actual computation for both components was performed in R (R Core Team 2015). Each analysis technique is described in detail in the following.

11.3.1 Discounted Cash Flow Analysis

The cash flow for any project is defined as the sum of all costs and revenue that accrue in a specified amount of time. For this assessment, the basis is the discounted cash flow accounting method described by Seider et al. (2009). Applying Seider's methodology to the in situ oil shale process described in Section 11.2, the cash flow at any time step t is

$$CF(t) = S(t) - C_V(t) - C_F(t) - T(t) - R(t) - C_{WC}$$
$$- C_{TDC} - C_L - C_S - C_{RIP} - C_P - C_{DC} - C_{WR} \qquad (11.3)$$

where
  CF is the annual cash flow
  $S(t)$ is the gross sales revenue
  $C_V(t)$ is the variable operating costs
  C_F is the fixed operating costs
  $T(t)$ is the total taxes
  $R(t)$ is the royalties on oil and gas production
  C_{WC} is the working capital
  C_{TDC} is the total depreciable capital costs (i.e., heating system, PSS, etc.)
  C_L is the capital cost of mineral leases and of land on which production
     facilities are built
  C_S is the capital cost of start-up
  C_{RIP} is the capital cost of royalties for intellectual property
  C_P is the capital cost of permitting
  C_{DC} is the capital cost of heating wells (drilling) and production wells
     (drilling and completion)
  C_{WR} is the capital cost of well reclamation.

Equation 11.3 is generalized so that it covers any time step t of any scenario. However, no time step t includes all of the terms listed and some terms are paid for over many time steps. Terms that are functions of time vary because the oil production rate varies. Each of the scenarios analyzed follows the same relative investment schedule outlined in Table 11.1.

Time steps in the model are tracked on a daily basis. Each scenario begins in the design phase where mineral rights are leased (C_L), all aspects of the project are permitted (C_P), and 25% of C_{TDC} is spent. The duration of the design phase is assumed to be equal to one-third of the construction phase. In the construction phase, all wells are drilled (C_{DC}) and the remainder of the total depreciable capital costs (C_{TDC}) is invested. The amount of time spent

TABLE 11.1

Capital Investment Schedule

Phase	C_{TDC}	C_L	C_P	C_{DC}	C_{WC}	C_{RIP}	C_S	C_{WR}
			Investment					
Design	X	X	X	—	—	—	—	—
Construction	X	—	—	X	—	—	—	—
Start-up	—	—	—	—	X	X	X	—
Production	—	—	—	—	—	—	—	—
Shutdown	—	—	—	—	X	—	—	X

in the construction phase is equal to the amount of time necessary to drill all wells in the scenario or 9 months (whichever is greater). Start-up occurs in the time step immediately following the completion of construction and concurrently with the beginning of the production phase. Working capital is invested (C_{WC}) and royalties for intellectual property (C_{RIP}) and start-up capital (C_S) are spent. The production phase begins after construction is complete. No further capital is invested during this period, but variable expenses and other costs that are functions of time (labor, taxes, royalties, etc.) are paid. The last step of the project, shutdown, occurs in the time step immediately following the end of the production phase, at which time the working capital is reclaimed, production is terminated, and well reclamation costs (C_{WR}) of $30,400 per well (Andersen et al. 2009) are paid.

To account for the time value of money, the cash flow for each year of the project is multiplied by a discount factor f, defined as

$$f_n = \frac{1}{\left(1+r_d\right)^n} \tag{11.4}$$

where r_d is the desired annual discount rate (i.e., interest rate) that the entity financing the project wishes to make each year, n, of a given project. Summing the discounted cash flows for each year of a project gives the net present value (NPV) of the project:

$$NPV = \sum f_n CF_n \tag{11.5}$$

Where CF_n is cash flow in year n. When Equation 11.5 equals zero (i.e., the NPV of the project is zero), the discount rate in Equation 11.4 is defined as the internal (or investors) rate of return (IRR). IRR is a common financial metric used to compare the value of different projects. Equation 11.5 can also be used to find the oil supply price, which is the oil price needed to produce an NPV of 0 at a given IRR. The IRR was selected as one of the parameters for this analysis. The IRR value range used in this analysis, with a mean of 15%, was selected based on recommendations from Seider (Seider et al. 2009) and from reported IRR values for conventional oil projects (Standard & Poor's 2011).

11.3.1.1 Capital Costs

The capital costs in Equation 11.3 were estimated using a combination of several techniques: vendor estimates, Williams six-tenth's rule (Williams 1947), statistical analysis of publicly available cost data, and Seider's capital costing method (Seider et al. 2009). Vendor estimates were used for the downhole heating system (McQueen et al. 2009) and electrical grid connections (Black & Veatch 2014). Williams six-tenth's rule was used for estimating the scaled costs of the PSS system. According to Williams, economies of scale in process equipment can be modeled using the equation:

$$C = C_o \left(\frac{Q}{Q_o} \right)^{0.6} \left(\frac{I}{I_o} \right)$$ (11.6)

where

    C is the cost

    Q is the material capacity (in this case, the oil production rate)

    I is an appropriate cost index or inflation index (the Consumer Price Index was used here)

    o refers to the base value of the subscripted variable.

As discussed in Section 11.2, the capital costs of drilling and completion were obtained from publicly available cost data (Utah DOGM 2015). Finally, the capital costs for all other terms in Equation 11.3 such as working capital, land, and permitting were estimated based on the capital costing model of Seider et al. (2009) as shown in Table 11.2.

11.3.1.2 Operating Costs

The operating costs in each scenario can be differentiated into variable (C_V) and fixed (C_F) costs based on whether they are functions of the in situ production operation. The variable costs for the process proposed in Section 11.2 are the costs of operating the PSS system and of buying electricity for the heater system. Electricity is purchased at $0.0607/kWh, which is the average retail price of electricity for industrial users in the state of Utah in 2014 (U.S. EIA 2015b). PSS operating costs reported in Section 11.2.4 are scaled linearly with the actual daily production for each scenario.

    The fixed costs for the process are the costs of labor, maintenance, property taxes, and insurance, all of which are estimated as suggested by Seider et al. (2009). Labor costs related to operations are estimated according to assumed

TABLE 11.2

Capital Costing Method

Category	Symbol and Definition
Total bare module investment (TBM)	C_{TBM} = sum of costs for heating and PSS systems
Cost of site preparation and service facilities	C_{SS} = 20% of C_{TBM}
Allocated costs for utility plants	C_{alloc} = cost of electrical grid connection
Total direct permanent investment (DPI)	$C_{DPI} = C_{TBM} + C_{site} + C_{serv} + C_{alloc}$
Cost for contingencies and contractor fees	C_{cont} = 15% of C_{DPI}
Total depreciable capital (TDC)	$C_{TDC} = C_{DPI} + C_{cont}$
Cost of mineral rights and land leases	C_L = 2% of C_{TDC}
Cost of permitting	C_P = \$0.10/bbl of oil produced
Cost of royalties for intellectual property	C_{RIP} = 2% of C_{TDC}
Cost of plant start-up	C_S = 10% of C_{TDC}
Cost of drilling and completing wells (DC)	C_{DC} = Estimated as part of DOE analysis
Total permanent investment (TPI)	$C_{TPI} = C_{TDC} + C_L + C_P + C_{RIP} + C_S + C_{DC}$
Working capital (WC)	C_{WC} = 5% of C_{TPI}
Total capital investment (TCI)	$C_{TCI} = C_{TPI} + C_{WC}$

Source: Modified from Seider, W.D. et al., *Product & Process Design Principles: Synthesis, Analysis and Design*, 3rd edn., John Wiley & Sons, New York, 2009.

hourly wages (\$30/h) and the number of operators required (three per shift following Seider's recommendations). The process is continuously manned during the production phase. Maintenance is estimated as 5% of C_{TDC} for all of the wages, salaries, and benefits paid to maintenance labor as well as the required materials, services, and overhead. Salaried labor costs, including process engineers for technical assistance and control laboratory staff, are assumed to be \$82,510 per person per shift per year (U.S. BLS 2014). Finally, the management, including accounting and business services, supervisors, human relations, and the mechanical department, is budgeted as operating overhead based on specific percentages of the total salaries, wages and benefits of the operators, maintenance personnel, lab personnel, and engineers. Property taxes and insurance are assumed to be 1% of C_{TPI}. These and other fixed costs are defined in Table 11.3.

11.3.1.3 Corporate Tax, Royalties, and Severance Tax

Oil and gas produced through in situ retorting (or any other method) is subject to a number of taxes and royalties. The first (and most straightforward to calculate) is royalty payments, defined as a percentage of the gross sales of the produced oil or gas. For most types of oil and gas production, the percentage collected is 12.5%. However, starting with the passage of the Energy Policy Act of 2005, royalty rates for oil produced from oil shale have been in political limbo. In its initial response to the Energy Policy act of 2005, the

TABLE 11.3

Fixed Costs Included in Scenario Analyses

Cost	Method of Calculation
Labor for operations	
Wages and benefits (LW)	LW = $30/operator-hour
Salary and benefits (LS)	LS = 15% of LW
Operating supplies and services	6% of LW
Technical assistance	$82,510/(operator/shift)/year
Control laboratory	$82,510/(operator/shift)/year
Maintenance (M)	5% of C_{TDC}
Wages and benefits (MW)	43.48% of M
Salary and benefits (MS)	10.87% of M
Materials and services	43.48% of M
Maintenance overhead	2.17% of M
Operating overhead	
General plant overhead	7.1% of (LW + LS + MW + MS)
Mechanical department services	2.4% of (LW + LS + MW + MS)
Employee relations department	5.9% of (LW + LS + MW + MS)
Business services	7.4% of (LW + LS + MW + MS)
Property tax	1.0% of C_{TPI}
Insurance	0.4% of C_{TPI}
General expenses	
Administrative expense	$200,000/(20 employees)/year
Management incentive compensation	1.25% of net profit

Sources: Modified from Seider, W.D. et al., *Product & Process Design Principles: Synthesis, Analysis and Design*, 3rd edn., John Wiley & Sons, New York, 2009; U.S. Bureau of Labor Statistics (BLS), Occupational employment statistics, May 2014 occupational employment and wage estimates, Salt Lake City, UT, http://www.bls.gov/oes/current/oes_ut.htm#17-0000, 2014.

U.S. Bureau of Land Management (BLM) proposed a starting rate of 5% for 5 years, followed by an increase of 1% per year up to 12.5%. Subsequent revisions to that proposal/decision have left royalty rates unclear. BLM's most recent Programmatic Environmental Impact Statement for oil shale identified a variety of different methods for setting royalty rates, such as determining them by public comment for each lease during the lease sale, using a sliding scale based on market prices for oil and gas, or establishing a minimum rate of 12.5% with an option for the Secretary of the Interior to increase the rate in the future (U.S. BLM 2013). For this analysis, a 12.5% royalty rate is used for both oil and gas.

The state of Utah collects severance taxes on oil and gas using a split rate system based on the market price of each product at the wellhead. The first $13/bbl for oil and $1.50/MCF for gas are taxed at a rate of 3%; any additional value above these thresholds is taxed at a rate of 5%. An additional 0.2% of the total value (TV) is taxed as a conservation fee (r_{cf}). This set of tax rules is implemented using the following equation:

$$ST = TV\left\{r_{cf} + \left[0.03\left(1 - f_{ST}\right) + 0.05 f_{ST}\right]\right\} \tag{11.7}$$

where
 ST is the severance tax due to the state on a dollar-per-barrel basis
 f_{ST} is the fraction of TV above the threshold value.

The results of this equation are then multiplied by the volume of oil or gas produced to find the total severance tax due for each product.
 Finally, corporate incomes taxes are calculated assuming the top rates of 5% and 35% at the state and federal levels, respectively, of taxable income. Taxable income (TI) is defined as

$$TI = P\left(S - C_V - d\right) - C_F - D - R - ST \tag{11.8}$$

where
 d is depletion
 D is depreciation

and all other variables are as defined previously. Cost depletion is used to determine d assuming that the cost depletion factor, p_t, is equal to the capital cost of land divided by the total planned oil production. The depletion charge in any given year is then the number of barrels of oil extracted that year multiplied by the depletion factor p_t. A 10-year modified accelerated cost recovery system method is used for calculating depreciation, with the first depreciation charge occurring at start-up. Since state corporate income taxes (T_S) are deductible from federal corporate income taxes (T_F), the total corporate tax liability is given by

$$T_S = t_s \cdot TI \tag{11.9}$$

$$T_F = t_f \cdot \left(TI - T_S\right) \tag{11.10}$$

where t_S and t_F are the respective state and federal corporate tax rates. Given that property taxes are accounted for as a fixed expense, the total tax liability used in Equation 11.3 is the sum of the severance tax (ST), the state corporate tax (T_S), and the federal corporate tax (T_F).

11.3.1.4 Model Outputs

Model outputs include the oil supply price, an itemized breakdown of capital and operating costs, and the estimated external energy ratio (EER).

EER, introduced in Chapter 10, is the ratio of the energy obtained from the produced oil and gas to the energy required for heating as defined in the following equation:

$$\text{EER} = \frac{v_o \rho_o ED_o + v_g \rho_g ED_g}{E_{in}} \tag{11.11}$$

where
E_{in} is the energy input required for retorting
ED_o and ED_g are the energy densities of oil and gas, respectively, as specified in Section 11.2.4.

The other parameters are as identified in Equations 11.1 and 11.2.

11.3.2 DOE Analysis

In the DOE analysis technique, the values of the input parameters are varied systematically to determine the contribution of each parameter to the overall system response. In this DOE analysis, the output response is the oil supply price. There are two types of input parameters. Those parameters marked as "well geometry" were varied for the in situ retort analysis discussed in Chapter 10 and are summarized here in Table 11.4. The parameters marked as "economic" were added in this chapter. Economic parameters are modeled as either normal or log-normal distributions and are summarized in Table 11.5.

Latin hypercube sampling (LHS) is a statistical method for selecting sets of parameter values that ensures that there is no overlap in any of the randomly selected parameter values. LHS was used to select 2000 unique combinations of the "economic" input parameters given in Table 11.5. A small fraction (49)

TABLE 11.4

Well Geometry Input Parameters and Value Ranges for DOE Analysis

Input Parameter	Range
Horizontal well spacing (H_{space}, ft)	16.70–81.89
Vertical well spacing (V_{space}, ft)	8.37–80.74
Offset angle (V_{angle}, degrees)	0.11–63.39
Location of 1st row of wells in formation ($V_{location}$, ft)	0.30–328.02
Well radius (r, in.)	4–6
Number of well rows (n_{row})	110
Number of wells (n_{well})[a]	5232

[a] The number of wells is calculated from the parameters for the number of rows and the well spacing.

TABLE 11.5

Economic Input Parameters, Distribution Types, Fit Parameters, and Percentiles for DOE Analysis

Input Parameter	Distribution	Mean	SD	P10	P90
Drilling time (t_{Drill}, days)	Log-normal	4.24	0.641	30	158
Drilling capital cost (C_{Drill}, $1e6/well)	Normal	1.88	0.781	$0.89	$2.9
Completion capital cost (C_{Compl}, $1e6/well)	Log-normal	14.6	0.577	$1.1	$4.7
Total well length (L, ft)	Log-normal	9.36	0.246	8497	15,962
Recovery mass fraction (x_r)	Normal	0.85	0.071	0.76	0.94
Gas mass fraction (x_g)	Normal	0.29	0.057	0.22	0.36
Wellhead natural gas price (gp, $/MCF)	Normal	3.80	0.832	$2.73	$4.86
IRR	Normal	0.15	0.025	11.8%	18.2%

Note: SD, standard deviation; P10, 10th percentile; P90, 90th percentile. Mean and SD values for log-normal distributions are the log-mean and log-standard deviation.

of these points was excluded for returning nonphysical values of at least one parameter (e.g., negative wellhead gas prices, negative gas fractions, etc.). In order to reduce the computational expense of finding an optimal set of LHS points, the randomly selected probability values for drilling costs were also applied to completion costs. For example, if the 40th percentile was selected for drilling costs, then the 40th percentile was also selected for completion costs.

Using the discounted cash flow methodology discussed in Section 11.3.1, each of these input parameter combinations was then used to calculate the oil supply price for all 242 designs in Chapter 10 (spanning the ranges of the "well geometry" input parameters). The result was 472,142 unique input parameter combinations (or scenarios) in the DOE analysis.

Next, linear regression was performed to fit the oil supply price results as the sum of all the input parameters using the following equation:

$$OSP = a \cdot H + b \cdot V + c \cdot V + d \cdot V + e \cdot r + f \cdot n + g \cdot n_{well} + h \cdot t_{Drill} + i \cdot C_{DC}$$

$$+ j \cdot L + k \cdot x_r + l \cdot x_g + m \cdot gp + n \cdot IRR \tag{11.12}$$

where
OSP is the oil supply price
Lower case variables a through n are all fitted coefficients

All other terms are the input parameters listed in Tables 11.4 and 11.5. Multiplying the fitted coefficients by the average value of each term then reveals the contribution (on average) of each term to the oil supply price.

11.4 Results and Discussion

11.4.1 Oil Supply Price Results

Raw results from the analysis of the 472,142 scenarios are presented in Figures 11.8 and 11.9. Figure 11.8 shows oil supply price versus each of the 14 input parameters. Figure 11.9 replicates Figure 11.8 but only shows the results from scenarios that have oil supply prices ≤ EIA's 2015 Annual Energy Outlook average wellhead oil price for the Rocky Mountain region between 2015 and 2040 under the high-oil-price forecasting assumption. This "economically viable" oil price is approximately $174/bbl (U.S. EIA 2015a). This limit was picked because historically EIA has tended to underpredict oil prices by as much as half of the actual price.

Two types of plots are shown in Figures 11.8 and 11.9. The first type is hexbinning, which counts the number of points located in a particular region of the plot and colors regions with higher numbers of points more darkly. This plot type is used for input parameters that are varied (more or less) continuously. The second type, a violin plot, is used for the input parameters that have discrete values, well radius (r) and the number of well rows (n_{row}). Only five values of r and ten values of n_{row} were considered. The violin plot shows the probability that a value y (oil supply price) will occur for each discrete x value of the input parameter. Thicker regions indicate greater numbers of results. While all of the plots show edges to the hexbin region, these edges are unlikely to be true limits. The DOE analysis conducted here only sampled a small portion of all the possible combinations of well geometry and economic input parameters in Tables 11.4 and 11.5 (0.0045% of all the unique combinations of 13 parameters if each parameter was sampled at the 10th, 20th, 30th, ... 90th percentiles). However, the actual distribution (i.e., shape) of the results is representative of the full parameter space.

The oil supply price results in Figures 11.8 and 11.9 display some general trends. Horizontal well spacing (H_{space}) between 20 and 40 ft is highly likely to lead to oil supply prices in the $100–$1000/bbl region; increasing horizontal spacing above 40 ft tends to substantially increase the maximum oil supply price results. Vertical spacing (V_{space}) of about 60 ft results in an order of magnitude increase in the oil supply price results, perhaps indicating that well geometries with that spacing are missing an important oil shale layer. From the violin plots for the number of rows (n_{row}), it is clear that at least two well rows are required to achieve low oil supply prices. The plot for the number of wells (n_{well}) shows that having either too few or too many wells leads to higher prices. In terms of the economic parameter set, shorter drilling times, lower drilling and completion costs, longer wells, higher recovery fractions, lower gas fractions, and lower IRR values all lead to lower oil supply prices. Gas prices appear to have negligible impact on oil supply prices.

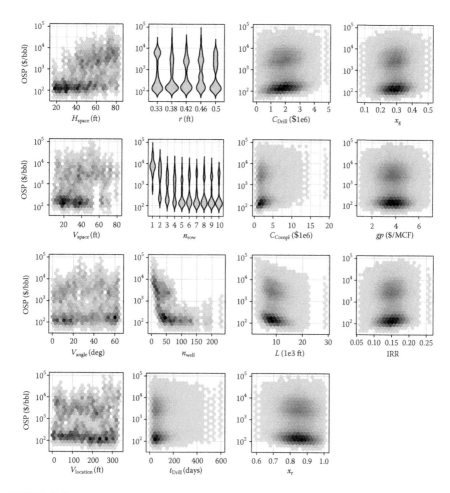

FIGURE 11.8

Hexbin *xy*-scatterplots and violin plots of oil supply price versus the input parameters in Tables 11.4 and 11.5. The *y*-axis for all plots is on a log-scale. The input parameters include horizontal well spacing (H_{space}), vertical well spacing (V_{space}), offset angle (V_{angle}), the location of the first row of wells in the formation ($V_{location}$), well radius (r), number of well rows (n_{row}), number of wells (n_{well}), drilling time (t_{Drill}), drilling capital cost (C_{Drill}), completion capital cost (C_{Compl}), total well length (L), recovery mass fraction (x_r), gas mass fraction (x_g), wellhead natural gas price (gp), and internal rate of return (IRR).

Many of the economic parameter trends are obscured by the impact of the in situ designs in Figure 11.8 but are more noticeable in Figure 11.9. As discussed in Chapter 10, the well geometry parameters play a large role in determining the in situ retort's EER. The EER, which collectively captures the impact of the well geometry parameters and the values of recovery mass fraction (x_r) and gas mass fraction (x_g), was calculated for each of the

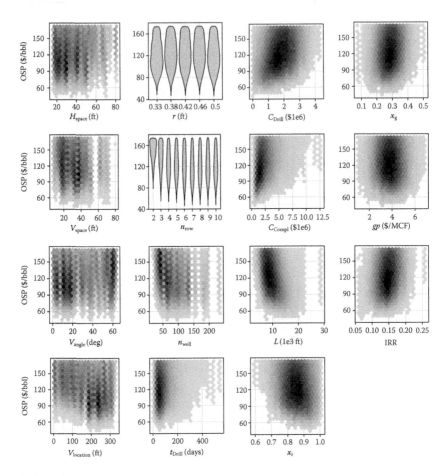

FIGURE 11.9
Hexbin xy-scatterplots and violin plots of oil supply price versus the input parameters in Tables 11.4 and 11.5 for scenarios with oil supply prices ≤ \$174/bbl. The y-axis is on a linear scale. The input parameters include horizontal well spacing (H_{space}), vertical well spacing (V_{space}), offset angle (V_{angle}), the location of the first row of wells in the formation ($V_{location}$), well radius (r), number of well rows (n_{row}), number of wells (n_{well}), drilling time (t_{Drill}), drilling capital cost (C_{Drill}), completion capital cost (C_{Compl}), total well length (L), recovery mass fraction (x_r), gas mass fraction (x_g), wellhead natural gas price (gp), and internal rate of return (IRR).

scenarios using Equation 11.11. The results are shown in Figure 11.10a for the full dataset and in Figure 11.10b for the economically viable dataset. Since most of the in situ designs have low EER values, there is a high density of oil supply price results around \$10,000/bbl visible in Figure 11.10a. A second cluster of results is visible in the region between EER values of 4 and 10, which results in oil supply prices between \$100–\$200/bbl. These two clusters are visible in all of the economic parameter plots in Figure 11.8.

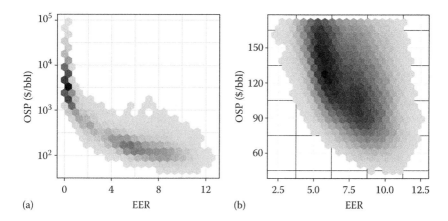

FIGURE 11.10

Hexbin xy-scatterplots of oil supply price versus EER for (a) all scenarios (note that the y-axis is on a log-scale) and (b) for scenarios with oil supply prices $\leq$ $174/bbl.

The economically viable set in Figure 11.9 effectively "zooms" in on the cluster in the EER range of 4–10.

The plot of the economically viable set in Figure 11.10b shows that given the selected LHS sample set, an EER value of at least 2.5 is required to achieve oil supply prices $\leq$ $174/bbl. Imposing this EER limit excludes 128 of the original 242 well geometry designs from consideration. Interestingly, the in situ designs with the highest EER do not necessarily have the lowest oil supply prices. The best in situ design from the perspective of EER values (EER = 12.5) produces approximately 40% less oil after 500 days than the in situ design with the lowest oil supply prices (EER = 10.1).

Another interesting way of looking at the data is to calculate the median value of each input parameter at different oil supply price thresholds and to observe how it changes. Table 11.6 shows these median values for the full-results dataset and for results less than three price thresholds: $174/bbl (the average EIA high oil forecast price), $100/bbl, and $75/bbl. The largest change in the input parameters occurs when moving from the full-results dataset to the economically viable dataset, which removes nearly half of the well geometry designs from consideration (although almost every economic parameter set is still included at this stage). At the $100/bbl and $75/bbl thresholds, the well geometry parameter set narrows further with the most substantial changes being the elimination of unfavorable economic parameter sets.

All of these trends are reflected quantitatively in the regression analysis results shown in Table 11.7. The regression analysis was performed with the full dataset. The relative impact of each input term, calculated by normalizing all the impact values by the largest impact term, is shown in Figure 11.11.

TABLE 11.6

Median Values of Input Parameters and of EER (Model Output) at Different Oil Supply Price Cutoff Thresholds

Item	Full Range	≤ $174/bbl	≤ $100/bbl	≤ $75/bbl
Number of well geometry parameter sets	242	134	112	76
Number of economic parameter sets	1,951	1,915	1,164	383
Percent of all LHS cases	100%	35.5%	8.4%	1.4%
Model input parameters				
Horizontal well spacing (H_{space}, ft)	49.0	34.7	30.5	30.1
Vertical well spacing (V_{space}, ft)	33.7	31.2	32.9	32.6
Offset angle (V_{angle}, degrees)	31.1	26.7	18.9	18.9
Location of 1st row of wells in formation ($V_{location}$, ft)	168.6	187.8	229.1	234.3
Well radius (r, in.)	5	5	5	5
Number of well rows (n_{row})	5	7	7	7
Number of wells (n_{well})	43	73	88	88
Drilling time (t_{Drill}, days)	69	67	62	58
Drilling capital cost (C_{Drill}, $1e6/well)	$1.89	$1.75	$1.41	$1.02
Completion capital cost (C_{Compl}, $1e6/well)	$2.23	$2.02	$1.57	$1.18
Total well length (L, ft)	11,617	12,325	13,690	15,268
Recovery mass fraction (x_r)	84.8%	85.5%	87.0%	89.5%
Gas mass fraction (x_g)	29.0%	28.4%	27.3%	26.3%
Wellhead natural gas price (gp, $/MCF)	$3.79	$3.80	$3.84	$3.85
IRR	15.0%	14.7%	14.1%	13.8%
Model output values				
EER	3.92	7.18	8.41	9.06

As in the aforementioned graphical analysis, well geometry input parameters have the largest impact. Increasing well spacing (in either the horizontal or vertical direction) leads to the largest increases in the oil supply price, while increasing well radius or the number of rows leads to the largest reductions in oil supply price.

All of these results can be explained by considering the physics of heat transfer in the oil shale deposit. Since the oil shale deposit is being heated conductively, the rate of heat transfer to any point in the deposit is proportional to the inverse square of the distance between the heater and that point. Therefore, increasing the well spacing dramatically increases the amount of time that must elapse before the heating zones of adjacent wells begin to

TABLE 11.7

Regression Results and Impact Analysis

Input Parameter	Fitted Coefficient	Median Value	Impact Factor
Horizontal well spacing (H_{space}, ft)	154.31	48.95	7554
Vertical well spacing (V_{space}, ft)	76.39	33.67	2572
Offset angle (V_{angle}, degrees)	12.36	31.13	385
Location of 1st row of wells in formation ($V_{location}$, ft)	1.35	168.57	227
Well radius (r, in.)	−401.47	5	−2007
Number of well rows (n_{row})	−1071	5	−5353
Number of wells (n_{well})	52.92	43	2276
Drilling and completion time (t_{Drill}, days)	0.08	69	5
Drilling cost (C_{Drill}, $1e6/well)	1.11E−04	1.89E+06	209
Completion cost (C_{Compl}, $1e6/well)	5.65E−05	2.23E+06	126
Well lateral/heating length (L, ft)	−0.16	11617	−1903
Recovery mass fraction (x_r)	−3888	0.85	−3298
Gas mass fraction (x_g)	2029	0.29	589
Wellhead gas price (gp, $/MCF)	−40.54	3.79	−154
IRR	3248	0.15	487

overlap, thus reducing the heating efficiency of the retort system. Increasing the well radius leads to larger amounts of surface area acting as a heat source. Since the retorting simulations specify a constant-temperature boundary condition, the total heat flux from a well will increase linearly with well radius. The higher number of well rows lead to more overlapping heating zones, which in turn improve efficiency.

Interestingly, increasing the number of wells (n_{well}) increases the oil supply cost. On the one hand, each additional well increases capital and operating costs, but each well also gives more production and at least one additional overlapping heating zone (depending on the number of well rows, n_{row}). Part of this result may be explained by the use of only first-order interactions in Equation 11.12, which excludes any possible interactions between input terms such as n_{well} and n_{row}. The first-order model has 15 terms, all statistically significant except for drilling time, and an R^2 of 0.58. If the model considered all second-order interactions (every possible combination of two input parameters multiplying each other, e.g., $a \cdot n_{well} \cdot n_{row} + b \cdot n_{row} \cdot C_{Drill} + \cdots + xyz \cdot gpIRR$), then the model would have 120 terms, 90 of which are statistically significant (i.e., p-value ≤ 0.05), and an $R^2 = 0.74$. Nevertheless, Equation 11.12 is more likely to be the better model because it is less prone to overfitting and it explains the majority of the variation in the oil supply price results.

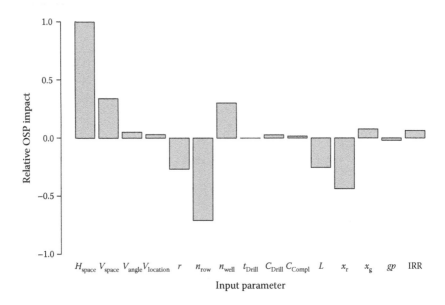

FIGURE 11.11

Relative oil supply price impact of each input parameter. Increases in parameters with positive impact values result in higher oil supply prices, while increases in parameters with negative impact values result in lower oil supply prices. The input parameters include horizontal well spacing (H_{space}), vertical well spacing (V_{space}), offset angle (V_{angle}), the location of the first row of wells in the formation ($V_{location}$), well radius (r), number of well rows (n_{row}), number of wells (n_{well}), drilling time (t_{Drill}), drilling capital cost (C_{Drill}), completion capital cost (C_{Compl}), total well length (L), recovery mass fraction (x_r), gas mass fraction (x_g), wellhead natural gas price (gp), and internal rate of return (IRR).

11.4.2 Detailed Breakdown of Costs

A detailed economic breakdown of both the capital costs and per-barrel costs for the economically viable well sets is shown in Figures 11.12 and 11.13, respectively. In these boxplots, the middle line represents the median value, the top and bottom of the box represent the 75th and 25th percentiles of the results, and the whiskers on the top and bottom represent the maximum and minimum values. Also note that the y-axis in Figure 11.12 is shown on a log-scale. The capital cost breakdown in Figure 11.12 clearly shows that the single biggest cost driver is well drilling and completion (C_{DC}), accounting for 42% of C_{TCI} on average for the economically viable scenarios. For comparison, the next largest cost driver, the allocated costs for utilities (i.e., the electrical line, or C_{alloc}), is 16% of C_{TCI} on average. It should also be noted that given the capital costing methodology outlined in Table 11.2, the only independently calculated capital costs are for the heating system, PSS, electrical grid connection (C_{alloc}), and well drilling and

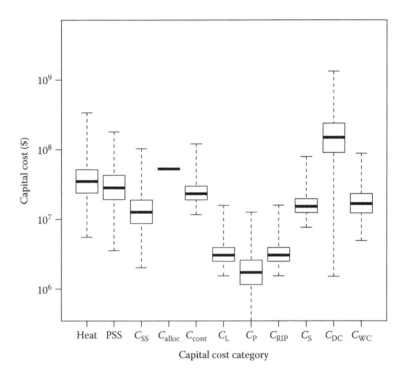

FIGURE 11.12
Capital cost breakdown for wells with oil supply price $\leq$ \$174/bbl. Capital cost categories are the heating system (Heat), production, separation and storage (PSS), site preparation and service facilities (C_{SS}), allocated costs for the electrical grid connection (C_{alloc}), contingencies and contractor fees (C_{cont}), mineral rights and land leases (C_L), permitting (C_p), royalties for intellectual property (C_{RIP}), start-up (C_S), drilling and completion (C_{DC}), and working capital (C_{WC}); see Table 11.2.

completion. All other capital cost categories (C_{ss}, C_{cont}, C_L, C_S, and C_{WC}) are defined as percentages of other terms.

The per-barrel cost breakdown in Figure 11.13 shows that taxes are the largest per-barrel cost. The amount of money paid in taxes is proportional to the amount of profit taken, as specified in each scenario by the IRR selection. Increasing the IRR results in higher per-barrel profits, which in turn results in higher taxes and royalty payments. After profits, taxes, and royalties, the next largest expense is the cost of capital (C_{TCI}). Electricity purchases, which are (on average) 91% of the variable operating costs (C_V), are the fourth biggest expense. However, as noted in Section 11.2.3, the cost of building and running a dedicated power plant would be even higher. Fixed operating costs are negligible as both the labor and maintenance required for an in situ operation are minimal.

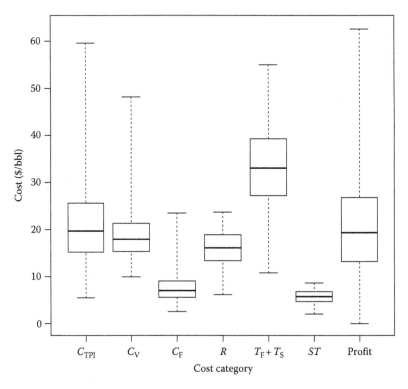

FIGURE 11.13

Costs on a dollar-per-barrel basis for wells with oil supply price $\leq$ \$174/bbl. The cost categories, defined in Section 11.3.1, are total permanent investment (C_{TPI}), variable costs (C_V), fixed costs (C_F), royalties (R), federal (T_F) and state (T_S) taxes, severance taxes (ST), and net profit (Profit). Profit is the net earnings on each barrel of oil necessary to produce each scenario's specified IRR.

11.5 Conclusions

In situ oil shale could be economically viable if oil prices recover. Of the scenarios tested in this work, 36% have oil supply prices less than \$174/bbl (average oil price between 2015 and 2040 under EIA's high oil price forecast). However, if oil prices only recover to \$90/bbl (average oil price between 2015 and 2040 under EIA's reference forecast), then only 5% of the scenarios would be viable. The primary driver of oil supply prices is the EER, which depends on well geometry parameters, particularly horizontal and vertical well spacing, well radius, and the number of well rows. However, even with ideal EER values, the profitability of oil shale projects is hindered from a financial perspective by (1) the longtime delay between the start of the project and the start of production and (2) the capital expense of drilling and completing wells, both of which make low oil supply prices difficult to achieve.

References

Andersen, M., R. Coupal, and B. White. 2009. Reclamation costs and regulation of oil and gas development with application to Wyoming. In *Western Economics Forum*. Laramie, WY: Western Agricultural Economics Association. http://ageconsearch. umn.edu/bitstream/92846/2/0801005.pdf. Accessed on October 2015.

Baker Hughes. 2015 (May 13). North America rotary rig count archive—U.S. monthly average by state through 2013. http://phx.corporate-ir.net/phoenix. zhtml?c=79687&p=irol-reportsother. Accessed on October 2015.

Bartis, J. T., T. LaTourrette, L. Dixon, D. J. Peterson, and G. Cecchine. 2005. *Oil Shale Development in the United States—Prospects and Policy Issues*. Santa Monica, CA: RAND Corporation.

Beer, G. L., E. Zhang, S. Wellington, R. Ryan, and H. Vinegar. 2008. Shell's in situ conversion process—Factors affecting the properties of produced shale oil. Paper presented at the *26th Oil Shale Symposium*, Golden, CO. http://www.ceri-mines. org/documents/28thsymposium/presentations08/PRES_3-2_Beer_Gary.pdf. Accessed on October 2015.

Bezdek, R. H., R. M. Wendling, and R. L. Hirsch. 2006. *Economic Impacts of U.S. Liquid Fuel Mitigation Options*. Pittsburgh, PA: National Energy Technology Laboratory.

Black & Veatch. 2014. Capital costs for transmission and substations. Western Electricity Coordinating Council. Retrieved from https://www.wecc.biz/ Reliability/2014_TEPPC_Transmission_CapCost_Report_B+V.pdf. Accessed on October 2015.

EPA Oil Shale Work Group. 1979. *EPA Program Status Report: Oil Shale—1979 Update*. Washington, DC: Environmental Protection Agency. https://books.google. com/books?id=ET9SAAAAMAAJ&source=gbs_navlinks_s. Accessed on October 2015.

Institute for Clean and Secure Energy. 2013. *A Market Assessment of Oil Sands and Oil Shale Resources*. Salt Lake City, UT: Institute for Clean and Secure Energy, University of Utah. http://www.icse.utah.edu/assets/for_download/pdfs/proj ects/2013OilShaleMarketAssessment.pdf. Accessed on October 2015.

INTEK Inc. 2009. *National Strategic Unconventional Resource Model—A Decision Support System*. Washington, DC: INTEK Inc.

INTEK Inc. 2011. *Secure Fuels from Domestic Resources—Profiles of Companies Engaged in Domestic Oil Shale and Tar Sands Resource and Technology Development*. Washington, DC: INTEK Inc. http://energy.gov/sites/prod/files/2013/04/f0/ SecureFuelsReport2011.pdf. Accessed on October 2015.

McQueen, G., D. Parman, and H. Williams. 2009. Enhanced oil recovery of shallow wells with heavy oil: A case study in electro thermal heating of California oil wells. *2009 Record of Conference Papers—Industry Applications Society 56th Annual Petroleum and Chemical Industry Conference*, PCIC 2009, Piscataway, NJ. http:// doi.org/10.1109/PCICON.2009.5297168. Accessed on October 2015.

R Core Team. 2015. *R: A Language and Environment for Statistical Computing*. Vienna, Austria: R Core Team. http://www.r-project.org/. Accessed on October 2015.

Ryan, R. C., T. D. Fowler, G. L. Beer, and V. Nair. 2010. Shell's in situ conversion process—From laboratory to field pilots. In *Oil Shale: A Solution to the Liquid Fuel Dilemma*, eds. O. I. Ogunsola, A. M. Hartstein, and O. Ogunsola, pp. 161–183. Washington, DC: ACS Symposium Series.

Seider, W. D., J. D. Seader, D. R. Lewin, and S. Widagdo. 2009. *Product & Process Design Priniciples: Synthesis, Analysis and Design*, 3rd edn. New York: John Wiley & Sons.

Standard & Poor's. 2011 (December). Is natural gas drilling economic at current prices? *CreditWeek*, 9–12. http://www.standardandpoors.com/spf/swf/oiland-gas/data/document.pdf. Accessed on October 2015.

STRAAM Engineers. 1979. *Capital and Operating Cost Estimating System Handbook: Mining, Retorting, and Upgrading of Oil Shale in Colorado, Utah, and Wyoming*. Irvine, CA: STRAAM Engineers.

U.S. Bureau of Labor Statistics (BLS). 2014. Occupational employment statistics, May 2014 occupational employment and wage estimates, Salt Lake City, UT. http://www.bls.gov/oes/current/oes_ut.htm#17-0000. Accessed on October 2015.

U.S. Bureau of Land Management (BLM). 2013. Secretary Salazar finalizes plan promoting responsible oil shale and tar sands research, demonstration and development. http://www.blm.gov/wo/st/en/info/newsroom/2013/march/nr_03_22_2013.html. Accessed on October 2015.

U.S. Department of Energy (DOE). 1979. *Oil shale data book*. Washington, DC: U.S. Department of Commerce, National Technical Information Service. https://www.ntis.gov/Search/Home/titleDetail/?abbr=PB80125636. Accessed on October 2015.

U.S. Energy Information Administration (EIA). 2010. Oil and gas lease equipment and operating costs 1994 through 2009. http://www.eia.gov/pub/oil_gas/natural_gas/data_publications/cost_indices_equipment_production/current/coststudy.html. Accessed on October 2015.

U.S. Energy Information Administration (EIA). 2013. Updated capital cost estimates for utility scale electricity generating plants. http://www.eia.gov/forecasts/capitalcost/pdf/updated_capcost.pdf. Accessed on October 2015.

U.S. Energy Information Administration (EIA). 2015a. Annual energy outlook 2015. http://www.eia.gov/forecasts/aeo/pdf/0383(2015).pdf. Accessed on October 2015.

U.S. Energy Information Administration (EIA). 2015b. Average retail price of electricity to ultimate customers. http://www.eia.gov/electricity/data.cfm#sales. Accessed on October 2015.

U.S. Energy Information Administration (EIA). 2015c. U.S. natural gas prices. http://www.eia.gov/dnav/ng/ng_pri_sum_dcu_nus_m.htm. Accessed on October 2015.

U.S. Energy Information Administration (EIA). 2015d. Utah electricity profile 2013. http://www.eia.gov/electricity/state/utah/. Accessed on October 2015.

Utah Division of Oil, Gas, and Mining (DOGM). 2015. Data research center. http://oilgas.ogm.utah.gov/Data_Center/DataCenter.cfm. Accessed on October 2015.

Wellington, S. L., I. E. Bercenko, E. P. de Rouffignac et al. 2005. In situ thermal processing of an oil shale formation to produce a desired product. U.S. Patent US6880633 B2.

Williams, R. 1947. Six-tenths factor aids in approximating costs. *Chemical Engineering* 54(12):124–125.

12

Oil Shale Development, Air Quality, and Carbon Management

Kerry E. Kelly, John C. Ruple, and Jonathan E. Wilkey

CONTENTS

Oil shale development will result in the emission of pollutants that could contribute to poor air quality. To understand how oil shale development could affect air quality and how air quality regulations may constrain oil shale development, this chapter begins with an overview of the air quality regulations and of existing air quality within Utah's Uinta Basin. It then discusses the air pollutant emissions that are anticipated to result from oil shale development and how the convergence of emissions and evolving regulations is likely to affect the nascent oil shale industry. Next, this chapter discusses emissions of greenhouse gases (GHG) and of the chemical precursors of ozone (O_3), the pollutants of greatest concern with respect to air quality within the Uinta Basin. The discussion summarizes the modeling of both types of emissions from the in situ oil shale scenarios presented in Chapters 10 and 11. Oil shale's carbon footprint and net energy ratio (NER) are then compared to the carbon footprint and NER of other liquid transportation fuels in order to understand emissions in the context of broader energy policy. This chapter concludes with a summary of rapidly developing carbon management policies.

12.1 Air Quality Regulation and the Existing Conditions within the Uinta Basin

12.1.1 Air Quality Policy

Under the Clean Air Act (CAA), the U.S. Environmental Protection Agency (EPA) establishes air quality standards that apply nationwide. These standards include National Ambient Air Quality Standards (NAAQS) for six criteria pollutants: particulate matter (PM), O_3, carbon monoxide (CO), sulfur dioxide (SO_2), nitrogen oxides (NO_X), and lead (42 U.S.C. §§ 7408, 7409 (2012)). Areas that exceed a standard are deemed "nonattainment" areas for that pollutant.

Individual states develop State Implementation Plans (SIPs) to maintain compliance with EPA-established NAAQS and to guide efforts to bring nonattainment areas into compliance with CAA requirements (42 U.S.C. § 7410(a) (2012)). SIPs are submitted to the EPA and are subject to Agency approval. Once approved, the state assumes primary responsibility for air quality monitoring, permitting, and enforcement. Indian tribes can request to be treated as states by the EPA and develop Tribal Implementation Plans, if they so choose. The EPA retains CAA jurisdiction within "Indian Country" (see Section 12.1.4) where federally recognized tribes choose not to assume CAA implementation authorization.

12.1.2 Air Quality in the Uinta Basin

Monitoring data suggest relatively low levels of pollutants regulated under the NAAQS within the Uinta Basin, except for O_3. Ozone results when volatile organic compounds (VOCs) and NO_X react with sunlight. High O_3 levels within the Uinta Basin are associated with O_3 precursors emitted by oil and gas development (Edwards et al. 2014, Rodriguez et al. 2009). Although wintertime O_3 formation is not well understood, VOCs are currently believed to be the critical (rate-limiting) species in O_3 formation in the basin (Environ 2015). Exceedances of the NAAQS for O_3 are more common in the Uinta Basin during winter inversion periods when the ground is covered by snow and stagnant pools of cold air are present. In 2013 and 2014, the Uinta Basin exceeded the O_3 NAAQS on 26 days and 7 days, respectively (UDAQ 2015). The highest 8-hour ozone concentration reported in 2013 was 126 ppb, nearly twice the federal NAAQS.

In 2012, the lack of long-term monitoring data that met EPA quality control requirements led the EPA to determine that the Uinta Basin was "unclassifiable" for O_3 attainment. Although this finding was contested in court, the U.S. Court of Appeals for the District of Columbia (D.C.) Circuit recently upheld the unclassifiable finding (Miss. Comm'n on Envtl. Quality v. E.P.A., 790 F.3d 138 (D.C. Cir. 2015) (per curiam)). The D.C. Circuit's opinion does not resolve the matter because the EPA will likely designate the Uinta Basin as a nonattainment area if past air quality trends continue, and this designation will trigger imposition of more stringent emission controls.

On an independent front, the EPA announced downward revision of the federal NAAQS for O_3 from 75 to 70 ppb, effective December 28, 2015 (U.S. EPA 2015b). The combination of elevated O_3 levels associated with recent increases in hydrocarbon development (Edwards et al. 2014, Rodriguez et al. 2009) and the tighter federal ozone standards (U.S. EPA 2015b) suggest more stringent future regulation of O_3 precursor emissions.

12.1.3 Oil Shale Development and Air Quality

In addition to NAAQS regulation, the CAA requires operating permits, specifying acceptable air pollutant emissions levels, for all major stationary air pollution sources and all hazardous air pollution sources (42 U.S.C. §§ 7411, 7412 (2012)). Permitting requirements for a new or modified source depends, in part, upon whether the source is located in an attainment or nonattainment area. Nonattainment area permitting requires that the permittee demonstrate an ability to offset its new emissions in order to advance the area's attainment potential. Attainment area permitting falls under the Prevention of Significant Deterioration program, which limits the degree of increased air pollution that can be permitted.

Notably, the EPA is in the process of developing New Source Performance Standards for the oil and gas industry (U.S. EPA 2015a). These standards are likely to address production, processing, transmission and storage, including well completions and hydraulic fracturing. Whether these standards would apply to oil shale production, and if so, what their potential impact might be, will become clearer with rule finalization.

Finally, the EPA, federal land managers, and states and tribes must also cooperatively develop and implement plans to preserve visibility. At present, visibility in the Uinta Basin is quite good, but impacts to downwind national parks, national monuments, and congressionally designated wilderness areas are a concern.

Separate from regulation under the CAA, the Bureau of Land Management (BLM), which manages most of the federal public lands within the Uinta Basin, must comply with the provisions of the Federal Land Policy and Management Act (FLPMA). FLPMA, among the act's other obligations, requires the BLM to protect "environmental, air, and atmospheric" resources (43 U.S.C. § 1701(a)(8) (2012)). The BLM is therefore likely to condition or restrict energy development activities, including activities related to oil shale development, in an effort to comply with FLPMA and to demonstrate protection of environmental, air, and atmospheric resources as well as CAA compliance.

As the preceding paragraphs make clear, oil shale development within Utah's Uinta Basin will take place against a highly regulated backdrop and in a region that is already challenged by high ground-level O_3 levels. Increases in hydrocarbon production within the Uinta Basin over the past decade have contributed to compromised air quality, increasing the risk that additional emissions will lead to downwind visibility impairment or other CAA violations. Thus, both the changing development and regulatory landscape are likely to make CAA and FLPMA compliance more difficult, which will in turn make oil shale development more difficult. As the Government Accountability Office (GAO) acknowledged recently, "air quality... appears to be particularly susceptible to the cumulative affect[sic] of energy development, and according to some environmental experts, air quality impacts may be the limiting factor for the development of a large oil shale industry in the future" (U.S. GAO 2010).

The BLM provides context for the GAO's conclusions by describing the scope of anticipated air quality impacts resulting from commercial oil shale development as follows:

> Temporary, localized impacts (primarily PM and NO_X, with some CO, VOC, and SO_2 emissions) would result from the clearing of the project area; grading, excavation, and construction of facilities and associated infrastructure; and mining (extraction) or drilling of the oil shale resource. Long-term, regional impacts (primarily NO_X and CO, with lesser amounts of PM, VOCs, and SO_2) would result from oil shale processing, upgrading, and transport (pipelines). Depending on site-specific locations, meteorology, and topography, NO_X and SO_2 emissions could cause regional visibility impacts (through the formation of secondary aerosols) and contribute to regional nitrogen and sulfur deposition. In turn, atmospheric deposition could cause changes in sensitive (especially alpine) lake chemistry. In addition, depending on the amounts and locations of NO_X and VOC emissions, photochemical production of O_3 (a very reactive oxidant) is possible, with potential impacts on human health and vegetation. Similar impacts could also occur from the additional coal-fired power plants that would be needed to supply electricity for in situ oil shale extraction. Localized impacts due to emissions of hazardous air pollutants

(HAPs) (particularly BTEX and formaldehyde) and diesel PM could also present health risks to workers and nearby residences.... During all phases of oil shale development, GHG emissions of CO_2 and lesser amounts of CH_4 [methane] and N_2O [nitrous oxide] from combustion sources could contribute to climate change. Depending on the situation, dust emissions could exert either a cooling or a warming effect. (BLM 2012).

In light of the existing air quality concerns, particularly those associated with O_3 and its chemical precursors, as well as the emissions that are likely to result from commercial-scale oil shale development, prospective oil shale developers, policy makers, and resource managers will all need to pay careful attention to protecting air-quality-related resources. While air quality protection will likely increase the cost of development, failure to plan proactively to protect these resources is certain to result in permitting delays and invite legal challenges. Moving forward, therefore, will require careful planning and cooperation between industry and regulators.

12.1.4 Indian Country Jurisdiction and the Need for Collaboration

Addressing air quality challenges requires working closely, and proactively, with multiple regulatory agencies. As noted earlier, states are required to develop SIPs, which provide for implementation, maintenance, and enforcement of primary and secondary air quality standards (42 U.S.C. § 7410(a)(1) (2012)). States also have the discretion to develop programs to regulate new stationary sources of air pollution (42 U.S.C. § 7411(c) (2012)) and hazardous air pollutant emissions (42 U.S.C. § 7412(l) (2012)).

States, however, cannot assume regulatory jurisdiction on Indian lands. An Indian tribe may petition the EPA to be treated as a state for purposes of virtually all sections of the CAA and may assume CAA regulation, provided that the tribe (1) is federally recognized, (2) has a governing body carrying out substantial governmental duties and powers, and (3) is capable of implementing the program consistent with the CAA and applicable regulations (40 C.F.R. §§ 49.1 through 49.11 (2014)). The EPA administers the CAA for Indian lands when tribes have not assumed CAA jurisdiction. The Ute Tribe of Indians has not assumed CAA regulatory jurisdiction; thus, the EPA will administer the CAA as it applies to Indian land within the Uinta Basin. Oil shale developers must therefore work with the EPA to obtain appropriate environmental permits for their operations.

The challenge presented by working in Indian Country is twofold. First, since SIPs do not apply to Indian lands and the EPA rarely has the resources to develop a Federal Implementation Plan (FIP) for Indian lands, regulated entities may be subject to different and less clearly defined regulatory requirements than those that would apply in areas subject to an SIP.

The lack of a FIP creates uncertainty with respect to what will be required by the EPA to maintain or achieve CAA compliance and what is expected of regulated entities.

Second, Indian Country jurisdictional boundaries can be difficult to define, particularly in the Uinta Basin (Tanana and Ruple 2012). "Indian Country" can include current and past Indian reservation lands, dependent Indian communities, and Indian allotments—any of which can extend well beyond current reservation boundaries. Indian Country within eastern Utah extends well beyond the current boundary of the Uintah and Ouray Indian Reservation and includes the majority of the region's oil shale and other energy resources. The EPA will therefore have primary regulatory juris-diction over air quality permitting for oil shale developments within Utah. Nevertheless, the State of Utah will almost certainly want to coordinate with the EPA to ensure that federal actions and authorizations harmonize with the state's ongoing efforts to protect and improve regional air quality. Prospective developers should therefore begin working with the EPA early on in order to understand applicable requirements and to minimize the risks associated with potentially confusing jurisdictional boundaries and differ-ent regulatory requirements.

12.2 Estimating GHG and Air Impacts from In Situ Oil Shale Development

The modeling performed in this chapter builds on the energy balance and economic evaluation of the in situ heating scenarios discussed in Chapters 10 and 11. The goal of the modeling in this chapter is to predict GHG and VOC emissions and the related impact that would result from in situ oil shale development in the Uinta Basin. When the kerogen in oil shale is heated in situ in the absence of air (pyrolysis), it produces both liquid- and gas-phase products; see Chapter 11 for details.

12.2.1 Methods

This chapter's modeling evaluation focuses on NER total GHG and VOC emissions, and normalized GHG and VOC emissions (emissions per unit energy) of the in situ scenarios considered in Chapters 10 and 11. The chapter considers a more comprehensive system boundary than those considered in Chapters 10 and 11. In addition to the production of electricity to heat the formation as discussed in Chapters 10 & 11, this evaluation also includes the energy and emissions from drilling and from separation/storage of the liquid product. This evaluation does not include the energy and emissions associated with site preparation, fugitive emissions associated with drilling into a formation, transport of the materials and equipment to the site, hydraulic fracturing of the wells, or site restoration as these processes are difficult to quantify with sufficient accuracy to provide meaningful results. They are also likely to have minimal impact on the results because they represent a small fraction of the project's emissions.

This evaluation limits the discussion to the most feasible cases: those with an oil supply price < \$174/bbl (see Chapter 11) and with an NER > 1. These constraints reduce the total number of scenarios from 472,142 to 169,340. The emissions in this chapter are presented in two ways: normalized to the energy content of the fuel (higher heating value [HHV]) and total annual emissions. Total annual emissions are estimated by assuming that the emissions are distributed equally over the project's 7-year lifetime.

12.2.1.1 Net Energy Ratio

In this chapter, NER is defined as

$$\text{NER} = \frac{E_{\text{liq}}}{E_{\text{req}}} \tag{12.1}$$

where
   E_{liq} represents the energy contained in the liquid (oil) product
   E_{req} represents the energy required to produce the liquid product (i.e., energy to drill the wells and to electrically heat the formation).

NER differs from the external energy ratio (EER) discussed in Chapters 10 and 11 by including the energy required to drill the wells in E_{req} and by considering the liquid product rather than both products (oil and gas). This definition of NER was selected to facilitate comparison with the normalized GHG emissions and with literature NER values. Normalized GHG emissions are typically reported on a basis of the energy content of the liquid product, rather than a total product basis, and NERs are typically reported for a single product, i.e., crude oil, natural gas, gasoline, etc. The energy required to generate oil is calculated by multiplying the total energy required to generate products (both oil and gas) by the ratio of the total energy produced in the form of crude oil to the total energy generated in both gas and oil products. Compared to the energy required to heat the formation, the energy required to drill the wells causes a negligible increase (less than 0.5% of the total energy) in the total energy requirement (E_{req}). Consequently, the NER results are similar to the EER results shown in Chapters 10 and 11.

12.2.1.2 Total and Normalized GHG and VOC Emissions

Several sources of GHG (N_2O, CH_4, and CO_2) and VOC emissions are considered in this analysis. First are the GHG and VOC emissions resulting from the combustion of fuel needed to drill the wells. Second are the GHG and VOC emissions from the power generation required to electrically heat the formation. Third are the VOC emissions from separating and storing the liquid fuels prior to sale. Drilling into oil shale formations, particularly those that are far from outcrops, could release fugitive CH_4 and other light hydrocarbon (VOC) emissions (Matta et al. 1977). However, this release is difficult to quantify due to lack of data. Consequently, the present analysis does not consider the release of fugitive (adsorbed) CH_4/VOC emissions. From an air-emissions standpoint, the majority of CH_4/VOC released during the in situ process would be captured as part of the product streams.

Table 12.1 summarizes the selected emission factors and energy consumption factors. Diesel fuel consumption for well drilling is estimated using the relationship between fuel consumption and well depth shown in the following equation:

$$V = 1003.26513 \times (1.00027178)^D \tag{12.1}$$

where
  V is the volume of diesel in gallons
  D is the well depth in feet.

The relationship was developed from analyzing the fuel consumption data available for 153 oil and gas wells drilled in the Uinta Basin since 2008 (UDOGM 2014). In determining the energy equivalent for the volume of fuel consumed, diesel's HHV of 128,450 BTU/gal is used. GHG and VOC emission factors for the combustion of diesel fuel are from Argonne National Laboratory (ANL 2012).

GHG emissions from electricity generation are assumed to be the average for the state of Utah, estimated from EPA's eGRID (EPA 2014). For 2010, the most recent year available, Utah's power was generated from approximately 81% coal, 15% natural gas, 3% hydropower and other renewables, and 1%

TABLE 12.1

Emission Factors and Energy Usage for Processes Considered in This Chapter

Process	Energy	CO_2 Equivalent	VOC
Drilling	See Equation 12.1	9980 g CO_{2e}/gal diesel fuel	0.26 g VOC/gal diesel fuel
Electricity generation	—	1830.2 g/kW h	0.0139 g/kW h

Note: kW h, kilowatt hour.

miscellaneous fossil-energy sources. EPA's eGRID does not provide VOC emissions, so VOC emissions are estimated from Utah's resource mix and emission factors for electricity generation from ANL (2012).

While total GHG and VOC emissions from oil shale development are important to consider, the increase in VOC emissions in the Uinta Basin itself is of particular concern due to the air quality issues discussed previously in this chapter. Only one of Utah's major electricity-generation facilities, the Bonanza Power Plant, is located within the Uinta Basin. The Bonanza Power Plant produces 10% of the state's electricity. For this analysis, it is assumed that electricity for in situ production of oil shale would be purchased from the state's grid, but it is unclear whether this power would come from the Bonanza Power Plant. If power were generated at the Bonanza Power Plant, this additional generation would result in increased VOC emissions within the Basin.

VOC emissions from oil separation and storage are estimated as 0.364 g of CH_4/bbl using an emission factor from California EPA for light crude (CAL EPA 2013), an assumed 95% reduction of VOCs based on the New Source Performance Standards for oil storage tanks (40 C.F.R. § 60.5395 (2015)), and the average ratio of VOCs to CH_4 of 0.16 for the Uinta Basin (Zhang et al. 2009). The emissions per unit of energy of product are based on an HHV of 42.55 MJ/kg for crude oil and of 758 BTU/scf for gas.

12.2.2 Results

This discussion focuses on the 169,340 scenarios that have an oil supply price <$174/bbl (see Section 11.4.1) and an NER > 1. Figure 12.1 shows a set of hexbin plots of the normalized GHG emissions in g CO_{2e}/MJ crude oil for selected input parameters. These plots count the number of points located in a particular region of the plot and color regions with higher numbers of points more darkly. Normalized GHG emissions correlate well with NER ($R^2 = 0.89$) and with normalized VOC emissions ($R^2 = 0.98$). Heating the formation requires the majority of the energy input (compared to well drilling or separation/ storage), so the emissions from the production of electricity dominate total GHG emissions and are a significant contributor to VOC emissions. Hence, GHG and VOC emissions tend to correlate with each other. In addition, lower normalized GHG emissions are associated with higher recovery fractions of the converted kerogen. None of the other parameters correlate with normalized GHG emissions ($R^2 \le 0.1$).

Another way of evaluating the data is to calculate the median value of each model input parameter at different values of normalized GHG emissions to observe how each parameter affects normalized GHG emissions. Table 12.2 shows these median values for the 169,340 scenarios and for results that correspond to the 25th, 50th, and 75th percentiles—22.3, 26.5, and 32.1 g CO_{2e}/MJ, respectively. The input parameters in the table are (1) geometry/design inputs (discussed in Chapter 10) including number of wells, rows of wells, and

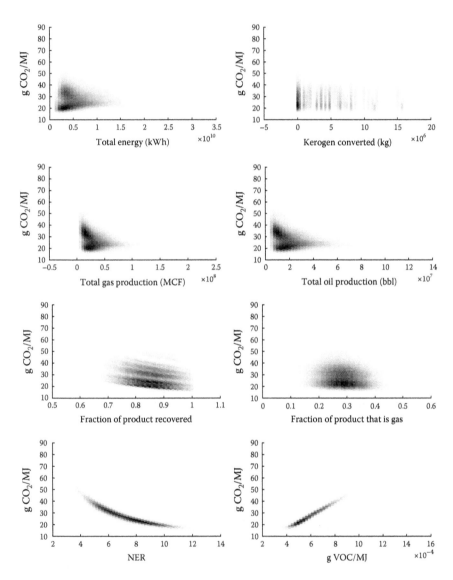

FIGURE 12.1

Hexbin plots of selected parameters versus normalized CO_{2e} emissions. The darker hexes indicate greater frequency. The parameters include total energy required to heat the formation, mass of kerogen converted to liquid and gaseous products, total gas production, total oil production, fraction of product recovered, fraction of the product that is gas, net energy ratio, and normalized volatile organic compound emissions.

TABLE 12.2

Median Values of Input Parameters and Model Output Values at the 25th, 50th, 75th, and 100th Percentile Cutoff Thresholds for Normalized CO_2 Equivalent Emissions

Item	≤22.3 g CO_{2e}/MJ	≤26.5 g CO_{2e}/MJ	≤32.1 g CO_{2e}/MJ	All
Percent of scenarios (%)	25	50	75	100
Model input parameters				
Number of rows	6	7	7	7
Number of wells	67	84	81	73
Length of production segment (ft)	8961	9084	9200	9473
Fraction of gas in the product (x_g)	0.294	0.290	0.288	0.284
Fraction of product[a] recovered	0.882	0.864	0.862	0.855
Model output values				
Energy needed to heat formation (kWh)	3.83E+09	4.66E+09	4.85E+09	4.61E+09
Mass of kerogen converted (kg)	3.16E+06	2.94E+06	2.94E+06	2.84E+06
Oil supply price ($)	105.51	109.40	115.28	123.32
Total oil production (bbl)	1.84E+07	2.06E+07	1.95E+07	1.71E+07
Total gas production (MCF)	2.90E+07	3.18E+07	2.99E+07	2.58E+07
NER	9.26	8.48	7.82	7.16
g CO_{2e}/MJ crude oil	20.4	22.3	24.2	26.5
g VOC/MJ crude oil	4.77E−04	5.10E−04	5.43E−04	5.81E−04
Total CO_2 (ton/year)[b]	5.11E+05	6.08E+05	6.34E+05	6.03E+05
Total VOC (ton/year)[b]	11.9	13.9	14.2	13.2

[a] Includes gas and liquid product that is generated from kerogen pyrolysis.
[b] Average over the 7-year project life.

production length and (2) economic inputs (discussed in Chapter 11) including the fraction of the pyrolyzed kerogen that is recovered and the fraction of the product that is gas. Also listed in Table 12.2 are the median values of the model outputs for the same set of GHG emissions cutoff points. These results show that lower normalized GHG emissions are associated with higher gas fractions, higher kerogen recovery fractions, lower oil supply prices, higher NERs, and lower normalized VOC emissions.

Table 12.3 shows the median normalized VOC emissions (g VOC/MJ) for the same set of model input parameters and output values listed in Table 12.2. Lower normalized VOC emissions are associated with higher gas fractions, higher recovery fractions, lower oil supply prices, higher NERs, and lower normalized GHG emissions. These findings are generally intuitive with the potential exception of lower normalized GHG and VOC emissions being associated with higher gas fractions. Higher gas fractions (alternatively lower oil fractions) lead to lower emissions from oil separation and storage.

TABLE 12.3

Median Values of Input Parameters and Model Output Values at the 25th, 50th, 75th, and 100th Percentile Cutoff Thresholds for Normalized Volatile Organic Compound Emissions

Item	$\leq 5.10 \times 10^{-4}$ g VOC/MJ	$\leq 5.81 \times 10^{-4}$ g VOC/MJ	$\leq 6.76 \times 10^{-4}$ g VOC/MJ	All
Percent of scenarios (%)	25	50	75	100
Model input parameters				
Number of rows	6	7	7	7
Number of wells	70	84	81	73
Length of production segment (ft)	8939	9095	9212	9473
Fraction of gas in the product (x_g)	0.306	0.294	0.290	0.284
Fraction of product[a] recovered	0.881	0.865	0.862	0.855
Model output values				
Energy needed to heat formation (kWh)	3.91E+09	4.65E+09	4.85E+09	4.61E+09
Mass of kerogen converted (kg)	3.04E+06	2.94E+06	2.94E+06	2.84E+06
Oil supply price ($)	107.39	110.07	115.58	123.32
Total oil production (bbl)	1.84E+07	2.04E+07	1.95E+07	1.71E+07
Total gas production (MCF)	3.06E+07	3.22E+07	3.00E+07	2.58E+07
NER	9.18	8.47	7.82	7.16
g CO_{2e}/MJ crude oil	20.4	22.3	24.2	26.5
g VOC/MJ crude oil	4.77E-04	5.10E-04	5.43E-04	5.81E-04
Total CO_2 (ton/year)[b]	6.03E+05	5.11E+05	6.08E+05	6.34E+05
Total VOC (ton/year)[b]	13.2	11.9	13.9	14.2

[a] Includes gas and liquid product that is generated from kerogen pyrolysis.
[b] Average over the 7-year project life.

In 2011, VOC emissions for Uintah and Duchesne Counties were 112,793 ton/year (Harper 2013). The median annual VOC emission for the feasible set of in situ oil shale scenarios is 13.2 ton/year (Table 12.3), suggesting that VOC emissions from most scenarios would have a minor, incremental effect on the Uinta Basin's VOC burden. In addition, the electricity generation associated with heating the formation is responsible for 8.3–11.9 ton/year of the total VOC emissions (the minimum and maximum of the 169,340 scenarios). The electricity could be generated in the basin at the Bonanza Power Plant, at a newly constructed power plant in the basin, or elsewhere in the state. Regardless, the location of the electricity generation would have a limited effect on the basin's VOC burden. The VOC emissions in the basin from drilling and separation/storage, ranging from 1.3-4.8 ton/year, would similarly have a limited effect on air quality. However, in light of the more stringent O_3 standards discussed, any increase in emissions of O_3 precursors, including VOCs, has the potential to raise permitting challenges.

For GHG emissions, the majority (>99%) of the emissions is associated with electricity generation.

12.3 Comparing Oil Shale's Net Energy Ratio and Carbon Footprint to Other Fuels

12.3.1 Wellhead GHG Emissions

For the 169,340 in situ oil shale production scenarios analyzed in this chapter, the median value of computed GHG emissions is 26.5 g CO_{2e}/MJ (Table 12.2). This value is approximately five times the U.S. average normalized GHG emissions at the wellhead of 5.1 g CO_{2e}/MJ for conventionally-produced crude oil (ANL 2014).

Tables 12.4 and 12.5 summarize the range of reported wellhead GHG emissions and NERs for recovery of crude oil from conventional sources, oil sands, and oil shale, including this chapter's results. Since most evaluations of oil sands and oil shale production methods report GHG emissions and NER on a basis of energy content of a refined fuel, the tables include values that were converted to a basis of energy content of crude oil using information from the publication as well as reported wellhead values.

A good deal of variability exists in the GHG emissions estimates. The range for the in situ oil shale scenarios analyzed in this chapter is 15.2–81.5 g CO_{2e}/MJ, a higher range than for oil produced conventionally or by the Shell in situ conversion process (ICP) method (Brandt 2008). One

TABLE 12.4

GHG Emissions from Conventional and Unconventional Oil Sources per MJ of Crude Oil Equivalent at the Wellhead

Crude Source	Low (g CO_{2e}/MJ)	High (g CO_{2e}/MJ)
Conventional	1[a]	17[a,b]
Oil sands—in situ	9[a]	41[a]
Oil sands—ex situ	8.7[c]	15[a]
Oil shale—Shell ICP	7.5[d,e]	12[d,e]
Oil shale in situ—this chapter	15.2[f]	81.5[f]

[a] TIAX (2009).
[b] Jacobs (2009).
[c] Cleveland and O'Connor (2011).
[d] Brandt (2008).
[e] Energy value of crude oil estimated by assuming a higher heating value of 5746 MJ/bbl for the crude equivalent.
[f] Minimum and maximum values from the 169,340 scenarios.

TABLE 12.5

Net Energy Ratio from Conventional and Unconventional Oil Sources
(Based on MJ of Crude Oil Equivalent at the Wellhead)

Crude Source	Low NER	High NER
Conventional	12[a]	20[b]
Oil sands—ex situ	4.7[c]	15[c]
Oil shale—ex situ	1.1[b]	1.8[b]
Oil shale—Shell ICP	1.6[d]	2[d]
Oil shale in situ—this chapter	2.49[e]	12.5[e]

[a] Murphy and Hall (2010).
[b] Cleveland and O'Connor (2011).
[c] Brandt et al. (2011).
[d] Brandt (2008).
[e] Minimum and maximum values from the 169,340 scenarios.

likely reason for the emission differences between this study and Brandt's study is the GHG intensity of the electricity required for formation heating. Brandt assumed that produced gas was combusted in a 45% efficient, natural gas combined cycle plant and that the remaining energy was purchased from the Colorado grid (72% coal and 24% natural gas). Both of these sources of electricity are less GHG-intensive than the Utah grid (81% coal). Also, Brandt assumed that the oil shale resource was homogeneous, while this study takes into account the detailed stratigraphy of the Green River Formation in the Coyote Wash core (see Chapter 10).

12.3.2 Life Cycle GHG Emissions

When comparing fuels derived from oil shale to other transportation fuel sources, it is important to consider the fuel's entire life cycle. Oil shale is considered an unconventional source of crude, as are oil sands and heavy oil. Figure 12.2 shows an example of GHG emissions from both well-to-pump (WTP) and well-to-wheel (WTW) fuel cycles for a conventional source of crude oil. The WTP cycle includes raw material extraction, transportation, processing (including upgrading), refining, and delivery to the pump. The WTW cycle extends the WTP cycle to include tailpipe emissions from fuel combustion.

Worldwide, WTP GHG emissions for producing a barrel of crude oil are likely to increase. The U.S. Energy Information Administration (EIA) projects that world production of unconventional liquid fuels, which require more energy to produce and generate larger WTP GHG emissions, will increase from 3.4 million barrels per day (BPD) in 2007 to 12.9 million BPD by 2035 (EIA 2015, NPC 2007), accounting for 12% of world liquid fuel supply. In addition, a good deal of variability already exists in the WTP GHG profiles of crude oils, depending upon resource origin, as seen in Figure 12.3.

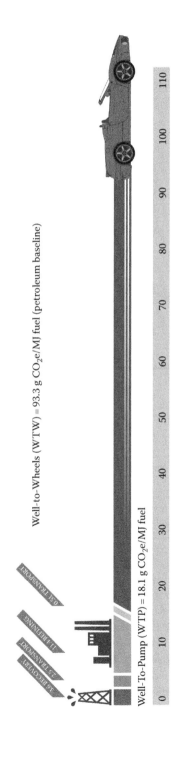

FIGURE 12.2
The GHG emissions in CO_2 equivalents include emissions of CO_2, CH_4, and N_2O. Units in the figure are given in g CO_2e/MJ. Well-to-pump and well-to-wheel GHG emissions for gasoline produced from conventional crude oil. (Data from the U.S. Department of Energy, National Energy Technology Laboratory, An evaluation of the extraction, transport, and refining of imported crude oils and the impact on life cycle greenhouse gas emission (DOE/NETL-2009/1362), National Energy Technology Laboratory, Pittsburgh, PA, 2009).

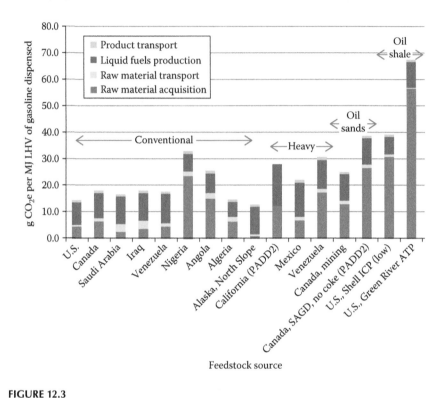

Feedstock source

FIGURE 12.3

LHV refers to the lower heating value of the fuel and PADD2 is the Midwest Petroleum Administration Defense District. Well-to-pump greenhouse gas emission profiles for gasoline produced from various sources of crude oil. (Data for this figure were obtained from U.S. Environmental Protection Agency, Endangerment and cause or contribute findings for greenhouse gases under Section 202(a) of the Clean Air Act; Final Rule, 74 Fed. Reg. 66,494 (December 15, 2009), 2009; Brandt, A.R., *Environ. Sci. Technol.*, 42(19), 7489, 2008; Brandt, A.R. and Farrell, A.E., *Clim. Chang.*, 84, 241, 2007; National Petroleum Council, Topic paper #22–Heavy oil, National Petroleum Council, Washington, DC, July 18, 2007, http://www.npchardtruthsreport.org/topicpapers.php; IHS CERA, Oil sands, greenhouse gases, and US oil supply–Getting the numbers right, 2012, http://www.api.org/policy-and-issues/policy-items/oil%20sands/greenhouse-gases; 2012; Charpentier, A.D. et al., *Environ. Res. Lett.*, 4(1), 14005, 2009.)

Canadian oil sands, Venezuelan bitumen, and California heavy oil all have greater WTP GHG emissions than U.S. conventional crude oil. However, high WTP GHG emissions are not limited to unconventional fuels; Nigerian crude has the fourth highest WTP GHG emissions profile in Figure 12.3, primarily due to the venting and flaring of nearly all of the coproduced natural gas (NETL 2009). The WTP GHG emission's estimate for Shell's ICP (Brandt 2008) is in the same range as Canadian oil sands and Nigerian crude, while that for the ex situ Alberta Taciuk process oil shale retort is substantially higher (Brandt 2009).

12.4 Carbon Management Policies

The EPA has found that GHG emissions endanger public health and welfare (U.S. EPA 2009), and this finding is driving new regulations. The EPA also requires certain fossil fuel suppliers, industrial gas suppliers, and direct GHG emitters to monitor and report their CO_{2e} emissions (40 C.F.R. pt. 98 (2014)).

In the absence of a national carbon policy, states have taken the lead in addressing liquid transportation fuels' contribution to GHG emissions. For example, the California State Air Resources Board adopted a number of regulations, including a low carbon fuel standard (LCFS) that requires the reduction of the carbon content of transportation fuels sold, supplied, or offered for sale in California (Cal. Code Regs. tit. 17, §§ 95480-90 [West, Westlaw through 10/16/15 Register 2015, No. 42]). California's LCFS resulted in several unsuccessful legal challenges, and the LCFS remains in force. Additionally, in 2009, the Oregon Legislature authorized the Oregon Department of Environmental Quality (ODEQ) to promulgate an LCFS. The ODEQ has issued draft regulations requiring a 10% reduction in life cycle GHG emissions from Oregon's transportation fuels over a 10-year period (Or. Admin. R. 340-253 (2015) (proposed rule)), but that rule will not take effect until finalized.

This carbon-management regulatory landscape is evolving, and it is unclear whether LCFSs or other carbon-management strategies will become more widespread. Since GHG emissions from the production of oil shale under almost any development scenario, including the in situ scenarios evaluated in this chapter, are significantly higher than the values reported for conventional oil (Figure 12.3; Table 12.4), the combination of oil shale's carbon footprint and burgeoning regulations may have significant regulatory consequences and may limit development.

12.5 Opportunities for GHG and Ozone-Precursor Emission Reductions

In laboratory- and pilot-scale tests of various oil shale production methods, the two main contributors to GHG emissions are the production of thermal and electrical energies to power the operation and the decomposition of carbonate minerals to form CO_2. For VOC emissions, separation (of the oil, gas, and water) and storage are important contributors to emissions.

For thermal and electrical energy production, lower-emitting sources have been proposed, including a high-efficiency, combined-cycle natural gas power plant (Thomas et al. 2009) as well as renewable sources such as wind and solar (Brandt 2008, Hanson and Limerick 2009). As pointed out by Brandt (2008), in situ thermal production methods would not likely be affected by the intermittency of renewables due to long heating times and the large thermal mass and high heat capacity of the oil shale. Carbon capture and storage has also been examined as a way to mitigate power-generation GHG emissions with recent studies looking at possible sequestration and enhanced oil recovery targets in saline formations in Utah's Paradox Basin (EGI 2016).

Other efficiency measures can reduce energy input requirements and the resulting emissions. For in situ production, Brandt surmises that a conductive heating method such as Shell's ICP would reuse waste heat from depleted production cells, thereby increasing efficiency (Brandt 2008). Efficiency could also be improved by introducing fractures into the formation, thus allowing for convective and not just conductive heat transfer. Both Chevron, Inc. and American Shale Oil, Inc. have proposed methodologies that allow for convective heating of an in situ retort created by fracturing (Burnham 2010, Crawford and Killen 2011. However, because of the uncertain role that fracturing may play in in situ production, as well as the uncertainty associated with related fugitive emissions, well fracturing was not considered in this analysis.

With respect to mineral decomposition, carbonate minerals in oil shale begin decomposing (and releasing CO_2) near 1049°F (565°C) for dolomite and 1148°F–1247°F (620°C–675°C) for calcite (Brandt 2009, Burnham 1993). Through understanding of the mineralogy of a particular oil shale resource and careful monitoring of process conditions/temperatures, the carbonate decomposition temperature window(s) can be avoided. However, the tradeoff for operating at lower temperatures is that residual carbon is left on the spent shale.

12.6 Conclusions

This chapter summarizes regulatory programs protecting air-quality-related values and existing air quality conditions within the Uinta Basin, considers the effect of oil shale development on GHG and VOC emissions in the Uinta Basin, and addresses how oil shale production may be affected by a rapidly changing regulatory environment. Oil shale operators may face air quality constraints due to the basin's likely designation as an O_3 nonattainment area as well as evolving GHG regulatory programs. VOC emissions from in situ oil shale production could place some additional burdens on an air-quality-constrained region, although the extent of the burden would depend on the location of the energy generation required to heat the formation. With respect to GHG emissions, crude oil produced from in situ oil shale processing would have a larger GHG footprint than either crude produced conventionally or by ex situ processing of oil sands, but its footprint would be in the range of values reported for crudes produced from in situ treatment of oil sands. In light of the evolving air quality and GHG regulatory landscape and the jurisdictional challenges inherent in working in the Uinta Basin, those involved in oil shale development should engage state and federal regulators at the outset of the development planning process in order to minimize the risk of costly and time-consuming delays.

References

Argonne National Laboratory (ANL). 2014. The greenhouse gases, regulated emissions, and energy use in transportation (GREET) model. https://greet.es.anl.gov/. Accessed on August 20, 2015.

Argonne National Laboratory (ANL). 2012. *Updated Greenhouse Gas and Criteria Air Pollutant Emission Factors and Their Probability Distribution Functions for Electric Generating Units* (ANL/ESD/12-2). Argonne, IL: Argonne National Laboratory.

Brandt, A. R. 2008. Converting oil shale to liquid fuels: Energy inputs and greenhouse gas emissions of the Shell in situ conversion process. *Environ. Sci. Technol.* 42(19):7489–7495.

Brandt, A. R. 2009. Converting oil shale to liquid fuels with the Alberta Taciuk processor: Energy inputs and greenhouse gas emissions. *Energy Fuels* 23:6253–6258.

Brandt, A. R., J. Boak, and A. K. Burnham. 2011. Carbon dioxide emissions from oil shale derived liquid fuels. In *Oil Shale: Solutions to the Liquid Fuel Dilemma*, eds. O. I. Ogunsola, A. M. Hartstein, and O. Ogunsola. Washington, DC: ACS Symposium Series.

Brandt, A. R. and A. E. Farrell. 2007. Scraping the bottom of the barrel: Greenhouse gas emission consequences of a transition to low-quality and synthetic petroleum resources. *Clim. Chang.* 84:241–263.

Burnham, A. K. 1993. *Chemical Kinetics and Oil Shale Process Design (UCRL-JC-114129)*. Livermore, CA: Lawrence Livermore National Laboratory.

Bureau of Land Management (BLM). 2012. Proposed land use plan amendments for allocation of oil shale and tar sands resources on lands administered by the Bureau of Land Management in Colorado, Utah, and Wyoming and final programmatic environmental impact statement, Vol. 2, Chapters 4 and 5. Washington, DC: Bureau of Land Management, pp. 4–53. http://ostseis.anl.gov/documents/peis2012/vol/OSTS_Volume_2.pdf.

Burnham, A. K. 2010 (April). Progress and plans on AMSO's RD&D lease tract. Paper presented at the *University of Utah Unconventional Fuels Conference*, Salt Lake City, UT.

California Air Resources Board (CARB). 2011. Low carbon fuel standard program. http://www.arb.ca.gov/fuels/lcfs/lcfs.htm.

California Environmental Protection Agency (CAL EPA). 2013 (October). 2007 oil and gas industry survey results–Final report (revised). http://www.arb.ca.gov/cc/oil-gas/draftreport.pdf.

Charpentier, A. D., J. A. Bergerson, and H. L. MacLean. 2009. Understanding the Canadian oil sands industry's greenhouse gas emissions. *Environ. Res. Lett.* 4(1):14005–14016.

Cleveland, C. J. and P. A. O'Connor. 2011. Energy return on investment (EROI) of oil shale. *Sustainability* 3:2307–2322.

Crawford, P. M. and J. C. Killen. 2011. New challenges and directions in oil shale development technologies. In *Oil Shale: Solutions to the Liquid Fuel Dilemma*, eds. O. I. Ogunsola, A. M. Hartstein, and O. Ogunsola. Washington, DC: American Chemical Society.

Edwards, P. M., S. S. Brown, J. M. Roberts et al. 2014. High winter ozone pollution from carbonyl photosynthesis in an oil and gas basin. *Nature* 514:351–354.

Elliott, K. 2008 (October). Examination of oil sands projects: Gasification, CO_2 emissions and supply costs. Paper presented at the *SPE International Thermal Operations and Heavy Oil Symposium*, Calgary, Alberta, Canada.

Energy & Geoscience Institute. (n.d.). Carbon science and engineering research, Paradox Basin, Southwest Regional Partnership on carbon sequestration. Retrieved March, 23 2016. http://co2.egi.utah.edu/projectsites/paradox/index.htm.

Environ. 2015. Final report–2014 Uinta Basin winter ozone study. Salt Lake City, UT: Utah Division of Air Quality. http://www.deq.utah.gov/locations/U/uintahbasin/ozone/docs/2015/02Feb/UBWOS_2014_Final.pdf.

Hanson, J. L. and P. Limerick. July 12, 2009. What every Westerner should know about oil shale: A guide to shale country. Report from the Center #10. Boulder, CO: Center of the American West, University of Colorado at Boulder. Retrieved October 11, 2011. http://centerwest.org/wp-content/uploads/2011/06/oilshale.pdf.

Harper, K. 2013 (August 14). *2011 Uinta Basin Oil and Gas Emissions Estimates.* Salt Lake City, UT: Utah Division of Air Quality.

IHS CERA. 2012. Oil sands, greenhouse gases, and US oil supply–Getting the numbers right. http://www.api.org/policy-and-issues/policy-items/oil%20sands/greenhouse-gases.

Jacobs Consultancy. 2009. *Life Cycle Assessment Comparison of North American and Imported Crudes (AERI 1747).* Calgary, AB: Alberta Research Energy Institute, File no. AERI 1747.

Matta, J. E., J. C. LaScola, and N. Kissell. 1977. *Methane Absorption in Oil Shale and Its Potential Mine Hazard (RI 8243).* Pittsburgh, PA: U.S. Department of the Interior, Bureau of Mines.

Murphy, D. J. and C. A. S. Hall. 2010. Year in review–EROI or energy return on (energy) invested. *Ann. NY Acad. Sci.* 1185:102–118.

National Petroleum Council (NPC). 2007 (July 18). Topic paper #22–Heavy oil. Washington, DC: National Petroleum Council. http://www.npchardtruthsreport.org/topicpapers.php.

Oregon Department of Environmental Quality (ODEQ). 2015. Oregon clean fuels program. http://www.deq.state.or.us/aq/cleanFuel/.

O'Donoghue, A. J. 2012. Uintah Basin ozone problem triggers lawsuit against EPA. Deseret News, July 23, 2012. http://www.deseretnews.com/article/865559426/Uinta-Basin-ozone-problem-triggers-lawsuit-against-EPA.html?pg=all.

Rodriguez, M. A., M. G. Barna, and T. Moore. 2009. Regional impacts of oil and gas development on ozone formation in the Western United States. *J. Air Waste Manage. Assoc.* 59:1111–1118.

Tanana, H. and J. Ruple. 2012. Energy development in Indian Country: Working within the realm of Indian law and moving towards collaboration. *Utah Environ. Law Rev.* 32(1).

Thomas, M. M., E. A. Alvarez, G. Ghurye et al. 2009. Responsible development of oil shale, ExxonMobil Upstream Research Company. Paper presented at the *29th Oil Shale Symposium*, Golden, CO.

TIAX. 2009. Comparison of North American and imported crude oil lifecycle GHG emissions. TIAX Case No. D5595.

U.S. Department of Energy, National Energy Technology Laboratory (NETL). 2009. An evaluation of the extraction, transport, and refining of imported crude oils and the impact on life cycle greenhouse gas emission (DOE/NETL-2009/1362). Pittsburgh, PA: National Energy Technology Laboratory.

U.S. Energy Information Agency (EIA). 2015. Market trends: Liquid fuels. http://www.eia.gov/forecasts/aeo/mtliquidfuels.cfm.

U.S. Environmental Protection Agency (US EPA). 2014. Emissions & generation resource integrated database. http://www.epa.gov/cleanenergy/energy-resources/egrid/.

U.S. Environmental Protection Agency (U.S. EPA). 2015a. New Source Performance Standards for the oil and gas industry, Proposed Rule, 80 Fed. Reg. 56,593 (September 18, 2015).

U.S. Environmental Protection Agency (U.S. EPA). 2015b. National Ambient Air Quality Standards for ozone, Final Rule, 80 Fed. Reg. 65,292 (October 26, 2015) (to be codified at 40 C.F.R. pts. 50, 51, 52, et al.).

U.S. Environmental Protection Agency (U.S. EPA). 2009. *Endangerment and Cause or Contribute Findings for Greenhouse Gases under Section 202(a) of the Clean Air Act*; Final Rule, 74 Fed. Reg. 66,494 (December 15, 2009). Washington, DC: U.S. Environmental Protection Agency.

U.S. Government Accountability Office (US GAO). 2010. *Energy-Water Nexus: A Better and Coordinated Understanding of Water Resources Could Help Mitigate the Impacts of Potential Oil Shale Development* (GAO 11-35). Washington, DC: U.S. Environmental Protection Agency.

Utah Division of Air Quality (UDAQ). 2015. Annual summary reports. http://www.airmonitoring.utah.gov/dataarchive/archo3.htm.

Utah Division of Oil, Gas and Mining. 2014. Live data search—Online oil and gas information system. http://oilgas.ogm.utah.gov/Data_Center/LiveData_Search/main_menu.htm.

Zhang, Y., C. W. Gable, G. A. Zyvoloski, and L. M. Walter. 2009. Hydrogeochemistry and gas compositions of the Uinta Basin: A regional-scale overview. *AAPG Bull.* 93:1087–1118.

Index

A

Absolute permeability, 188, 195, 227–228, 235–236
Air quality regulation
  air quality policy, 308
  GHG and VOC emissions
    carbon management, 323
    electricity generation, 318–319
    emission factors and energy consumption factors, 314
    hexbin plots, 315–316
    median values, 315, 317–318
    NER, 313, 319–322
    normalized, 315–319
    sources, 314
    total, 315
  Indian country jurisdiction and collaboration, 311–312
  oil shale development
    attainment area permitting, 309
    BLM, 310–311
    EPA, 309
    FLPMA, 310
    nonattainment area permitting, 309
  in Uinta Basin, 308–309
Argonne Premium Coal Sample Bank, 90
Aurasource, Inc., 7, 24
Axial deviatoric stress *vs.* axial strain
  nonzero axial deviatoric stress, 231
  radial strain, 231
  stress–strain–time data, 229–230
  testing chronology and sample behavior, 229–231
  transverse isotropy, 232
  triaxial testing, 231–233

B

Black Sunday, 32, 44, 53
Bureau of Land Management (BLM), 4, 6–7, 16, 20–21, 310–311

C

Carbon management policies, 323
Central Utah Project, 52
Chemical percolation devolatilization (CPD) model
  aliphatic labile bridge, 146
  Bethe lattice, 144
  black liquor pyrolysis, 145
  bridge-breaking scheme, 146–147
  bridge variables calculation, 149–150
  char bridges, 149–150
  chemical structure parameters, 145–147
  detached clusters, 144–145
  labile bridges, 149–150
  liberated fragments, 146
  light gas formation, 148
  molecular weight distributions, 147–148
  percolation lattice, 144
  rate coefficients, 147–149
  tar and char yields calculation, 147–149
Chevron Oil Shale Company, 6
Core-scale oil shale pyrolysis
  ex situ processing, 158–159
  GR1, GR2, and GR3 samples
    elemental analysis, 161
    multiscale pyrolysis experiments, 166, 179, 181–184
    TGA experiments, 160–161, 163, 168–169, 171
  ICP, 158–159
  kerogen decomposition, kinetic model, 158
  physical phenomena, 158
  product analysis, 166–167
  Red Leaf's EcoShale® in-capsule process technology, 158
  S1 and S2 oil shale samples
    elemental analysis, 159–160
    multiscale pyrolysis experiments, 164–166, 170–180, 183–185

Printed and bound by CPI Group (UK) Ltd, Croydon, CR0 4YY

01/11/2024

01782617-0010